Handbook of Fuels

Edited by
Barbara Elvers

Further Reading

O. Ocic

Oil Refineries in the 21st Century

Energy Efficient, Cost Effective, Environmentally Benign

2005

ISBN 978-3-527-31194-1

B. Kamm, P. R. Gruber, N. Kamm (Eds.)

Biorefineries – Industrial Processes and Products

Status Quo and Future Directions

2 Volumes

2006

ISBN 978-3-527-31027-2

G. A. Olah, A. Goeppert, G. K. S. Prakash

Beyond Oil and Gas: The Methanol Economy

2006

ISBN 978-3-527-31275-7

F. Shahidi (Ed.)

Bailey's Industrial Oil and Fat Products

6 Volumes

2005

ISBN 978-0-471-38460-6

Handbook of Fuels

Energy Sources for Transportation

Edited by
Barbara Elvers

WILEY-VCH

WILEY-VCH Verlag GmbH & Co. KGaA

The Editor

Dr. Barbara Elvers
Hennebergstr. 15
22393 Hamburg
Germany

Library of Congress Card No.:
applied for

British Library Cataloguing-in-Publication Data
A catalogue record for this book is available from the British Library.

Bibliographic information published by the Deutsche Nationalbibliothek
The Deutsche Nationalbibliothek lists this publication in the Deutsche Nationalbibliografie; detailed bibliographic data are available in the Internet at <http://dnb.d-nb.de>.

Typesetting Thomson Digital, India
Printing Strauss GmbH, Mörlenbach
Binding Litges & Dopf GmbH, Heppenheim

Printed in the Federal Republic of Germany
Printed on acid-free paper

ISBN: 978-3-527-30740-1

Table of Contents

Handbook of Fuels: Energy Sources for Transportation. Edited by Barbara Elvers

ISBN: 978-3-527-30740-1

Preface

We are living in an era that relies on mobility and transportation and have become used to – and expect – the fast transport of all types of goods even to remote locations, with personal mobility becoming a key issue in the everyday life of many people, at least in industrialized countries. However, transportation is not possible without usage of energy. When looking back in history, oats were grown to feed the horses, which had to carry riders and pull coaches, steam had to be produced from coal and water to drive steam engines, and the movement of ships and boats required wind and sails or a rowing crew.

Today's means of transportation mainly rely on fuels that are combusted in an engine or turbine. The vast majority of fuels are still petroleum-based products, utilizing the entire product slate of oil refining, extending from propane/butane for automotive fuels through gasoline, diesel fuel and kerosene, used in automobile engines and aviation, to the bottom-of-the-barrel residual fuels combusted in marine diesel engines. To gain a deeper insight into the complexity of a refinery and fuels production, this handbook will begin with a chapter on oil refining, providing the basis for the chapters on automotive, aviation turbine and marine fuels.

However, in an era facing dwindling oil reserves and growing environmental concerns, new alternative fuels and means of transportation have to be sought and exploited. To give an overview on current research and development in this area, two separate chapters are devoted to hydrogen fuel and fuel cells, while the alternative fuels bioethanol, natural gas, fatty acid methyl ester, synthetic fuels (coal/biomass/gas to liquids) and liquefied petroleum gas are treated in the chapter on automotive fuels. The application of methanol in fuel cells is described in the corresponding chapter, while a description of the use of methanol as a carburetor fuel has been omitted intentionally. The automobile industry is more than reluctant to develop engines utilizing this toxic liquid, which, in order to guarantee supply, has to be available and distributed at gasoline stations.

The combustion of fossil fuels results in the emission of sulfur and nitrogen oxides, unburned hydrocarbons, particulates and carbon dioxide gas. Measures to lower emission values are described in the chapters on automotive exhaust control and fuel cells. Also, combustion of bioethanol, fatty acid methyl esters and hydrogen is anticipated to result in a reduction of carbon dioxide emissions and is dealt

Handbook of Fuels: Energy Sources for Transportation. Edited by Barbara Elvers

ISBN: 978-3-527-30740-1

with in the corresponding chapters. Hybrid concepts, consisting of a combination of a fuel and a non–fuel–the battery–will not be described in this handbook.

Until a decisive breakthrough has been achieved with one of the alternative fuels we will certainly have to live with a fuel mix, exploiting the advantages of the different types of fuel on the one hand and minimizing their drawbacks on the other.

Hamburg, July 2007 *Barbara Elvers*

List of Contributors

Gunter Alfke
Sassenburger Weg
Hamburg
Germany

Geoff J. Bishop
Shell Aviation Ltd.
London
United Kingdom

H. P. Calis
Shell Global Solutions
The Hague
The Netherlands

Werner Dabelstein
Deutsche Shell AG (retired)
Hamburg
Germany

Rula Deeb
Malcolm Pirnie, Inc.
Emeryville
California 94608
USA

Ingo Drescher
Volkswagen AG, Section 3.8.3.3
38436 Wolfsburg
Germany

Matthias Duisberg
UMICORE AG & Co. KG
Hanau-Wolfgang
Germany

Nigar Frachet
World LP Gas Association
Rue Anatole de La Forge
75017 Paris
France

Lutz Guderjahn
Südzucker Ethanol GmbH
Gottlieb-Daimler Strasse 12
68165 Mannheim
Germany

Walther W. Irion
Deutsche Shell AG
Hamburg Germany

Michael Kavanaugh
Malcolm Pirnie, Inc.
Emeryville
California 94608
USA

Harald Klein
Linde Gas and Engineering
Höllriegelskreuth
Germany

Handbook of Fuels: Energy Sources for Transportation. Edited by Barbara Elvers

ISBN: 978-3-527-30740-1

Thomas Kreuser
OMG AG & Co.
Hanau
Germany

Maryline Laughier
Malcolm Pirnie, Inc.
Emeryville, CA
94608
USA

Gerhard Lepperhoff
FEV Motorentechnik
Aachen
Germany

Marco Lopez
UMICORE AG & Co. KG
Hanau-Wolfgang
Germany

Wolfgang Lüke
Shell Global Solutions
(Deutschland) GmbH
Hamburg
Germany

Torsten Mundt
Shell Global Solutions GmbH
Hamburg
Germany

Otto S. Neuwirth
UK Wesseling (retired)
Wesseling
Germany

Franz Nierlich
Oxeno Olefinchemie GmbH
Marl
Germany

Udo Peters
Oxeno Olefinchemie GmbH
Marl
Germany

Harald Ranke
Linde Gas and Engineering
Höllriegelskreuth
Germany

Klaus Reders
Deutsche Shell AG (retired)
Hamburg
Germany

Arno Reglitzky
Deutsche Shell AG (retired)
Hamburg
Germany

Gary Robertson
Shell International Oil Products
B.V.
The Hague
The Netherlands

Michael Sakuth
Huls Infracor GmbH
Marl
Germany

Kuno Schädlich
VEBA OEL AG (retired)
Gelsenkirchen
Germany

Peter Schug
VEBA OEL AG (retired)
Gelsenkirchen
Germany

Ekkehard Schulte-Körne
Oxeno Olefinchemie GmbH
Marl
Germany

Andrea Schütze
Deutsche Shell AG
Hamburg
Germany

Katharina Seitz
UMICORE AG & Co. KG
Hanau-Wolfgang
Germany

Gerhard Sextl
Zeochem AG
Uetikon
Switzerland

Martti Suominen
Fortum Oil and Gas Oy
Kilpilahti
Finland

Stephen M. Thompson
Shell International Petroleum
Maatschappij B.V.
The Hague
The Netherlands

Martin Votsmeier
OMG AG & Co.
Hanau
Germany

1 Introduction

Klaus Reders

The history and development of the automotive fuels gasoline and diesel together with their corresponding engines began more than 100 years ago and is still on going. During this entire period, engine and fuel technologies were mutually dependent on each other. However, with subsequent engine development, the fuel requirements changed and, likewise, improvements in engine technologies were often only possible through specific progress in fuel technologies.

In the early days of both the spark-igniting Otto and the self-igniting diesel engine, the mere starting and running of the engines, i.e., the preparation of a combustible air – fuel mixture, was the main problem. The most important quality criterion for *gasoline* was that it had to evaporate easily: a finger dipped into the fuel had to dry rapidly in the open air.

Diesel fuel had to ignite, with little delay, at the end of the compression stroke of the engine, as a result of increases in heat and pressure. In contrast to gasoline volatility, the ignition quality of a fuel was much more difficult to determine and it took RUDOLF DIESEL and other inventors considerable time and experimental effort to establish that lamp kerosene and gas oil were the most suitable fuels.

Even after the initial breakthrough, both engines suffered from fuel-related problems, the spark-ignited engine from uncontrolled early "self-combustion" of the air – fuel mixture, which caused engine damage and severely reduced engine efficiency, and the diesel engine from the problem of exact metering of the small amounts of fuel at high pressure and high speed, as is required for fuel-injection systems.

It was about 50 years after the invention of the spark-ignited engine that a gasoline additive (tetraethyllead) was found that reduced the uncontrolled combustion, known as engine knock, and thus allowed construction of engines with much improved efficiency.

Small engines suitable for use in trucks became available about 30 years after the invention of the diesel engine, when high-pressure fuel-injection pumps had been developed, but it took another 10 years before Mercedes Benz installed the first diesel engine in a passenger car.

Handbook of Fuels: Energy Sources for Transportation. Edited by Barbara Elvers

ISBN: 978-3-527-30740-1

Table 1.1 Engine fuels, a highly complex mixture of various components produced to international standards, meeting engine, customer and environmental requirement

Raw materials	Targets for fuel properties	Engine requirements
Crude oil– aromatics, olefins, paraffins, naphthenes (alcohols, ethers)	high energy content optimum combustion properties safe fuel handling no deposits, no corrosion	high engine performance low fuel consumption optimum drivability long engine life
Additives based on organic chemistry	acceptable environmental properties and toxicity	low emissions
Alternative fuels– natural gas biomass/ethanol vegetable oils or esters		
Solar energy/electricity		

Subsequently, prolonging engine life and safe operability under all climatic and driving conditions and increasing the yield of engine fuels from a given amount of crude oil by new refinery processes became important targets for fuel development.

Over the last two decades the reduction of emissions during production, transportation, storage and, of course, the use of fuels in engines has become the most important objective and although the availability of engine fuels from crude oil can be guaranteed for many decades to come, considerable effort is being spent on the development of alternative fuels, with the aim of also reducing emissions.

Engine fuels have meanwhile been optimized to a complex mixture of a large number of components meeting international standards and quality criteria. They guarantee maximum possible safety in the complete chain from production to their final usage, optimum engine operation over the whole engine lifetime and, at the same time, minimum possible effects on the environment (Table 1.1).

1.1 History of the Spark Ignited "Otto" Engine and of Gasoline

The development of motorized road transport was triggered by the dissatisfaction of Nikolaus Otto with the performance of the gas engines built in his factory in Deutz, near Cologne, which was founded in 1872. These heavy stationary engines used town gas as a fuel and operated at atmospheric pressure. They worked reliable and sold well in the beginning. However, all efforts to improve the incredibly low efficiency of these engines [3 horsepowers (hp) could be gained from an engine 4 m in height] failed, because a combustion process without compression was applied.

Nikolaus Otto therefore restarted tests that had been abandoned in 1862, that is: to develop a piston engine which combusts a gas – air mixture produced outside the engine after compression by the piston. This process would lead to much higher pressures in the combustion chamber and consequently to higher engine performance. This time he succeeded. In autumn 1876 the first four-stroke engine was being operated and a year later the German "Reichspatent" 532 was issued. Overnight all previous gas engines were outdated because of the new "high-efficiency" engine. However, initially the new Otto engine, which produced 20 hp, was still monstrous, weighed 6.8 t and was still fueled by town gas.

The technical director of the "Gas Engine Factory Deutz AG", Gottlieb Daimler, was not satisfied with the new engine. He envisaged a small high-speed four-stroke engine that could be built into a vehicle powered by a fuel which could be transported at ambient pressure. The Deutz factory was not interested in this concept and consequently Daimler left the company and moved to Bad Cannstatt near Stuttgart, where he began his own studies, together with the engineer Wilhelm Maybach. Right from the beginning Daimler intended to use a light mineral oil fraction, known as "ligroin", as the fuel, which up to that point had predominantly been used as a cleaning agent.

Within a very short period Daimler and the ingenious designer Maybach developed patentable basic technologies enabling their small four-stroke engine to run on a light gasoline fraction:

- the "uncontrolled glow tube ignition", a heated platinum tube pointing into the combustion chamber between the valves to ignite the compressed air – fuel mixture
- the "curve groove steering" to open and close the exhaust valve
- the "float carburetor" as a further development of the surface carburetor in which the intake air was sucked in through the "ligroin" (the light gasoline fraction)

The light gasoline, which had a density of well below 0.700 g/mL and a final boiling point of 85 °C, was sufficiently volatile to produce an ignitable air – fuel mixture in these early and simple carburetors. The float carburetor had a considerable size as it was also used as a fuel reservoir, nevertheless it operated reliably. In 1885 the first vehicle from the workshop of Daimler and Maybach, the "Petroleum Reitwagen" (riding car) motorcycle (Fig. 1.1), was on its "maiden trip" from Bad Cannstatt to Stuttgart-Untertürkheim. A year later the gas engine manufacturer Carl Benz surprised the public with a three-wheeled "patented motor vehicle" (Fig. 1.2) and in the same year Daimler installed his "small, lightweight, high-speed engine" in a four-wheeled motor carriage (Fig. 1.3).

Meanwhile, four-cylinder in-line engines had been constructed and in 1893 Maybach developed the first spray-nozzle carburetor, which replaced the large float carburetor. However, at that time it was not realized that this novel carburetor could also prepare ignitable air – fuel mixtures with gasolines of higher density and higher final boiling point and also of lower volatility. Therefore, the low-density light gasoline used up till then remained the preferred fuel for many years to come.

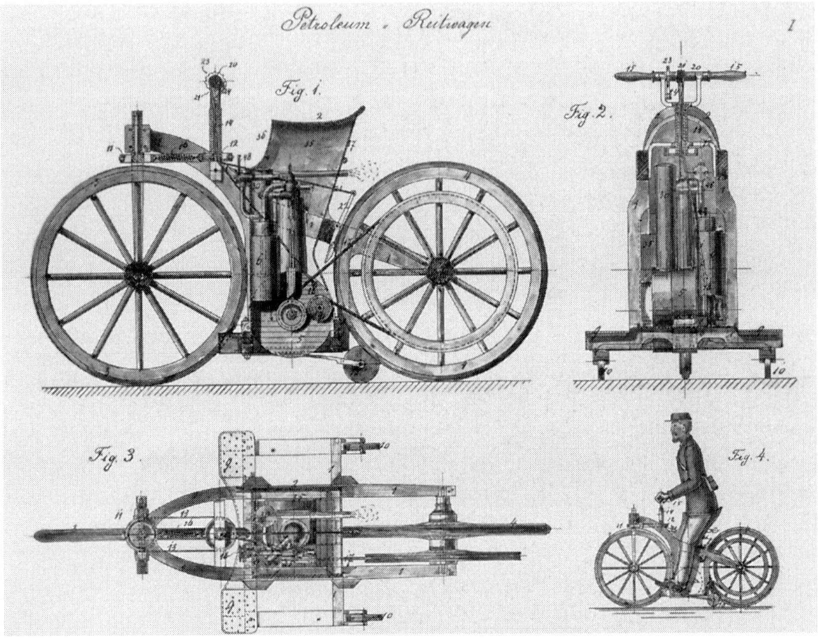

Fig. 1.1 Drawing of the Petroleum Reitwagen taken from the patent dated August 29, 1885. Technical data: air-cooled single-cylinder four-stroke engine, bore 58 mm, stroke 100 mm, swept volume 0.264 L, maximum output 0.5 hp, engine speed 600 rpm, weight 90 kg, vehicle speed 12 km/h.

The first recorded gasoline specification known to us originates from 1907. The US Navy specified a "high-grade refined gasoline, free from all impurities" with a Baumé density of 80°Be (about 0.673 g/mL). Motorists bought their gasoline at the pharmacy (the only place where it was available) and usually checked the density with a hydrometer.

When it was recognized that engines with spray-nozzle carburetors could also use heavier distillate gasolines, the fuel density was no longer the main specification parameter. It was discovered that the boiling characteristics of a gasoline was by far a better measure of its evaporation properties and hence new specifications were based on distillation properties.

Gasolines with a wider distillation range were also important from the supply point of view. With the rapidly increasing number of vehicles – in 1913 more than one million motor cars were registered worldwide – it became more and more difficult for the existing refineries to meet the demand for gasoline with the necessary quality (Figs. 1.4–1.7). In the USA, for example, the required amount of gasoline was higher than the available quantity of distillates (straight-run products even by 1912). It was therefore imperative to produce more gasoline suitable for motor

Fig. 1.2 The Daimler single-cylinder engine, patented in 1885.

vehicles from the available crude oils, independent of their natural composition. This was even more important for Europe than for the USA, who at that time had nearly unlimited crude oil resources.

To solve the supply problem, a new refinery technology, thermal cracking, known in principle since 1869, was used from 1913 onwards to increase the production of gasoline from a given amount of crude oil. This technology rapidly increased in significance. However, unfortunately, it produced not only suitable light gasoline components from heavy crude oil constituents but also unstable olefins, which led to oxidation problems (gum formation and deposits). Hence oxidation inhibitors became the first implied gasoline additives.

The spray-nozzle carburetor and gasolines of suitable volatility ensured more precise mixture preparation and solved the early starting and running problems. However, the desire to increase the power output for a given engine size required

Fig. 1.3 Daimlers "small, lightweight, high-speed engine" powering the first four-wheeled passenger car, "the motor carriage" from 1886; a boat (1886); a fair train, (1887); a truck, (1889).

Fig. 1.4 The "Gasoline Factory Rhenania" at Düsseldorf, Germany, in 1903. This was the first refinery of the Royal Dutch Petroleum Maatschappij, which merged with the Shell Transport and Trading Co. in 1907 to become the Royal Dutch/Shell group.

Fig. 1.5 Pump room and filling of cans and drums with gasoline at the "Gasoline Factory Rhenania" in 1905.

higher cylinder pressures, which could be effected, for example, by increasing the compression ratio. This led to a new problem called engine knock. Although the phenomenon was not understood for some time, the results of the pinking combustion noise were obvious: overheating of pistons and valves causing severe engine damage.

SIR HARRY RICARDO investigated the knock phenomena and by 1910 he had discovered that addition of benzene prevented knock and engine damage. In his

Fig. 1.6 The chemical laboratory of the Rhenania gasoline works near Hamburg, Germany, founded in 1910.

Fig. 1.7 1924 Shell began a "modern" service station network in Germany. A gasoline pump at the road side in Landau, in the upper Rhine area.

further studies he investigated the knock performance of gasolines with different compositions, their components and also that of specific hydrocarbons, alcohols and other oxygenates. In general he found that branched and cyclic hydrocarbons, and also straight-chain hydrocarbons with olefinic double-bonds, produced less knock and would allow higher engine compression ratios than the straight-chain *n*-paraffins, which were predominantly present in distillate straight-run gasoline. These results were another reason for the development of refinery conversion technologies.

As the engine knock problem could not be solved at that time by selection of gasoline components and refinery technology, research was directed towards anti-knock additives. In 1918 it was found that aniline had antiknock performance and by 1921 the General Motors researchers MIDGLEY and BOYD, after testing a large number of substances, found that tetraethyllead had a very strong antiknock effect. This highly toxic material and its derivatives were used for about half a century as gasoline additives. Rising environmental concerns and the introduction of exhaust gas catalyst after-treatment systems eventually required unleaded gasolines to be introduced.

In the following decades, new refinery processes to increase the gasoline output and to raise the antiknock performance (octane quality) were developed:

- *Catalytic cracking* (1935) had, compared with thermal cracking, the advantage of produce less unstable olefins.
- *Isomerization, alkylation* (1938) increased the yield of branched paraffinic hydrocarbons with better octane quality.
- *Cyclization* (with simultaneous dehydrogenation) (reformate 1939, platformate 1949) produced aromatics with high octane quality (modern catalytic reformers can minimize the benzene concentration).

With new refineries and the new upgrading processes it was possible to meet the worldwide increasing demand for gasoline in both quantity and quality. However, owing to political and other reasons there were local differences. Germany, for instance, had little access to crude oil, but a large coal production. During World War I, a gasoline consisting of 70 % benzene, produced as byproduct of coal coking, and 30 % ethanol from biomass was developed. In the 1920s blends of either gasoline or benzene with alcohols were used and also gasoline – benzene mixtures.

By order of the German government, from 1930 onwards gasolines had to contain 2.5% ethanol. Under pressure from farmers, this fraction was increased to 10% in 1932. However, as a result of the increasing motorization the necessary amounts of ethanol could not be supplied and therefore from 1936 onwards mixtures of ethanol – methanol were also used increasingly. It is interesting that even at that time, problems of corrosion of aluminum alloys and zinc-coated iron, caused by methanol, were being reported.

In spite of the significant substitution, a considerable amount of crude oil/gasoline had to be imported. During this period the first coal liquefaction plants were erected. The first unit working according to the Bergius – Pier coal hydrogenation method went on stream in 1936. The plant produced a gasoline with an octane quality (RON, Research Octane Number) of about 70. In comparison, the Fischer – Tropsch hydrocarbon synthesis supplied fuel with an RON of only 50 to 58. Therefore, it was still necessary to use alcohols and benzene to achieve an acceptable octane quality.

To control the gasoline quality problems in Germany at that time, in 1938 the motor industry was advised to limit the octane requirement of the engines to 74 RON. In May 1939 it was decided to add tetraethyllead to gasoline. The maximum concentration was 0.42 g/L and during World War II the minimum octane number of German gasoline was 74 RON.

At the end of the Second World War nearly all refineries and units for the gasoline production in Germany had been destroyed. An office for mineral oil, founded shortly before the war, continued to work after the war to organize the short-term supply of oil products. After this office was closed in 1951 and international oil companies had built new refineries according to modern and international standards, the gasoline quality in Germany improved and reached that of the remainder of the western world in about 1960. The requirements of motorists and the motor industry, together with strong competition between the oil companies, led to an ever increasing gasoline quality, which was achieved through new components and with gasoline additives. The aim was perfect engine operation under all climatic conditions, a long engine life, high power output and high fuel economy. Milestones in the additive development were, apart from alkyllead compounds and antioxidants, anti-icing additives to prevent carburetor icing, phosphorus-containing additives to reduce spark plug fouling brought about by highly leaded gasoline and various generations of detergents to keep carburetors and later the complete inlet system, including the fuel injection nozzles, free from deposits.

When it was recognized that the increasing use of leaded gasoline caused worldwide lead contamination of air, water and soil and that lead found its way into plants, animals and humans, it was decided to reduce the maximum lead concentration in gasoline stepwise and to compensate the loss in octane quality by the addition of refinery components, however, not by benzene this time! In Germany the maximum lead concentration was reduced to 0.4 g/L in 1972 and to 0.15 g/L in 1976. The introduction of unleaded regular grade (research octane number, RON 91) in Germany in 1985 allowed the motor industry to introduce vehicles with a catalytic system for treating exhaust gas. At that time unleaded regular grade was the most commonly used unleaded gasoline in the USA (the author remembers that the supplies to the first service stations started with blends prepared in the research laboratory). A year later the unleaded so called Euro premium (RON 95), a new grade, followed. The octane quality of 95, agreed between the oil and the motor industries, did not require investment in the refineries, because the composition was similar to that of leaded premium (RON 98), but of course without the lead. In 1989 the unleaded Super plus (RON 98) was introduced in small quantities. One of the components used as a lead replacement in this grade was methyl *tert*-butyl ether (MTBE). Vehicles without a catalyst could only use unleaded gasoline when their engines were equipped with sufficiently hard exhaust valve seats. The motor industry supplied the relevant information and the oil industry installed telephone hot lines to inform their customers accordingly.

The sale of leaded regular gasoline stopped in 1988 and of leaded premium in 1996. Long before that date the motor industry had equipped all new engines with hardened exhaust valve seats and the oil industry had developed additives (e.g., potassium based) to protect the exhaust valve seats of old engines so that leaded gasolines were no longer required.

For a long time the quality of gasoline was standardized locally by national standards. The political and commercial moves towards a united Europe also led to a

European quality standard for gasoline, EN 228. While most physical properties are identical all over Europe, allowance is made for different climates (northern and southern Europe, winter and summer).

With the introduction of unleaded gasoline a strong movement towards more environmentally acceptable gasolines began. Over the years the benzene content was reduced to below 1 %, sulfur, which reduces the efficiency of exhaust gas catalysts, was reduced to 150 mg/kg and further to 50 mg/kg or 10 mg/kg (sulfur-free) in 2005. Also the vapor pressure was reduced to minimize evaporative losses and upper limits of aromatics and olefins have been introduced to make the very small amounts of unburnt hydrocarbons, which remain in the exhaust gas after the catalytic converter, more acceptable to the environment. All these measures were and are taken in close conjunction with the motor industry and often joint test work is the basis for future specifications. Also, new fuel developments are often tested jointly with the motor industry to ensure that all important aspects, environmental acceptability, performance and economy are optimized in future fuels.

The latest development in Europe is the addition of bioethanol in row concentration to gasoline to reduce CO_2 emission. The consumption of gasoline in Europe is decreasing, partly because of more efficient engines but mainly because of an increasing number of diesel vehicles.

1.2
History of the Diesel Engine and of Diesel Fuel

For a long time the production and the processing of crude oil was dominated by the demand for gasoline, because the spark-ignited Otto engine was, apart from a few exceptions, the only engine available for mass motorization. When RUDOLF DIESEL experimented with his self-igniting engine at the Maschinenfabrik Augsburg Nürnberg (MAN) factory in Augsburg, Germany, nothing was known about the fuel requirements of his new engine and it seemed only natural that he tried gasoline as a fuel.

During the development of his engine, DIESEL's aim was an engine operating as closely as possible to the Carnot cycle, with high gas temperatures after compression, which gives the maximum theoretical efficiency. This concept required higher compression ratios than used in the spark-ignited Otto engine to enable self-ignition of the fuel towards the end of the compression stroke, which is followed by a controlled flame propagation during the expansion stroke, provided a suitable fuel is used.

The first ignition of DIESEL's engine using gasoline on 10th August 1893 was so vehement that his device to measure the cylinder pressure was destroyed. This proved two things: firstly, the principle of self-ignition worked, and, secondly, gasoline was unsuitable as a fuel for a DIESEL combustion process. Obviously the self-ignition properties of gasoline were inadequate, causing ignition delay, which was followed by unacceptably steep pressure rises.

Apart from finding a suitable fuel there was also the problem of precise high-pressure fuel injection into the combustion chamber. Adequate pumps and nozzles

Fig. 1.8 The first DIESEL engine from MAN, a single cylinder of 19.6 L capacity, weighing about 5 t. It operated on a mixture of gasoline and lamp oil with a hitherto unknown high efficiency of 26.2 %, producing 17.8 hp at less than 200 rpm.

were not available. RUDOLF DIESEL solved the problem by blowing the liquid fuel together with compressed air into the combustion chamber. This system worked reasonably well, but the air compressor reduced the engine efficiency and made the engine less suitable for use in vehicles.

Employing a mixture of gasoline and lamp oil and the air pressure injection system, DIESEL finally succeeded and he got his engine (Fig. 1.8) to run.

After 1897, experiments with other fuels started, which had little success at first. Later it was found that so-called middle distillates from crude oil, with a boiling range between 250 and 350°C, were ideal fuels for the DIESEL combustion process. These middle distillates were blended in low concentrations into town gas, which explains the name "gas oil" for this mineral oil fraction.

However, it took more than 25 years until the diesel engine found its way into vehicles, after the development of reliable high-pressure liquid-fuel injection systems. In 1923–1924 MAN and Daimler – Benz introduced commercial vehicles with diesel engines and the first passenger car, a Mercedes 260 D followed in 1936, powered by a 2.6-L 4 cylinder pre-chamber engine producing 45 hp at 3000 rpm.

The most important fuel property, which was required to guarantee safe starting and smooth running of the engines without any harsh noises and steep pressure rises, was adequate ignition quality to ensure a short ignition delay after fuel injection into the hot and compressed air in the combustion chamber. Ideal fuels are those which are unsuitable for the spark-ignited Otto engine, i.e., straight-chain paraffins with a boiling range above that of gasoline. This widened the availability of fuels from crude oil without the use of conversion processes. Furthermore, for a long time the typical diesel fuel was a straight-run distillate, being a comparatively cheap fuel. The low cost and abundant supply of the diesel fuel and the high efficiency of the diesel engine were three important factors in the use of diesel engines in commercial road transport vehicles. The disadvantages of the diesel engines at that time, i.e., noise, smoke, a limited engine speed range and less power output for a given engine size and weight, were tolerated in heavy-duty machinery and vehicles.

Another important fuel property was cleanliness: the fuel had to be free from foreign matter particles and to be stable at high temperatures to protect the injection system against wear and deposits, which caused rough running, smoke and power loss. Another important quality criterion was cold flow filterability and pumpability. Unfortunately the long-chain paraffins, with their excellent ignition properties, form wax crystals at low temperatures, which limit their use in winter periods.

During the Second World War, the development of diesel powered vehicles slowed down. It was too difficult to supply the troops with two different fuels. Because of the crude oil shortage in Germany, diesel fuel also had to be produced from alternative sources, namely coal. Fortunately the Fischer – Tropsch process supplied a component with an even higher ignition quality than straight-run diesel fuel from crude oil. This component was blended with low quality tar oil to obtain a diesel fuel of reasonable quality. The steep increase of motorization in Germany after the war changed the situation completely. The consumption of diesel fuel rose from about 700 000 t per annum in 1942 to 4.8×10^6 t per annum in 1960. The demand for diesel fuel and light heating oil (which has about the same boiling range) rose further and could no any longer be met by straight-run products. The refineries had to invest in cracking units for the production of middle distillates. While the crack components improved the cold-flow performance, it became more difficult to obtain an adequate ignition performance of the final diesel fuel blend.

Over the years there was a continuous struggle to improve the harsh combustion of diesel engines by optimizing the mixture preparation. Application of pre-chambers, swirl chambers, pintle nozzles, increased injection pressures, or spraying the fuel against the piston surface were some of the measures taken. Pre-chamber and swirl chamber engines, often in combination with turbocharging, finally reached such a developmental stage that diesel engines could compete with Otto engines and increasingly found their way into passenger cars. This meant a more widespread use of diesel fuel and although the engines behaved in a more "user-friendly fashion", for some time, diesel fuel kept its smelly and "sticky" appearance. This poor image was made even worse by the way diesel fuel was offered at the service stations, as it was intended for heavy-duty customers. The pumps were often hidden in a remote corner, the surroundings were scruffy and the supply nozzle sometimes too large for a passenger car. In some areas of the world, e.g., in the USA, the diesel fuel that was available was unsuitable for the new high-speed passenger car engines. Also, inadequate housekeeping at service stations caused problems. Water at the bottom of the tanks, attack of fungi and dirt in the fuel caused problems in the early diesel-powered passenger cars.

Technical requirements, environmental considerations and customers needs demanded distinct improvements. This was partly achieved by passenger car friendly fuel pumps at the service station forecourts. However, the product also had to be improved through refinery technology and also by the use of new additives.

One of the important aspects was the reduction of the chemically bound sulfur. In 1955 a maximum sulfur concentration of 1 % was specified, which was reduced, stepwise, to 0.2 % in 1988, 0.05 % (500 mg/kg) in 1993, 350 mg/kg in 2000 and 50 mg/kg or 10 mg/kg (sulfur-free) in 2005. At higher sulfur levels, sulfur dioxides get into the crankcase, via the crankcase ventilation system, resulting in depletion of the lubricant additives, causing engine wear. The reduction in sulfur dioxide emissions and lower particulate emissions require diesel fuels of very low sulfur content and this was the reason for the later reductions. Hydrotreatment of the diesel fuel components to reduce the sulfur content also reduced the smell. Unfortunately, the natural antiwear properties of sulfur were reduced along with lowering of the sulfur content and antiwear additives had to be developed and added to protect fuel pumps and nozzles.

The cetane number (i.e., the unit describing the ignition quality of a fuel), originally a minimum of 45–49 in 1993, was raised to 51 in 2000 to reduce ignition delay, improve combustion and reduce exhaust emissions. An important step in this direction was the introduction of electronically controlled injection allowing a stepwise high-pressure injection of the fuel into the combustion chamber. This breakthrough in technology made direct fuel injection sufficiently smooth and offered additional reductions of emissions so that the highly efficient direct-injection diesel engines were suitable for passenger cars, and they have since replaced the previously used swirl and pre-chamber engines. However, low emissions and smooth engine running can only be achieved with high-quality fuels and recent tests have shown that synthetic diesel fuels with ultra-high cetane numbers can reduce emissions further. This type of fuel is, of course, more costly than the straight-run distillate used in the past.

Diesel fuel additives have been mentioned, but their role is much wider and they are being used successfully in increasing varieties. They offer the oil companies possibilities to differentiate and to market fuels of higher performance than those of the competitors.

Being liquids, *cetane improvers* such as ethyl hexyl nitrates (EHN) are safe to handle and increase the ignition performance (working as mild and controlled explosives). This gives the refineries more flexibility, especially in winter periods when the cold-flow requirements limit the use of high-cetane paraffins. An important group of additives are *cold-flow improvers* and wax *antisettling additives.* Both hinder the paraffins in the formation of large wax crystals and guarantee sufficient low-temperature fuel pumpability and filterability. The two additives are often used in combination during winter periods. Together with vehicle fuel-system heating-systems they are responsible for safe vehicle operation without the fuel filter blocking in cold climates and allowed the oil industry to guarantee safe winter operability below −20 °C.

Another type of additive is *detergents*, which keep injector nozzles clean. They help to keep the exhaust emissions in the as-new condition for a long time.

Anticorrosion and *antiwear additives* (so called lubricity additives) protect not only the vehicle but also the fuel distribution system.

Reodorants, which were popular a few years ago, are generally no longer required because the low sulfur fuels have lost their bad smell; however, *antifoam additives*, which can also be regarded as fuel "cosmetics", remain important as they reduce foaming when vehicle tanks are refilled at service stations, preventing spillage and overfill.

Compared with the old days, diesel fuel is now a complex blend of hydrocarbon streams with a wider range of additives than in gasoline. The quality of diesel fuel is standardized Europe-wide in EN 590. Important performance aspects brought about by additives such as lubricity additives have been included and the standard also contains compositional changes required to ensure low exhaust emissions.

The further development of the diesel engine to an even more efficient and environmentally acceptable prime mover with complex mixture preparation systems, such as high-pressure common-rail injection, requires high-quality diesel fuels. New refinery technologies, synthetic fuels or components, new additives and to some extent fuel from biomass will help to solve further development tasks. Close cooperation between the oil and motor industries either bilaterally or industry-wide, such as the recent auto/oil programs in the USA and in Europe, ensures the necessary progress towards more environmentally acceptable diesel engines of the future.

To reduce carbon dioxide emissions it was decided to add low concentrations of fatty acid methyl esters produced from biomass as diesel fuel components.

The engine and fuel developmetns alone have led to an increasing proportion of diesel-powered passenger cars.

2
Oil Refining

Gunter Alfke, Walther W. Irion, Otto S. Neuwirth

2.1
Introduction

2.1.1
History

Crude oil or petroleum, and a limited number of oil derivatives such as asphalt, pitch, and naphtha have been known and sporadically used since medieval times in the Orient, Europe, and America. However, it was not until the 19th century that significant oil processing started.

The first commercial oil operations can be dated back to the 1850s, with production commencing in Russia, Romania, and the United States [1]. The first successfully purpose-drilled oil well in Pennsylvania in 1859 was the beginning of the modern oil industry, as it stimulated rapid development in the search, production, and refining of oil.

Subsequently, small and simple oil refineries were built and operated in the direct vicinity of the Pennsylvanian oil fields and on the east coast of the United States, where former coal distillation processes were modified for oil processing. These first distilling units worked in *batch operation* and produced a small number of oil fractions.

Batch distillation was later combined with a thermal cracking step in order to increase the distillate yield. In this crack distillation, the heavier part of the oil was heated further and cracked until coke remained as residue in the still. The lower sections of the vessels were equipped with fired heating tubes to improve heat transfer.

Important progress in oil processing was achieved in Russia with the *continuous crude distillation* process which was introduced in a refinery in Baku in 1875. The continuous principle worked with a battery of stills, arranged in series, with stepwise gravity overflow of the oil throughout the battery. The stills were heated up to successively increasing temperatures, so that each still could produce a distillate

Handbook of Fuels: Energy Sources for Transportation. Edited by Barbara Elvers

ISBN: 978-3-527-30740-1

fraction corresponding to its temperature level. This continuous process system offered the possibility for mass production at reduced manpower requirements, and utilization of the residual heat for preheating the crude oil feed.

During the first five decades of oil refining the predominant product was kerosene (lamp oil), and only small amounts of other products, such as fuel oil, pitch, and lubricating oil, were marketable. The lubricating oils were produced by vacuum distillation (United States, in the 1870s) and steam distillation (Russia, in the 1880s). The lightest and heaviest fractions of the crude oil distillation, namely gasoline and residue, were not used at that time and had to be burnt.

From the beginning of the 20th century, a complete shift in product demand occurred because of the increasing expansion of the use of electricity and motorization. The availability of electricity greatly diminished the demand for lamp oil, whereas the increase in motor vehicles increased the need for gasoline, which soon became the priority product.

The increasing gasoline demand initiated the development of *thermal cracking processes*. The Burton process, which came into operation ca. 1912, produced up to 30 % of cracked gasoline from paraffinic gas oil feedstocks.

During World War I the requirements for a variety of oil products, in particular gasoline and fuel oils, rose dramatically, and the demand continued to increase in the 1920s and 1930s. Gasoline again showed the highest growth rates. In the same period the bitumens (asphalts) became a desired and important product group, owing to the expansion of road construction.

By the mid-1920s a new generation of *improved thermal cracking processes* emerged, of which the Dubbs process gained supreme importance. The gas oil or residue feed was heated in a cracking furnace, and the cracking reactions were completed in reaction chambers in which coke was formed as the residual product. The yield of cracked gasoline could be as high as 50 %. However, with the advent of the *Houdry catalytic cracking process* during the second half of the 1930s, the thermal cracking capacities were largely replaced by catalytic cracking units, which took the lead in gasoline production in the next two decades.

The catalytic cracking process proved that the octane quality of cracked gasolines that contain unsaturated hydrocarbons was superior to that of distilled (straight-run) gasolines. Hence, the straight-run naphthas were increasingly processed in *thermal reformers* at high cracking temperatures, to add a higher portion of unsaturated hydrocarbons to the motor gasoline.

In the late 1930s the first *catalytic reforming processes* were developed with improved gasoline quality by simultaneous desulfurization and increase in octane number. These *hydroforming* processes used molybdenum-on-alumina catalysts in the presence of hydrogen at high temperature and pressure.

With the introduction of *platinum catalysts* in the late 1940s, the catalytic reforming process was greatly improved: high quality gasolines were produced in higher yields, and hydrogen was formed as a byproduct.

A desulfurization step (*hydrotreating*) was installed upstream of the reformer, using cobalt – molybdenum catalysts and hydrogen from the catalytic reforming process in order to remove sulfur from the gasoline feedstock. Sulfur would otherwise poison the noble metal catalyst in the reforming process. Owing to their quality and yield advantages, catalytic reforming units replaced the thermal reforming capacities in a relatively short period of time.

In the 1930s the United States and the Soviet Union were the main oil producers and supplied sufficient gasoline products for their own needs. The consumer countries in the rest of the world had to rely on the oil supplies from a number of other producing countries, chiefly from South and Central America (Venezuela, Mexico) and the Middle East (Iraq, Saudi Arabia).

The largest expansion of the oil industry took place within the three decades after World War II. Between 1950 and 1980, a five-fold increase in oil demand occurred worldwide. The formation of political and economic blocs in the "eastern" and "western" hemispheres and the foundation of the OPEC (Organization of the Petroleum Exporting Countries) in 1960 were significant for international oil trade and distribution.

Oil refining underwent a rapid development initially in the highly industrialized regions of North America, Western and Eastern Europe, Eastern Asia, and Australia, and later, in the wake of increasing populations and energy demands, also in many developing countries in Latin America, Africa, and Asia. Finally, large refinery plants were also built in the OPEC countries for the export of products at the expense of crude oil exports.

The two worldwide crude oil price increases in 1973–1974 and 1979–1980 prompted changes in oil consumption, which caused the oil industry to develop processes with increased yields and a reduction of refinery capacities, especially in North America and Western Europe. The refining structures became more complex with the installation of higher capacities of residue conversion processes, such as *hydrocracking*, which increased yields of transportation fuels while crude oil consumption could be reduced.

2.1.2
General Aspects of Oil Refining

Crude oils are complex mixtures of a multitude of — mostly hydrocarbon-type — constituents with a wide variety of properties, hence they cannot be directly used for industrial or commercial applications. Therefore, crude oils must be treated in a refinery by suitable processes to produce an assortment of products that can be marketed according to certain quality requirements.

These products are predominantly used as transportation fuels (gasolines, kerosene, diesel) and for heating purposes (LPG, fuel oils), but also as feedstocks for the petrochemical industry (naphtha, LPG) and special manufacturing routes (lubricating oils, greases, bitumen), as well as specialty products (solvents, wax, coke). All these products consist of hydrocarbon mixtures, and there are only a few materials that are produced as individual chemical compounds, e.g., propane, sulfur, or benzene – toluene – xylene (BTX) aromatics.

The crude oils are usually transported by tankers and/or pipelines to the refinery site, and stored in tanks before being processed. Refineries were preferably built in coastal regions near industrial consumer centers, but from the late 1950s on, they were also built in inland regions for area-oriented supplies. In the 1970s and 1980s large refinery complexes were also erected on the coasts of oil exporting regions.

The processing structure of a refinery depends on the market requirements for the products to be sold and on the available type(s) of crude oil. Both factors must be considered to avoid excessive production of unmarketable or otherwise undesirable product(s) and to determine the refinery integration. Typical integrated refinery structures are dealt with in Section 2.5.

Modern refineries are operated continuously and with a high degree of automation. The processing capacities are high compared with other chemical industries. A medium-sized refinery may have a crude capacity of 10 000 t per day [corresponding to ca. 73 000 barrels (bbl) per day]. The crude distillation determines the refinery capacity. A breakdown of worldwide refining capacities for 1997 and 2004 is given in Table 2.1, demonstrating the doubling of capacities for India and China,

Table 2.1 World oil refining capacities 1997 and 2004, capacity figures are given in 10^6 t/a (derived from [2])

	Refining capacity, 10^6 t/a	
Country/region	**1997**	**2004**
United States	795.3	838.7
Canada	92.6	100.9
Subtotal North America	**887.9**	**939.6**
Brazil	83.2	96.0
Mexico	76.0	84.2
Venezuela	58.9	64.1
Argentina	33.3	31.3
Cuba	15.1	15.1
Other Latin America	129.7	122.4
Subtotal Latin America	**396.2**	**413.1**
Italy	122.7	116.0
Germany	109.2	115.5
France	93.3	97.6
United Kingdom	91.3	91.3
Spain	64.7	63.6
Netherlands	59.4	61.4
Belgium	31.5	40.2
Romania	27.1	25.8
Sweden	21.4	21.7
Greece	19.6	20.1
Poland	18.3	17.5
Bulgaria	16.2	16.2
Norway	15.4	15.5
Portugal	15.2	15.2
Other Europe (excl. FSU)	104.4	129.6
Subtotal Europe	**809.7**	**847.2**
Russia	343.6	271.6
Ukraine	62.4	44.0
Belarus	27.2	24.7
Other FSU	75.6	60.6
Subtotal Former Soviet Union	**508.8**	**400.9**

Japan	248.4	235.3
China	148.4	232.5
South Korea	127.1	128.8
Saudi Arabia	82.6	87.3
Iran	67.9	73.7
Singapore and Malaysia	76.4	94.0
India	54.3	112.7
Indonesia	46.5	49.6
Kuwait	44.3	44.5
Taiwan	38.5	61.0
Turkey	35.4	35.7
Thailand	35.2	35.2
Iraq	17.4	29.9
Philippines	16.2	16.2
Other Asia (excl. FSU)	87.6	78.7
Subtotal Asia (excl. FSU)	**1126.2**	**1314.1**
Egypt	27.3	36.6
Algeria	24.2	22.5
South Africa	23.3	24.5
Nigeria	21.9	21.9
Libya	17.4	19.0
Other Africa	32.7	78.7
Subtotal Africa	**146.8**	162.3
Australia	38.1	78.7
Other AUS/Oceania	4.6	4.6
Subtotal AUS/Oceania	**42.7**	**42.3**
Total world	**3918.3**	**4119.4**

while all others remain rather constant. In this table the countries with oil refining capacities are allocated to seven world regions: North America, Latin America, Europe (excluding Former Soviet Union), Former Soviet Union (FSU), Asia (excluding FSU), and Australia/Oceania.

A selection of the following process units may be found in modern integrated refineries, depending on the process routes to be installed.

Process	**Purpose/Products**
Crude distillation	crude oil fractionation into gas, gasoline, kerosene, gas oils, and atmospheric residue
Crude desalter (as required)	crude oil desalination
Vacuum distillation	separation of atmospheric residue into vacuum gas oils and vacuum residue
Naphtha hydrotreater	gasoline desulfurization
Kerosene hydrotreater	kerosene desulfurization
Gas oil hydrodesulfurizer	diesel and light heating oil desulfurization
Catalytic reformer	conversion of desulfurized gasoline into high-octane gasoline, LPG, hydrogen, and fuel gas
Gas treating and separation	purification of LPG, fuel gas (e.g., removal of hydrogen sulfide)
Catalytic cracker	conversion of vacuum gas oils or residues into fuel gas, LPG, gasoline, and gas oils

(Continued)

Process	Purpose/Products
Visbreaker	thermal cracking of vacuum residue into fuel gas, gasoline, gas oil, and heavy fuel oil
Hydrocracker	conversion of vacuum gas oils or residues into fuel gas, gasoline, kerosene, and gas oils
Coker	thermal cracking of vacuum residue into fuel gas, gasoline, gas oil, and coke
Deasphalting	solvent extraction of vacuum residue for deasphalted oil and asphalt
Alkylation (C_3/C_4)	production of alkylate [a]
Polymerization (C_3/C_4)	production of polymerate [a]
Isomerization (C_5/C_6)	production of isomerate [a]
Aromatics plant	solvent extraction of high-octane reformate to produce BTX aromatics
Lubricating oil plant	extraction, dewaxing, and hydrofinishing of vacuum gas oils to produce lubricating oils
Bitumen blowing	oxidation of vacuum residue to bitumen grades
Sulfur recovery	conversion of hydrogen sulfide (from gas treating) into sulfur
Hydrogen plant	hydrogen manufacture
MTBE/ETBE/TAME-plant	production of these motor gasoline components

[a] Gasoline components

A photograph of a typical refining complex is shown in Fig. 2.1.

Fig. 2.1 Aerial photograph of a refining complex.

2.2 Crude Oils and Products

2.2.1 Crude Oil Constituents

Crude oils are composed of a large variety of *hydrocarbons*. Hydrocarbons (HC) with one to four carbon atoms (methane – butane) are gaseous at room temperature, hydrocarbons with more carbon atoms are liquid at room temperature (pentane, hexane, etc.). Hydrocarbons with more than 17 carbon atoms per molecule are solid at room temperature as pure compounds, but these are usually dissolved in the liquid portion of the crude oil.

Three types of molecular structures are possible for hydrocarbons, i.e., straight-chain, branched-chain, and ring structures. Three possible structures of hexane are given as an example:

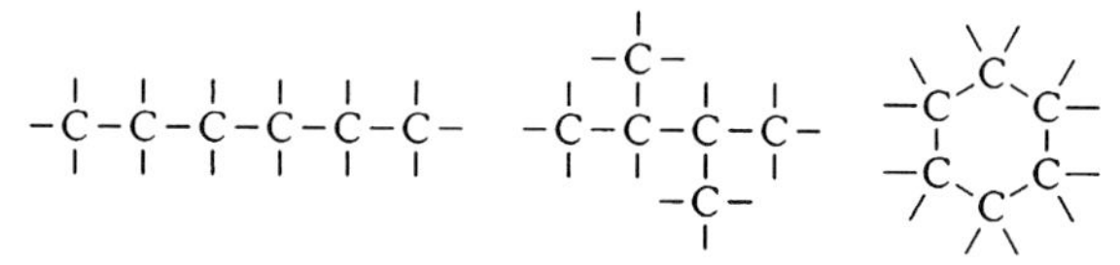

n-hexane (straight-chain structure) isohexane (branched-chain structure) cyclohexane (ring structure)

The higher and more complex hydrocarbon molecules consist of various combinations of these structures.

Saturated hydrocarbons are known as paraffins or alkanes (the IUPAC name). They can be further classified as normal, iso-, and cycloalkanes. Cycloalkanes are also known as naphthenes.

Unsaturated hydrocarbons (olefins or alkenes) are not normally contained in crude oils, but they are formed in various refining processes, particularly during cracking and dehydrogenation.

Crude oils also contain varying amounts of *aromatics*: simple aromatics such as benzene, condensed polynuclear aromatics, and aromatic ring systems with various paraffinic or olefinic side chains. Aromatics are also formed in various conversion processes.

There are also a number of heteroorganic compounds present in crude oils in relatively small concentrations. Examples are sulfur-, nitrogen-, and oxygen-containing organic compounds. Metal compounds, e.g., vanadium, iron, and nickel compounds occur in trace amounts.

Most of these components, i.e., their non-HC-elements, are undesirable because they either cause problems in crude oil processing or deteriorate the quality of the products. Therefore, they must be removed by suitable refining processes.

Sulfur compounds such as hydrogen sulfide, thiols (mercaptans), sulfides (thioethers), disulfides, polysulfides, and thiophenes, may cause corrosion of equipment, bad odor in products, and catalyst poisoning. Moreover, sulfur compounds in fuel products contribute to air pollution by formation of sulfur dioxide and trioxide when burnt, and thus have to be removed from oil products as far as possible. This is achieved by means of desulfurization processes and subsequent conversion of hydrogen sulfide into elemental sulfur as a marketable product.

Nitrogen compounds are more stable than sulfur compounds, and are only partially removed in the desulfurization processes. Specific denitrogenation may be desirable for catalytic cracking or hydrocracking feedstocks to avoid catalyst deterioration, and for fuel oils to reduce the formation of nitrogen oxides in burning.

Oxygen compounds may be present as naphthenic acids and phenols. *Naphthenic acids* (mostly mono- or dicyclic cycloalkane carboxylic acids, e.g., cyclopentane carboxylic acid) are constituents of naphthenic crude oils. Being corrosive, they have to be removed from crude oils or distillates by alkali extraction in the distillation process. The separated naphthenic acids are marketed as acidic extractants for metals; naphthenates are used as metallic soaps.

Phenols (which cause odor problems) are formed and recovered in the catalytic cracking process.

Metals act as catalyst poisons and can be removed by a demetallization step upstream of the catalytic residue conversion processes. In catalytic cracking metal passivation additives are applied where buildup of vanadium and nickel in the catalyst bed poses a problem.

2.2.2 Classification of Crude Oils

The quality of crude oils can be classified either according to hydrocarbon type, product fractions, or sulfur content. Although rather unspecific, these classification methods are useful for the selection of suitable crude oils for given refining structures and product demand patterns.

Classification According to Hydrocarbon Types Paraffin-based crude oils consist predominantly of paraffinic hydrocarbons. The low and medium molecular mass paraffins are suitable for all types of catalytic conversion into gasolines and middle distillates, and for chemical feedstocks. The high molecular mass paraffins are usually good raw materials for the manufacture of lubricating oils and paraffin waxes.

Naphthene-based crude oils contain mainly naphthenic, aromatic, and asphaltic hydrocarbons. The low and medium molecular mass ranges can be used for the production of valuable gasoline components, solvents, and feedstocks for aromatics manufacture. The high molecular mass ranges find use in special types of lubricants and in the production of bitumen.

Mixed-based crude oils have a more or less even distribution of paraffinic and naphthenic/aromatic hydrocarbons over all ranges of molecular mass. They constitute the largest class of crudes.

Classification According to Product Fractions This characterization commonly distinguishes between “light”, “medium”, and “heavy” crudes, i.e., the classification is based on the proportions of distillates and residue. This quality distinction takes into account the wide compositional varieties of crude oil fractionation. The percentages of the main fractions of some well-known crudes are listed in Table 2.2.

The *light crudes*, which are rich in gasoline and gas oil are, for example, represented by Norwegian and Nigerian crudes with correspondingly low residue contents. The *heavy crudes*, such as Venezuelan or Alaskan crude oils, contain relatively small amounts of distillates but high portions of residue, thus requiring higher conversion capacities than lighter crudes. A *middle group*, e.g., Saudi Arabian or British crude, has an advantageous (i.e., well-balanced) distribution of fractions.

Table 2.2 Yields of distillation fractions of selected crude oils

Type of crude, origin	Gas, LPG, vol %	Gasoline, vol %	Kerosene, vol %	Gas oil, vol %	Atmospheric residue, vol %
Nigerian light, Nigeria	3	26	14.5	27.5	29
Ekofisk, Norway	3.5	31	13.5	21.5	30.5
Forties, United Kingdom	4	22.5	12	22	39.5
Arabian light, Saudi Arabia	1.5	20.5	12	21	45
Alaska, United States	0.5	13.5	11.5	21.5	53
Tia Juana, Venezuela	—	1.5	3.5	14.5	80.5

Table 2.3 Sulfur contents of selected crude oils and their atmospheric residues

Type of crude, origin	Sulfur content of crude, wt %	Sulfur content of atmospheric residue, wt %
Ekofisk, Norway	0.1	0.3
Forties, United Kingdom	0.3	0.6
Es-Sider, Libya	0.5	0.9
Alaska, United States	1.0	1.5
West Siberia, Russia	1.5	2.6
Isthmus, Mexico	1.6	3.0
Kuwait, Kuwait	2.5	3.8
Arabian light, Saudi Arabia	3.0	4.5

Classification based on percentages of product fractions may assist in the crude selection for a refinery according to product priorities, taking into consideration the refinery conversion structure.

The lighter crude types are usually more expensive than those with high residual portions.

Classification According to Sulfur Content An important crude property is the sulfur concentration, which determines the processing complexity of a refinery to a considerable extent. The higher the sulfur content in the crude and its various fractions, the greater the need for adequate desulfurization and sulfur recovery capacities. Table 2.3 lists the sulfur contents of a number of crudes and their residues.

The sulfur content is an important factor worldwide in the pricing of crude oil; the group of low-sulfur crudes with <1 wt % of sulfur constitute the most desirable but generally most expensive qualities. Crude oils with 1.0–1.5 wt % of sulfur are classified as medium-sulfur, and those having above 1.5 wt % as high-sulfur crudes.

2.2.3 Oil Products

In general ca. 80–90 % of oil products are used as transportation and combustion fuels. The most important products, their applications, and some typical specifications are briefly described below.

2.2.3.1 Gas Fuels

Refinery gas consists of gas streams that contain mostly methane and ethane, which are burnt as fuel gas in refinery furnaces.

Liquefied petroleum gas (LPG, Chap. 3, Section 2.3.8.1) consists of propane, butanes, and mixtures thereof. It is used for heating purposes and special products. In some countries, e.g., Japan, the United States, and The Netherlands, LPG is also used to a considerable extent as a motor fuel. LPG specifications are based on national standards. The specified characteristics mainly concern vapor pressure, density, and contents of impurities (e.g., elemental sulfur, COS sulfur, hydrogen sulfide, vaporization residue, hydrogen, nitrogen, oxygen, and methane).

2.2.3.2 Liquid Fuels

Motor gasolines are blended products with a boiling range between 40 and 200 °C and are used for motor car engines (spark-ignition motors) (Chap. 3, Automotive Fuels). Typical gasoline components are light and heavy cracker and reformer gasolines, alkylates, isomerates, polymerates, pyrolysis gasoline, and LPG. Alcohols and ethers are added as nonhydrocarbon products. Gasolines have to comply with a large variety of specifications, among which the knock rating (octane number), volatility, boiling characteristics, density, oxidation stability, and lead contents are of prime importance. Typical specifications for unleaded gasoline (regular and premium grades, EN 228 from 2004, corrected 2006) are:

Unleaded gasoline	
Knock ratings	
RON	min. 95.0
MON	min. 85.0
Density (at 15 °C)	720 kg/m^3
max. 775 kg/m^3	
Vapor pressure	nationally specified according to applicable volatility classes (table in EN 228)
Boiling characteristics	nationally specified according to applicable volatility classes (table in EN 228)
Lead content	max. 0.005 g/L
Benzene content	max. 1.00 vol %
Olefins	max. 18 vol %
Aromatics	max. 35 vol %
Sulfur content	max. 50 mg/kg or 10 mg/kg
Oxygen content	max. 2.7 mass %
Oxidation stability	min. 360 min

The maximum addition rates of oxygen-containing gasoline components (*oxygenates*, e.g., methanol, ethanol, isopropanol, 2-methylpropanol, and ethers) inside the European Union are specified by EU Directive 85/536/EEC and have also been included in EN 228 (see below):

Methanol	max. 3 vol %
Ethanol	max. 5 vol %
Isopropanol	max. 10 vol %
tert-Butanol	max. 7 vol %
Isobutanol	max. 10 vol %
Ethers containing five or more carbon atoms per molecule	max. 15 vol %
Other oxygen-containing components	max. 10 vol %

In addition to this European standard for gasoline EN 228 (which is termed "Premium" or "Super" grade at the pump) member states are allowed to market a "Premium Plus" or "Super Plus" grade with octane ratings of RON/MON 98/88 and a regular grade with knock ratings of 92.5/81 (in Germany), respectively. These grades are included in national appendices to the EN 228.

While the specification of quality characteristics and specification data originates from market requirements, an increasing political influence has demanded severe changes to transport fuel characteristics for environmental reasons. The European motor gasoline standard EN 228 is based to a large extent on the European Directive 98/70/EC of October 13, 1998 and its amendment 2003/17/EC. By influencing product quality policy makers are responsible for a large part of the post-1970 refinery changes, which will be discussed.

Kerosenes are tailor-made products in the boiling range between 150 and 250 °C.

Lamp kerosene (lighting oil, cooking oil) has the longest tradition of oil products and is still widely used in most areas of the world. Smoke point, flash point, and volatility are important specifications.

Jet fuels, or aviation turbine fuels (Chap. 8), which are used worldwide in commercial aviation, have to undergo stringent quality screening. Control of density, freezing characteristics, viscosity, thermal stability, conductivity, corrosion behavior, water content, aromatics, and sulfur content is essential. A selection of Jet A-1 specifications is given below (based on relevant ASTM methods) (see also Chap. 8, Section 8.2.).

Density (at 15 °C)	775–840 kg/m^3
Distillation	
10 vol % recovered at	max. 205 °C
End point	max. 300 °C
Residue/loss	max. 1.5 vol % each
Flash point	min. 38 °C
Acid number	max. 0.015 mg KOH/g
Aromatics content	max. 22.0 vol %

Mercaptan sulfur content	max. 0.003 mass %
Sulfur content	max. 0.30 mass %
Freezing point	max. −47 °C
Viscosity (at −20 °C)	max. 8.0 mm^2/s
Specific energy	min. 42.8 MJ/kg
Smoke point	min. 25 mm
Copper corrosion (2 h /100 °C)	max. 1
Thermal stability (at 260 °C)	
Tube deposit rating	< 3
Filter pressure drop	max. 25.0 mm Hg
Electrical conductivity	50–450 pS/m

Further specifications are given for contaminants and additives. Additional domestic specifications have to be observed in various countries.

Gas oils are light and heavy gas oil fractions, and blends thereof, from straight-run and cracked origins, in a boiling range between 200 and 350 °C. The gas oils are predominantly used as automotive diesel fuels (Chap. 3) and as domestic heating fuels. Typical specifications for automotive diesel fuel and domestic fuel oil are listed below.

Automotive diesel fuel (specifications for moderate climate according to EN 590 from 2004, corrected 2006):

Density (at 15 °C)	min. 820 kg/m^3 max. 845 kg/m^3
Kinematic viscosity (at 40 °C)	min. 2.0 mm^2/s max. 4.5 mm^2/s
Distillation	
Recovery at 250 °C	max. 65 vol %
Recovery at 350 °C	min. 85 vol %
Recovery at 360 °C	min. 95 vol %
Cetane number	min. 51
Cetane index	min. 46
Flash point	min. 55 °C
Cold filtration test (CFPP)	classified between +5 °C and −20 °C for regional specifications and interseasonal qualities
Polycyclic aromatic hydrocarbons	max. 11 mass %
Sulfur content	max. 50 mg/kg or 10 mg/kg
Water content	max. 200 mg/kg
Ash content	max. 0.01 mass %
Carbon residue	max. 0.30 mass %
Oxidation stability	max. 25 g/m^3
Lubricity	max. 460 μm
Fatty acid methyl ester (FAME)	max. 5 vol%

Domestic fuel oil (specifications for extra light heating oil according to German standard DIN 51603–1 from 2003 as an example):

Density (at 15 °C)	max. 860 kg/m^3
Kinematic viscosity (at 20 °C)	max. 6.00 mm^2/s
Distillation	

Recovery at 250 °C	max. 65 vol %
Recovery at 350 °C	min. 85 vol %
Flash point	min. 55 °C
Cold properties depending on filterability and cloud point	
Calorific value (net)	min. 42.6 MJ/kg
Carbon residue	max. 0.3 mass %
Sulfur content for the standard grade	max. 0.20 mass %
Sulfur content for the low sulfur grade	max. 50 mg/kg
Water content	max. 200 mg/kg
Ash content	max. 0.01 mass %

Heavy fuel oils consist of various grades of mixtures of residual oils from distilling and conversion processes. High density, viscosity, and to some extent sulfur content of very heavy oils can be corrected by blending with gas oils. A low-sulfur quality with max. 1.0 wt % of sulfur has gained a considerable market share. Marine fuels (bunker fuels), power stations, and industrial furnaces are major outlets for heavy fuel oils. Specifications of various grades of marine distillate fuels and marine residual fuels are given in ISO 8217.

Important characteristics are density, kinematic viscosity, flash point, pour point (mainly defined for summer and winter qualities), carbon residue, ash, water, and sulfur content, and sediments.

2.2.3.3 Nonfuel Applications

Nonfuel applications are as follows:

1. LPG and light naphtha fractions are used as feedstocks for the petrochemical industry.
2. Special boiling-range naphthas are produced for various grades of solvents used in the chemical and pharmaceutical industry.
3. Paraffin waxes are used in the paper, chemical, and textile industries, and in medicine.
4. Heavy distillates are employed for the production of lubricants and greases.
5. Heavy residues are raw materials for the manufacture of road bitumens and asphalts.
6. Petroleum coke and high-grade coke are used for various industrial applications and electrode manufacture, respectively.

2.3 Oil Refining Processes

2.3.1 Crude Oil Distillation

The fractional atmospheric distillation of crude oil is performed in continuously operated fractionation columns. The necessary heat for the fractionation is introduced

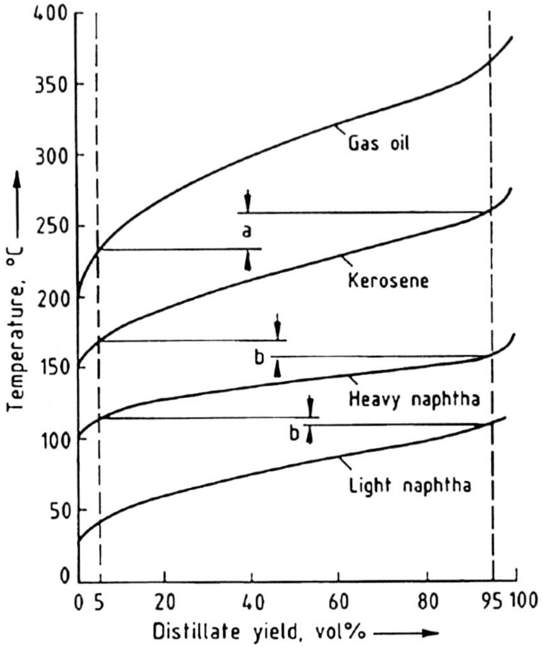

Fig. 2.2 Temperature – yield diagram of ASTM boiling ranges of crude oil fractions a) Boiling overlap; b) Boiling gap.

through fired tube furnaces. A great variety of trays or decks are applied as column interiors. Examples are grid trays, bubble cap trays, box trays, and valve trays which have been developed with the aim of improving the stage-to-stage vapor – liquid separation in the columns. Special types of trays are designed for distributing the incoming product and for collecting the draw-off product.

Figure 2.2 shows the temperature – yield diagram for the boiling curves of four fractions of a crude fractionation, i.e., light and heavy gasoline (naphtha), kerosene, and gas oil. Their boiling ranges between the 5 and 95 vol % yield points are ca. 45–105 °C, 110–155 °C, 165–260 °C, and 240–360 °C, respectively. In this example there is obviously a boiling "overlap" between kerosene and gas oil; boiling "gaps" exist between heavy gasoline and kerosene and between the two gasoline fractions. The overlap reveals a fairly inferior fractionation, whereas the gaps indicate good fractionation in the respective column sections.

Hydrocarbon fractions boiling above 360 °C will not be recovered by atmospheric distillation because cracking of the heavier products starts above this temperature. Therefore, the atmospheric residue must be subjected to vacuum distillation. Under the reduced pressure in vacuum columns, the higher-boiling oil fractions can be vaporized under similar temperature conditions as in the gas oil section of atmospheric fractionation. The final boiling point of a full-range vacuum gas oil (feedstock for catalytic cracker) would correspond to ca. 550 °C at atmospheric conditions.

2.3.1.1 **Atmospheric Distillation**

Atmospheric distillation is the first step in crude oil processing. The crude oil is separated into fractions of different boiling ranges. The distillation columns are usually operated continuously. Some major properties of the final products are established by the fractionation, for example, the boiling ranges of automotive fuels, the flash points and viscosities of lubricating oils, or the penetration (viscosity) of heavy fuel oil.

Modern fractionation columns are designed to yield one or several side streams in addition to a top and a bottom fraction. Figure 2.3 shows a typical atmospheric crude oil distillation unit, combined with a crude desalter and vacuum distillation. The crude oil is pumped from storage through the desalter (see Section 2.3.1.3) and is preheated in a series of heat exchangers (see Fig. 2.4) to ca. 220–250 °C before entering the furnace (crude heater). Here it is heated up to the required

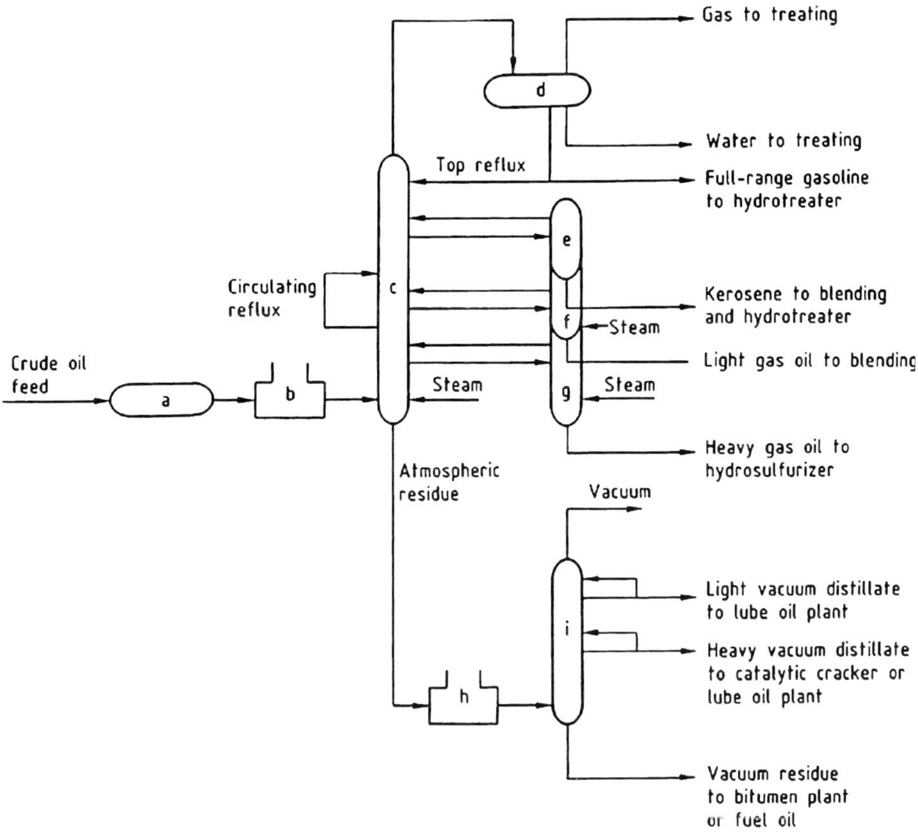

Fig. 2.3 Schematic representation of crude oil distillation including crude desalting and vacuum flashing a) Crude desalter; b) Crude heater; c) Main fractionator; d) Overhead accumulator; e) Kerosene stripper; f) Light gas oil stripper; g) Heavy gas oil stripper; h) Vacuum heater; i) Vacuum flasher.

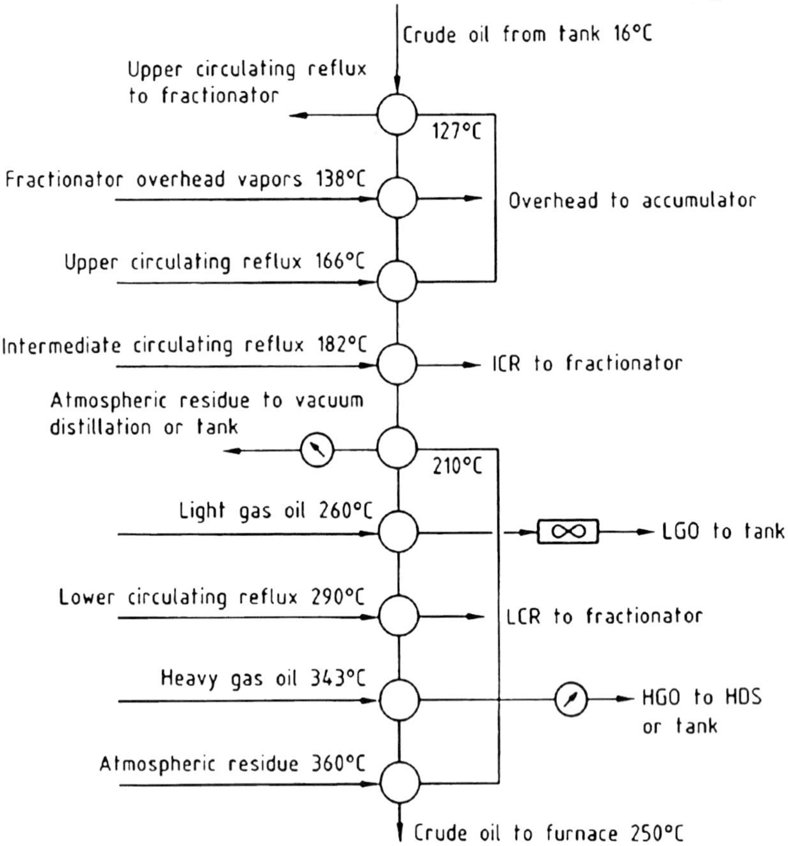

Fig. 2.4 Crude preheat train of a crude oil distilling unit.

column inlet temperature which is in the range of 360–380 °C. In the main fractionator the partially vaporized crude is separated into a top product, three side products, and a bottom product.

Top Product The column top vapors are condensed and separated into gas and full-range gasoline in the overhead accumulator. Part of the *liquid gasoline* is pumped back to the column as top reflux; the remainder is fed to the naphtha hydrotreater (see Section 2.3.3.1). The *gas stream* (uncondensable $C_1 - C_4$ hydrocarbons and H_2S) from the overhead accumulator is processed further in a gas treatment plant.

Process water originating from the stripping steam, which is applied to the fractionator and stripper columns, is withdrawn from the sump of the overhead accumulator and treated further in a wastewater treatment plant.

Side Streams The side streams comprising the middle distillates, i.e., kerosene, light and heavy gas oil, are transferred to their respective stripper columns where the lighter constituents are stripped off by injection of superheated steam.

The *kerosene stream* from the stripper column is pumped off either to a hydrotreater for the production of jet fuel, or via a heat exchanger and cooler directly to storage if it is to be used as a blending component for diesel fuel.

The *light gas oil* fraction is also either transferred to a desulfurizing unit or pumped to storage via a heat exchanger and cooler and is used for diesel and light heating oil blending.

The *heavy gas oil* fraction is fed to the hydrodesulfurization unit when it is used for the production of diesel fuel and light heating oil, or, if the sulfur content should be low enough, it is (partially) pumped via a heat exchanger and cooler to storage for direct blending.

One or more liquid side streams of the main fractionator column can be used as circulating reflux (CR). These circulation streams deliver heat to the crude oil via heat exchangers and contribute to the fractionation sharpness by returning cooled product back to the column.

Bottom Product The bottom product, namely atmospheric residue, can be supplied to a vacuum fractionator for the production of lubricating oil fractions (see Fig. 2.3) or to a vacuum flasher for the preparation of catalytic cracker feedstock (see Section 2.3.1.2.). Alternatively, the residue can serve as feedstock for a variety of other cracking processes, e.g., thermal and catalytic cracking, hydrocracking, or hydroconversion.

2.3.1.2 Vacuum Distillation

Generally two versions of vacuum distillation are applied — the vacuum flasher and the vacuum fractionator.

The *vacuum flasher* has a relatively simple design with a limited number of trays, because it serves only for the separation of atmospheric residue into one or two wide-boiling vacuum gas oils or waxy distillates which are used as catalytic cracker (CC) feedstock and a vacuum residue for bitumen manufacture or visbreaker feed. No specific separation severity is required for preparing the CC feed. Figure 2.5 shows a typical vacuum flasher. The vacuum is produced by a set of steam ejectors as shown in Fig. 2.6.

The *vacuum fractionator* is essential for the preparation of lubricating oil fractions. The vacuum fractionator separates the vacuum gas oil portion of the

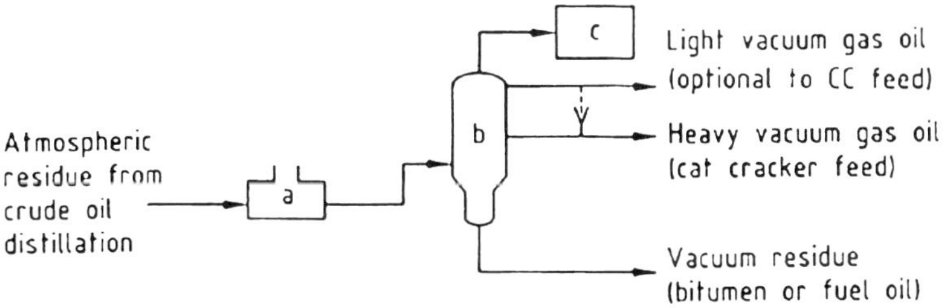

Fig. 2.5 Vacuum flasher for catalytic cracker feedstock a) Residue heater; b) Vacuum flasher; c) Vacuum system.

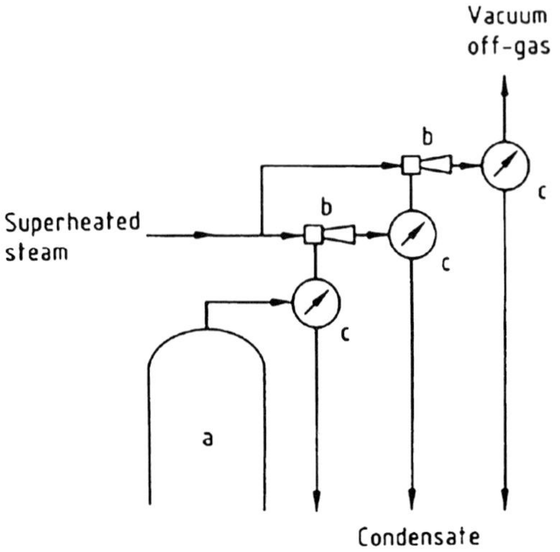

Fig. 2.6 Vacuum ejector system a) Vacuum column; b) Steam ejectors; c) Condensers.

atmospheric residue into light and heavy spindle oils, light and heavy machine oils, and cylinder oil. To achieve this high degree of fractionation the vacuum distilling section is often split into two fractionator columns. Care has to be taken to avoid carryover of residue components (asphaltenes) into the side streams.

2.3.1.3 Crude Oil Desalting

Although crude oils delivered from production fields are desalted, a small portion of water and salt always remains in the crudes that reach the refineries. Salt contents (sodium, calcium, and magnesium chlorides and sulfates) can have detrimental and costly effects on oil processing equipment. Examples are: corrosion in the fractionator overhead system due to the formation of hydrochloric acid; erosion or sedimentation of salt in lines, furnace tubes, pumps, and valves; and impaired heat transfer due to salt deposits in heat exchangers. It is therefore advisable to install a desalting unit in the crude feed line to the distillation unit.

The most common desalting unit is the *electrical desalter.* An electrical field is applied in the desalting vessel, which breaks up saltwater – oil emulsions by coalescence of the water droplets and segregation of the water from the oil phase. A typical electrical desalter is shown in Fig. 2.7.

2.3.2 Catalytic Cracking (Cat Cracking)

The commissioning of the first catalytic cracking plant in 1936 (Houdry process, United States) initiated a new era in petroleum refining. Until then, thermal cracking of distillates had been the governing conversion process, the role of which was very early on taken over by catalytic cracking. This development was particularly

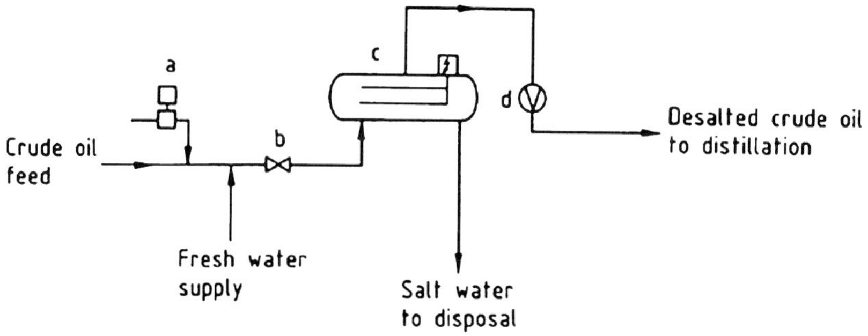

Fig. 2.7 Electrical crude oil desalter a) Chemical injection; b) Mixing valve; c) Crude desalter; d) Crude booster pump.

influenced by the increased demand for high-quality transportation fuels during World War II.

Four major advantages of catalytic over thermal cracking supported this development:

1. Lower production of C_1 and C_2 hydrocarbons to the credit of higher yields of C_3 and C_4 hydrocarbons (Fig. 2.8), which can be used for the manufacture of isoparaffins for motor gasoline via alkylation and polymerization processes.
2. Lower cracking temperatures, which result in a reduced production of diolefins, thus improving the oxidation stability of the gasoline fractions.
3. Higher octane ratings of gasoline by increased formation of branched alkanes, naphthenes (cycloalkanes), and aromatics.
4. Higher cracking reaction rates, allowing smaller dimensioning of plant equipment.

The conversion reactions of oil distillates in the catalytic cracking process occur mainly in the vapor phase at elevated temperatures in the presence of a cracking

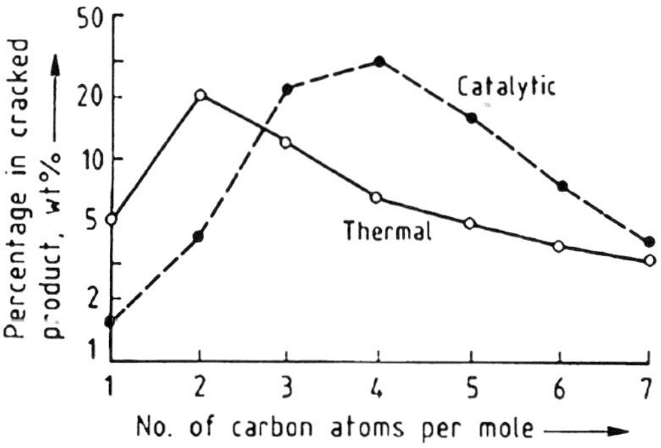

Fig. 2.8 Product distribution of C_1–C_7 hydrocarbons from thermally and catalytically cracked C_{16} key component [3].

catalyst (originally of the montmorillonite type, currently synthetic crystalline zeolites). The cracking reactions occurring at the active sites of the catalysts proceed via a carbenium ion mechanism, which predominantly affects the formation of olefins, isomeric components, and aromatics (the last via intermediate formation of cycloolefins). The formation of low-boiling olefins, branched alkanes, and aromatics favors the production of gasolines with high octane levels.

A considerable portion of the cracker feedstock is converted into gaseous components that are separated in a gas plant into fuel gas (containing hydrogen, methane, ethane, ethylene, and hydrogen sulfide) and LPG fractions, i.e., propane – propene and butane – butene. After treatment, these LPG streams are predominantly processed in alkylation and polymerization units for the production of high-octane gasoline components (see Sections 2.3.7.1. and 2.3.7.2.). Part of the propane and butanes is also used for LPG sales or as chemical feedstock.

The *cracked gasoline* (research octane number >90) must be treated to remove undesirable impurities, such as hydrogen sulfide, mercaptans, phenols, and nitrogen compounds, because they are responsible for unacceptable odor and corrosion.

The *middle distillates,* i.e., light and heavy gas oils, serve as blending stocks for light heating and diesel oils and heavy fuel oil, respectively. The cat cracked gas oils are traditionally referred to as light and heavy "cycle oils" as they have been widely used as recycled products to maximize gasoline production in the process.

An important byproduct of catalytic cracking is coke which originates from heavy carbonaceous material. Coke forms deposits on the catalyst particles and thus impairs catalyst activity. Therefore, the process must include a regeneration step where the carbon is burnt off the catalyst for reactivation before entering the reaction step.

Distillate fractions from crude oil have been used for about four decades as catalytic cracking feedstocks – chiefly *heavy gas oils* (vacuum gas oils) and to some extent *deasphalted oils.* Residual materials could not be processed because of their heavy metals content (nickel and vanadium) as this caused irreversible poisoning of the cracking catalysts. In addition, the metals and the asphaltene contents of the residues led to increased coke deposition on the catalysts. However, the introduction of metal-resistant catalysts and installation of feed pretreatment facilities since the mid-1970s have also made atmospheric residues accessible for catalytic cracking processes (see Section 2.3.6.4.2).

The development of catalytic cracking units gave rise to many improvements to the process within a relatively short time span. The first Houdry design was a fixed-bed process that required cyclic operation (switching of a series of vessels between the reaction and regeneration steps). A new continuous type of process, the Thermofor Catalytic Cracking (TCC), then emerged. This version of the process used a furnace (Thermofor kiln) as regenerator for coke burning, and mechanical catalyst transport by elevators between reactor and furnace and vice versa. This moving-bed system was later replaced by gas-flow, and even later by gas-lift transport.

The introduction of powdered catalyst types made way for the development of fluid-bed catalytic cracking (FCC) in 1942 (United States). Currently, FCC represents the most commonly applied catalytic cracking process.

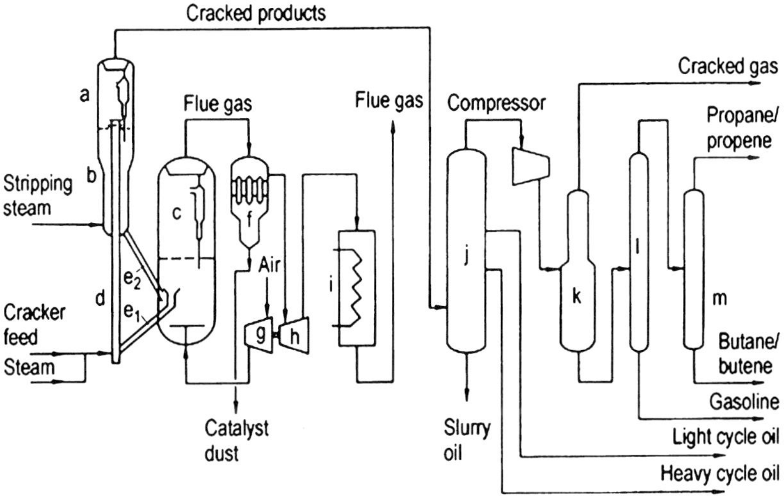

Fig. 2.9 Typical FCC unit a) Reactor; b) Stripper; c) Regenerator; d) Riser; e_1) Regenerator standpipe; e_2) Stripper standpipe; f) Cyclone vessel; g) Air blower; h) Flue gas expander; i) Waste-heat boiler; j) Fractionator; k) Absorber; l) Debutanizer; m) Depropanizer.

Significant improvements in the quality of the cracking catalyst were also achieved. The change from the natural clay materials (montmorillonite), which were originally used, to synthetic amorphous silica – alumina catalysts in the 1940s and the introduction of synthetic crystalline zeolites in the 1960s produced remarkable gains in activity, stability, and selectivity of the catalysts. The addition of highly active zeolitic material to the silica – alumina catalysts resulted in a strong selectivity gain, i.e., increased yields of gasoline and less gas production.

Process Description The flow diagram of a typical FCC process is shown in Fig. 2.9. Hot catalytic cracker feed and dispersion steam enter the process at the riser foot (d) where they are mixed with the hot reactivated catalyst flowing down the regenerator standpipe (e_1). The feed is vaporized by the hot catalyst, and the mixture is lifted up the riser into the reactor (a), where the catalyst is segregated from the vapors. The cracking reactions usually start in the riser and are completed in the reactor. The vapors (cracked products) pass through cyclone separators to eliminate any remaining catalyst particles before they enter the fractionator section.

The catalyst in the reactor bed, now loaded with coke, drops into the stripper (b) where any remaining hydrocarbons are stripped off by the addition of steam. From there the catalyst flows down via the stripper standpipe (e_2) into the regenerator bed (c), where the coke is burnt off by blowing combustion air into the vessel. The regenerated catalyst flows down through the regenerator standpipe to the riser foot. The flue gas leaves the regenerator via cyclone separators (f) to remove catalyst fines to avoid erosion of the internals of the downstream expander by catalyst particles. The gas then enters the heat recovery section (waste-heat boiler i).

The air blower (g) is driven by an expansion turbine (h), which utilizes the pressure difference of the flue gas. The heat of the flue gas is utilized in a waste-heat boiler for steam generation.

The fractionation and gas separation section consists of four columns – the fractionator (j), absorber (k), debutanizer (l), and depropanizer (m). The cracked products from the reactor are separated in these columns into fuel gas (C_2 and lighter), propane – propene, butane – butene, gasoline, light and heavy cycle oils (middle distillates), and slurry oil. The slurry oil contains some catalyst dust (carried over with the reactor product) and is normally recycled to the process.

The cracking sections of four different catalytic cracking processes are depicted in Figs. 2.10–2.13 in more detail. The downstream fractionation and gas separation sections are not shown.

Fluid-Bed Catalytic Cracking Unit (Fig. 2.10) [4–7] The partition of the regenerator vessel into a combustor and a disengaging section provides particularly efficient air – catalyst contact in the combustor section and low catalyst loadings in the regenerator cyclones.

Riser Reactor FCC (Fig. 2.11) [5–9] The cracking reactions proceed to completion in the riser, whereas the reactor serves as a catalyst – product disengaging vessel. Special flexibility is given by the two feed inlets at different elevations on the riser, and by the possibility of rapid conversion changes between maximum gasoline and maximum gas oil operation.

Figures 2.12 and 2.13 show the *riser-bed* and *transfer line flexicrackers*, respectively [6–8]. The transfer line configuration is applied for operations that require a short reaction time. The cracking reactions are interrupted at the top of the transfer line where fast separation of catalyst and cracked products occurs.

2.3.3 Hydrotreating

Until the 1940s there was little incentive for the petroleum industry to improve product quality by means of hydrogen treatment. Moreover, hydrogen was not available in refineries before the introduction of catalytic reforming (see Section 2.3.4.). Since the early 1950s, however, several influences have triggered the development of various hydrotreating processes. The increased production of high-sulfur crude oils and consequently the need to remove sulfur compounds from oil fractions, more stringent product specifications because of environmental requirements, and the production of increased quantities of cracked material from conversion processes boosted the installation of adequate hydrotreating capacities in refineries. Hydrotreating plants represent the highest capacities of all secondary refining processes worldwide.

The most important function of hydrotreating is the removal of sulfur components (hydrodesulfurization, HDS) by reaction with hydrogen in the presence of a suitable catalyst, to form hydrogen sulfide. Hydrogen sulfide is removed from the

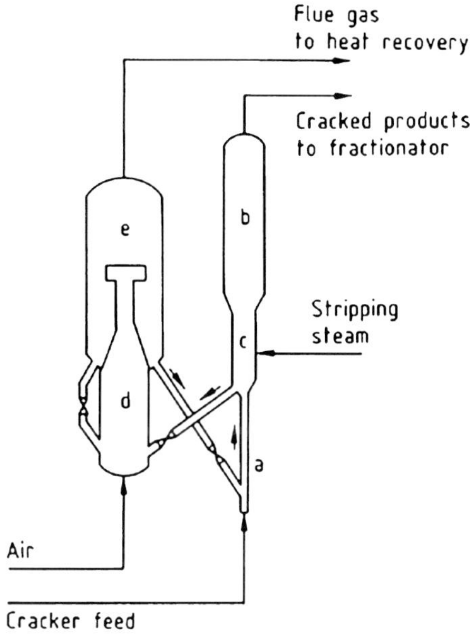

Fig. 2.10 Fluid-bed catalytic crackinga) Riser; b) Reactor; c) Stripper; d) Combustor; e) Disengaging vessel.

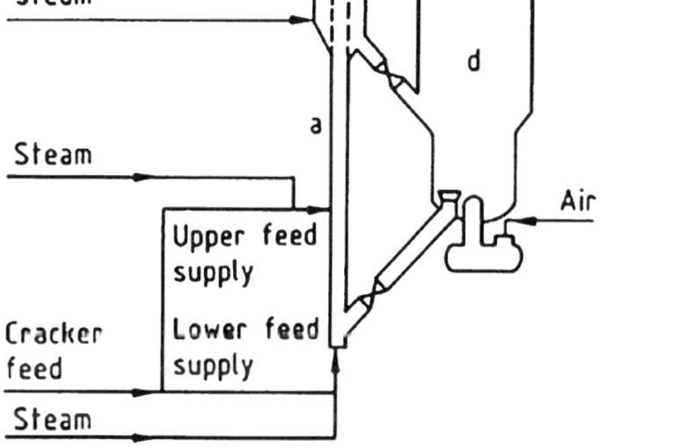

Fig. 2.11 Fluid-bed catalytic cracking with riser reactor a) Riser reactor; b) Cyclone vessel; c) Stripper; d) Regenerator.

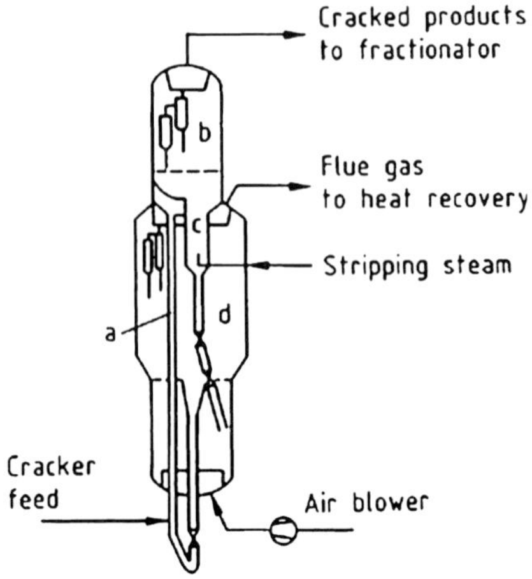

Fig. 2.12 Flexicracking with riser-bed a) Riser; b) Reactor; c) Stripper; d) Regenerator.

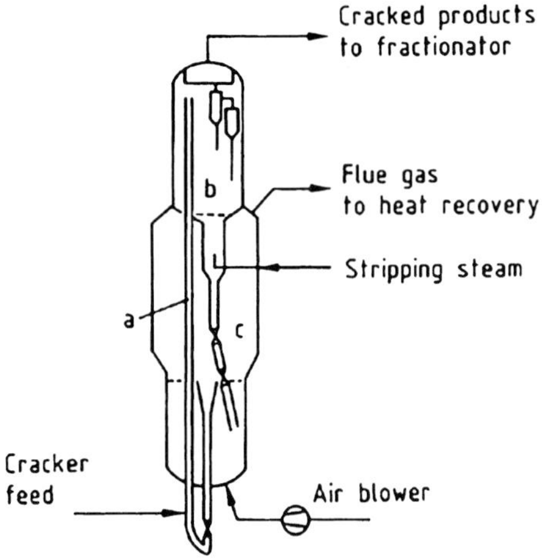

Fig. 2.13 Flexicracking with transfer line reactor a) Transfer line reactor; b) Cyclone vessel – stripper; c) Regenerator.

process gas stream using a solvent (e.g., amine) wash and is then converted into elemental sulfur in a Claus plant.

Hydrotreating processes are applied to achieve the following objectives:

1. Removal of sulfur, nitrogen, oxygen, and arsenic compounds from *reformer feedstocks* (naphthas) to prevent poisoning of the platinum catalysts in the reformers. Hydrogenation of the unsaturated hydrocarbons of cracked naphthas to minimize coking of the reformer catalysts.

2. Increase in the thermal stability of *jet fuels*. Additionally, an improvement of the smoke point can be aimed at by (partial) hydrogenation of aromatics.
3. Removal of sulfur compounds from *middle distillates* (gas oils) to achieve blending qualities for diesel fuels and light heating oil. Unsaturated hydrocarbons in cracked gas oils are hydrogenated to improve the color stability and reduce the coking tendency of the products.
4. Removal of sulfur, nitrogen, and metal compounds from *catalytic cracking feedstocks* (vacuum gas oils, atmospheric residues) to limit the catalyst deactivation by nitrogen and metals, and to yield lower sulfur contents in the cracked products.
5. Reduction of the sulfur content of *heavy fuel oils*.
6. Hydrogenation of diolefins in *pyrolysis gasolines* (originating from ethylene production plants) to avoid the formation of gum, which deteriorates the stability of these gasolines.
7. Improvement of odor, color, and oxidation stability of *lubricating oils* by hydrofinishing (mild hydrotreating). Dearomatization and removal of sulfur, nitrogen, and oxygen compounds by deep hydrogenation (high-pressure hydrotreating) of lubricating oils.

2.3.3.1 **Distillate Hydrotreating**

Two process types are used – the vapor-phase process for *hydrotreating of gasoline* (where the feed is vaporized in the reaction stage) and the trickle phase process for *hydrotreating and desulfurizing of straight-run middle distillates* or vacuum gas oils (where the major portion of the feed is liquid in the reaction stage).

The process scheme for both hydrotreating process types is identical. The feed stream is mixed with hydrogen-rich gas (usually fresh gas from catalytic reforming plus recycle gas from the hydrotreating process) and heated before it is passed through the catalyst bed in the reactor. The reactor product is cooled and separated into a gas and a liquid phase in two consecutive separation stages (high-pressure and low-pressure separators). The hydrogen-rich recycle gas from the high-pressure separator is recombined with the feed. The low-pressure off-gas that contains hydrogen sulfide is processed further in a gas treatment plant for removal of H_2S. The liquid reactor product is stabilized in a debutanizer (gasoline hydrotreater) or stripper column (gas oil hydrodesulfurizer).

The two process types are shown schematically in Fig. 2.14 (gasoline hydrotreating) and Fig. 2.15 (gas oil hydrodesulfurizing), respectively. The catalyst types employed and the main operating criteria of distillate hydrotreatment processes are given in Table 2.4.

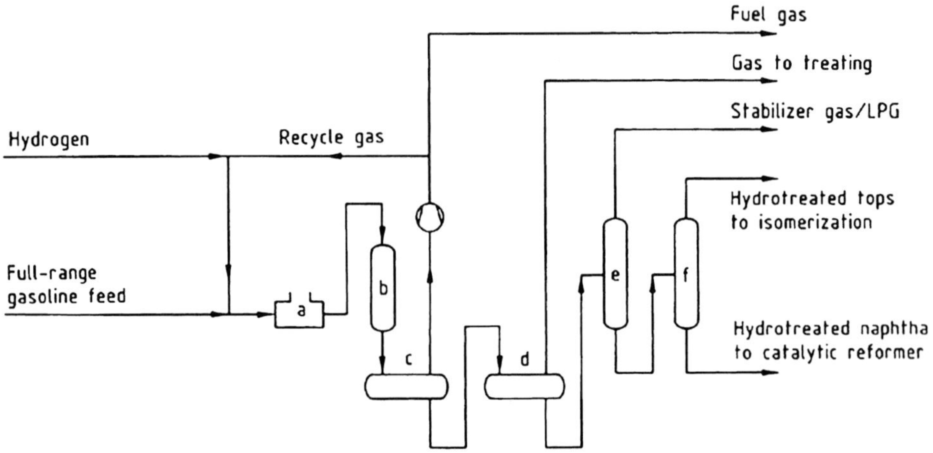

Fig. 2.14 Schematic representation of gasoline hydrotreater
a) Process heater; b) Reactor; c) High-pressure separator;
d) Low-pressure separator; e) Stabilizer; f) Gasoline splitter.

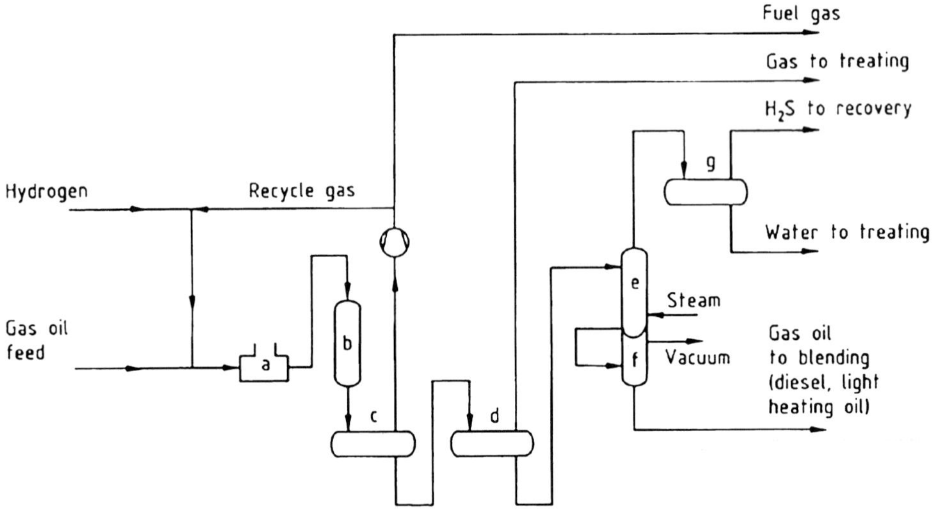

Fig. 2.15 Gas oil hydrodesulfurizer a) Process heater; b) Reactor; c) High-pressure separator; d) Low-pressure separator; e) Gas oil stripper; f) Gas oil dryer; g) Stripper overhead drum.

A large variety of commercial distillate hydrotreating processes are described in [10], [11], for example:

for straight-run distillates

Distillate HDS & SPI (IFP)
Distillate HT/Gulfining (Air Products – Gulf) [12]
Hydrofining (BP) [13]

Hydrofining (Exxon) [14]
Smoke point improvement (Shell)
Thermal naphtha HT (ARCO)
for vacuum distillates (FCC feedstocks)
HGO – HDS (ARCO)
GO-Fining (Exxon) [15]
for highly unsaturated distillates
DPG – HT (Lummus Crest)
Pyrolysis distillate hydrogenation (IFP) [16]
Pyrolysis gasoline hydrogenation (BASF – VEBA)
Hydroisomerization/HPN IVB (Engelhard) [17].

Hydrotreating (hydrodesulfurization, hydrodemetallization) processes for residues are dealt with in Section 2.3.6.3.1.

Table 2.4 Catalyst types and main operating criteria of distillate hydrotreating processes [18]

Process	Catalyst type	H_2 partial pressure, MPa	LHSV[a], vol/vol
HT [b] of naphthas	Co – Mo	0.5–0.8	5–8
HT [b] of kerosene	Co – Mo	0.8–1.2	4–6
HDS [c] of gas oils	Co – Mo	1.0–2.0	4–6
HDS [c] of VGO [d]	Co – Mo	2.5–4.0	1–3
HDS[c]–HDN [e] of naphthas	Ni – Mo	2.0–3.0	2–3
HDS [c] – HDN [e] of gas oils	Ni – Mo	2.0–3.0	1.5–2.5
Smoke point improvement of kerosene			
Moderate	Ni – W	2.0–3.0	1–3
High	Ni – W	3.0–4.0	1–1.5
HDS [c] of unsaturated naphthas and gas oils (coker, visbreaker, FCC)	Ni – Mo Ni – W	2.0–3.0	1.5–3

[a] LHSV = liquid hourly space velocity (volume of product per volume of catalyst).
[b] HT = hydrotreating.
[c] HDS = hydrodesulfurization.
[d] VGO = vacuum gas oil.
[e] HDN = hydrodenitrification.

2.3.3.2 **Pyrolysis Gasoline Hydrotreating**

This widely applied process can be used in a single-stage and a two-stage version, depending on the destination of the products. The feedstock is usually a pyrolysis gasoline from olefin plants (byproduct of naphtha steam cracking).

In the *first stage* the unsaturated and aromatics-rich material is selectively hydrotreated so that the conjugated double bonds of the diolefins and styrene are hydrogenated to produce a stable gasoline blending component. The main stability criteria are a reduction of the diene value (maleic anhydride value) from a range of 30–70 to 1–2, accompanied by a reduction of the bromine number from a range of 50–90 to 30–60. The induction period and potential gum content are also greatly reduced.

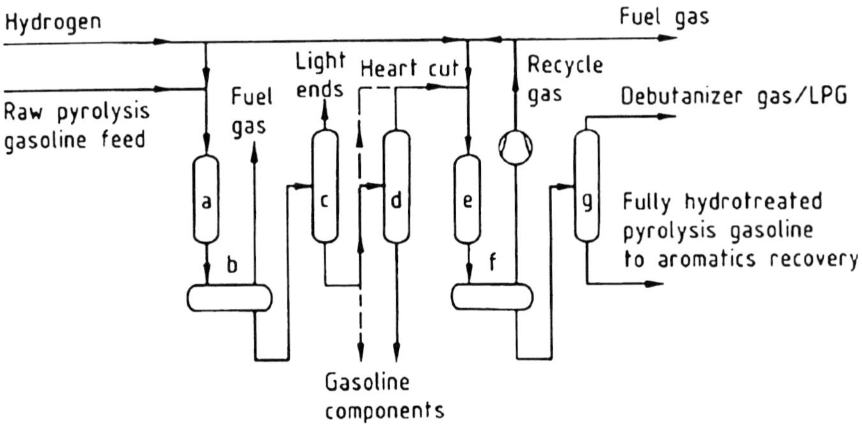

Fig. 2.16 Two-stage pyrolysis gasoline hydrotreater a) First stage reactor; b) First stage separator; c) Depentanizer; d) Gasoline (heart cut) column; e) Second stage reactor; f) Second stage separator; g) Debutanizer.

In the *second stage*, the first stage product or a fraction thereof ("heart cut") is hydrotreated further to achieve complete hydrogenation of the olefins and to accomplish desulfurization. Thus, an aromatics-rich product is yielded that is suitable for BTX recovery.

The main quality characteristics of the second stage are a reduction of the bromine number to < 1 and a sulfur removal from 1000–5000 ppm to <1 ppm.

Figure 2.16 shows a typical two-stage unit.

2.3.3 Desulfurizing by Adsorption

Hydrotreating of FCC gasoline usually reduces the octane rating due to saturation of olefinic double bonds. A new process by Philips Petroleum Co. (S-Zorb) avoids this disadvantage. The gasoline is contacted with an adsorptive material in a fluidized bed in the presence of hydrogen. The sulfur on the sorbent is oxidized to SO_2, which can either be reacted with H_2S from other sources in a Claus plant or be reduced to H_2S. The resulting gasoline sulfur level is below 10 mg/kg with minimal loss in octane number. The application of this technology for gas oils is very promising. Pilot plant data indicate that sulfur levels below 10 mg/kg can be achieved without changing the HC pattern [17a].

2.3.4 Catalytic Reforming

2.3.4.1 Introduction

In the reforming processes, gasoline fractions (naphthas) with a low octane number are converted into a high-octane reformate, which is a major blending product for motor gasolines (Chap. 3.4). The most valuable byproduct is hydrogen, as the catalytic reformer is the only indigenous hydrogen source of the refinery. The other important byproduct is liquefied petroleum gas.

Straight-run or possibly hydrocracked naphthas usually have octane numbers of only 35–65. Therefore these products are unsuitable for direct gasoline blending. Reformer feedstocks must be pretreated to adjust them to the required specifications (e.g., sulfur content, see Section 2.3.3).

With the development and introduction of noble metal catalysts at the beginning of the 1950s, the catalytic process displaced other (mainly thermal) reforming processes, because the catalytic conversion produces higher liquid yields – together with higher octane numbers – and hydrogen. Catalytic reforming is chiefly based on the catalytic conversion of normal paraffins and cycloparaffins into aromatics and isoparaffins.

The catalysts applied are either of the single metal platinum-on-alumina type, or bimetallic species where platinum is used in combination with a second metal, e.g., rhenium. The bimetallic catalysts have a better operational stability, but are more sensitive toward poisoning (e.g., by sulfur) or deficient regeneration. The catalyst metal is dispersed on the porous carrier material (alumina). The various commercial catalyst types contain 0.25–0.8 wt % platinum and up to 1 wt % of a halogen, usually chlorine. In these "bifunctional" catalysts the metal promotes hydrogenation and dehydrogenation reactions, while the chlorine catalyzes isomerization and cracking reactions.

The chemical reaction mechanisms of the catalytic reforming process can be characterized as follows:

Dehydrogenation of C_6-Naphthenes to Aromatics Example:

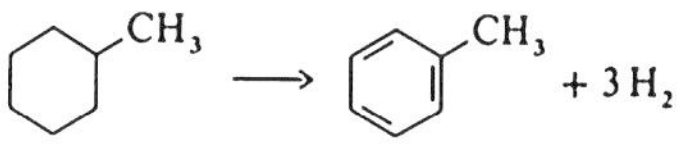

(1)

Conversion of methylcyclohexane into methylbenzene (toluene) with formation of 3 mol hydrogen per mole of converted hydrocarbon. The research octane numbers (RON) of methylcyclohexane and toluene are 75 and > 100, respectively.

Dehydroisomerization of Alkyl-C_5-Naphthenes to Aromatics Example:

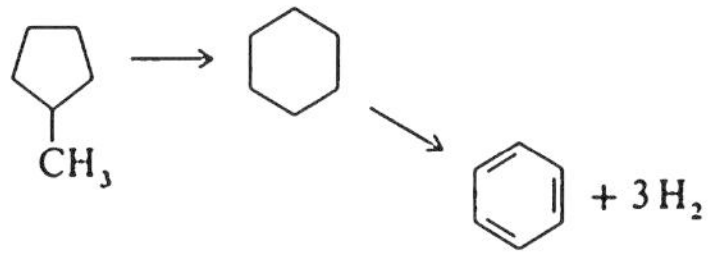

(2a)

(2b)

In the first step, methylcyclopentane is isomerized to cyclohexane (2a).

According to Eq. 1, cyclohexane is then dehydrogenated in a second step to form benzene with formation of 3 mol hydrogen, Eq. 2b.

The RONs of methylcyclopentane, cyclohexane, and benzene are 91, 83, and >100, respectively.

Dehydrocyclization of Paraffins to Aromatics Example:

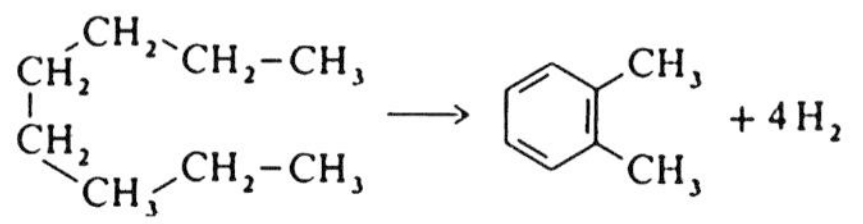
(3)

n-octane is converted into 1,2-dimethylbenzene (*o*-xylene) with the formation of 4 mol hydrogen. The RONs of these components are 0 and >100, respectively.

Hydrocracking of Paraffins and Naphthenes to Smaller Paraffin Molecules Example:

$$CH_3CH_2CH_2CH_2CH_2CH_2CH_3 + H_2 \longrightarrow CH_3CH_2CH_3 + CH_3CH_2CH_2CH_3 \quad (4)$$

n-heptane is split into propane and normal butane with the consumption of 1 mol hydrogen per mole of paraffin (2 mol hydrogen per mole of naphthene). This reaction does not affect the RON.

Isomerization of *n*-Paraffins to Isoparaffins Example:

$$CH_3CH_2CH_2CH_2CH_3 \longrightarrow CH_3CH_2CH(CH_3)CH_3 \quad (5)$$

Conversion of *n*-pentane into isopentane. The respective RONs are 62 and 92.

Hydrogenation of Unsaturated Hydrocarbons This reaction is also important because unsaturated hydrocarbons can act as coke precursors. Coke deposits would deactivate the catalyst.

The dehydrogenation reactions (Eq. 1) and (Eq. 3) are both endothermic and require appropriate heat input (i.e., 32 and 38 kJ/mol, respectively, at the reforming temperature level). However, because the reaction in Eq. 3 is considerably slower than that of Eq. 1, higher temperatures must be applied to achieve the desired reaction rate. The isomerization reactions (Eq. 2a) and (Eq. 5) are thermally almost neutral, whereas the hydrocracking reaction (Eq. 4) is exothermic (ca. −16 kJ/mol at the reforming temperature).

As the permissible benzene concentration in motor gasoline has been reduced to 1.00 vol % the dehydrogenation reactions (Eq. 2 and Eq. 3) are no longer desired. To prevent these reactions, benzene precursors methylcyclopentane and cyclohexane are normally removed from the reformer feedstock by distillation.

Hydrocracking (Eq. 4) is favored at high temperatures and high hydrogen partial pressures. These reactions are usually undesirable in catalytic reforming, because they consume hydrogen and produce gaseous hydrocarbons from liquid ones, thereby decreasing the yield of liquid reformate. This influence is illustrated in Fig. 2.17, where reformate yields are related to reactor pressure at various octane levels.

At a given pressure, the octane number can be increased by applying higher reactor temperatures; this, however, results in yield loss by increased hydrocracking.

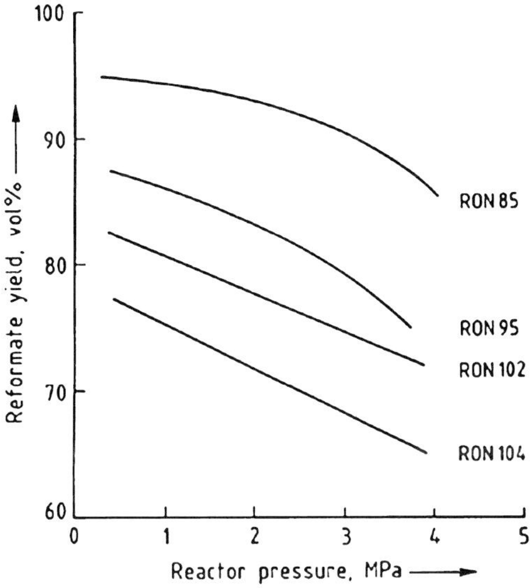

Fig. 2.17 Reformate yield and reactor pressure at various octane levels.

As a result, catalytic reformers should generally be operated at low pressure to achieve high liquid yields. However, the hydrogen partial pressure must be high enough to avoid the formation of unsaturated compounds that may polymerize and cause increased coke deposition on the catalyst, and hence its deactivation.

Catalytic reformers are usually available in three different versions, namely the semiregenerative, the fully regenerative, and the continuously regenerative reformer. Common to each process type is a reactor section and a product workup section. The fully and continuously regenerative process versions are also equipped with a catalyst regeneration section.

Various licensed reforming processes are described in [19] and [20]. Examples are Platforming (UOP), Magnaforming (Engelhard), Catalytic Reforming (IFP), Powerforming (Exxon), and Rheniforming (Chevron).

2.3.4.2 **Semiregenerative Reformer**

This conventional catalytic reforming process is shown in Fig. 2.18. The plant has three or four reactors that are arranged side by side at ground level. After addition of hydrogen-rich recycle gas to the naphtha feed, the mixture is heated up in the charge – product heat exchanger (a) and brought to reaction temperature in the charge heater (b). The vaporized feed is then successively passed through the catalyst beds of the reactors (c, e, and g). Between the reactors, makeup heat is provided in the intermediate heaters (d and f) to compensate for heat losses after each (endothermic) reaction step. The effluent from the last reactor is cooled in the heat exchanger and aftercooler (i). It is then separated in the product separator (h) into hydrogen-rich gas and liquid reaction product.

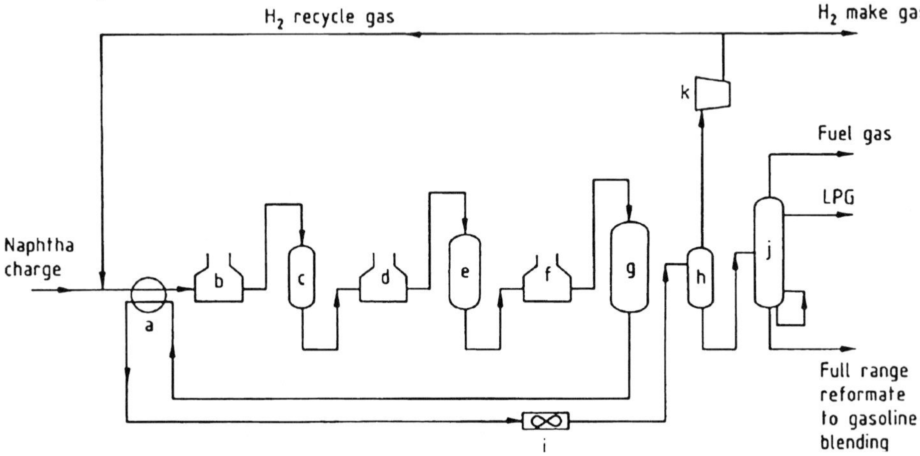

Fig. 2.18 Semiregenerative catalytic reformer a) Charge – product heat exchanger; b) Charge heater; c) First reactor; d) First intermediate heater; e) Second reactor; f) Second intermediate heater; g) Third reactor; h) Product separator; i) Product cooler; j) Stabilizer; k) Recycle gas compressor.

The H_2 gas is recompressed in the recycle gas compressor (k) and part is returned to the feed stream; the net H_2 "make gas" is supplied to hydrogen-consuming units, such as hydrotreaters and hydrodesulfurizers.

The liquid reactor product is fractionated in a stabilizer column (j) to separate the C_4 and lighter hydrocarbons from the reformate. The LPG stream (C_3 and C_4) is usually split in the LPG recovery plant for propane and butane production; (see Section 3.8.1.3. Liquefied Petroleum Gas – Production and Processing); the stabilized reformate can be used directly for gasoline blending or is fractionated further into light and heavy gasoline components before blending.

The catalyst is gradually deactivated during the operation period; this can be counteracted by stepwise increase in the reactor temperatures to keep the product quality at the desired level. Once the recommended ultimate temperature is reached, the plant has to be shut down and the catalyst is regenerated. This procedure is usually performed in three successive steps, i.e., carbon burn-off, metal reduction, and metal redispersion. The plant is then restarted for its next operation period. When the operation cycles between regenerations become unacceptably short, the catalyst inventory is replaced by a new batch. The spent catalyst batch is worked up for metal recovery (platinum, rhenium).

2.3.4.3 Fully Regenerative Reformer

The fully-regenerative, or cyclic, reforming processes have been devised to increase the on-stream time of the reforming plants by introducing the "swing reactor" principle. A typical example is the Power forming process [21]. In a system of four reforming reactors, three reactors are on-stream, in the fourth the catalyst is regenerated. After completion of the regeneration, this reactor is brought on-stream, while another reactor is switched to regeneration. The cyclic process requires a complex switch system with automatically operated valves. However, with this method a continuity of operation at high reformation severities can be maintained over a long period, with catalyst lifetimes of up to five years.

2.3.4.4 Continuously Regenerative Reformer

Special moving-bed techniques were developed for reforming units with continuous catalyst regeneration. The *IFP catalytic reforming process* which employs a conventional side-by-side arrangement of reactors uses a system of lift pots to transport the moving catalyst in portions from the bottom of a reactor to the top section of the next reactor [22], [23]. The catalyst portions obtained from the last reactor are lifted to the regenerator, where carbon burn-off and reconditioning of the catalyst take place. The catalyst is then reinjected into the first reactor bed.

A sophisticated catalyst moving-bed system was realized in the UOP Platforming process as illustrated in Fig. 2.19 [24], [25]. The reactors are arranged in a "stacked" construction. The catalyst trickles through the system from the top of the first reactor to the bottom of the third reactor, where it is collected and lifted in portions to the regenerator section.

"Sluices" are incorporated into the catalyst transport system, i.e., the spent catalyst from the third reactor is depressurized, freed from hydrogen, and then lifted in an inert gas stream to the regenerator; in turn, the regenerated and reconditioned catalyst has to be flushed free of combustion gases and pressurized with hydrogen before it is reinjected to the first reactor.

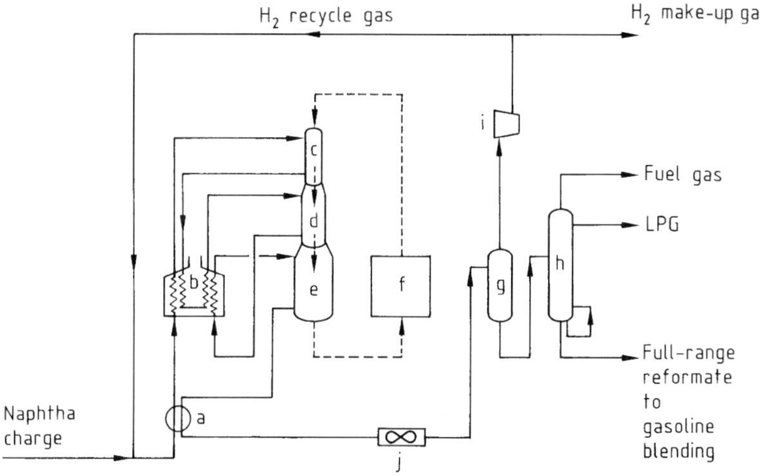

Fig. 2.19 Continuously regenerative catalytic reformer (UOP Platforming) a) Charge – product heat exchanger; b) Process furnace (charge heating cell, first intermediate heating cell, and second intermediate heating cell); c), d), e) Reforming reactors; f) Catalyst regeneration section; g) Reactor product separator; h) Stabilizer; i) Recycle gas compressor; j) Product cooler.

With continuous catalyst reactivation severe reforming conditions can be applied to achieve RON values of well above 100 in the reformate product. Moreover, operation at low reaction pressures of 7–10 bar results in favorable yield structures, i.e., high reformate and hydrogen quantities at low gas production.

2.3.5
Hydrocracking

The hydrocracking (HC) process was developed to produce high yields of distillates with better qualities than can be obtained by fluid-bed catalytic cracking. Hydrocracking feedstocks are typically flashed distillates from vacuum distillation, from cat and thermal cracking, or deasphalted oils. Operating conditions in the reactor section of hydrocrackers are usually about 400 °C and 8 –15 MPa. The reactions take place on metal sulfide catalysts in the presence of hydrogen. Major advantages of the process include the full-scale production of high-quality products, i.e., no low-grade fuel oil remains as residue, and the high flexibility towards product yields, i.e., possible gasoline maximization, or gasoline – kerosene, or kerosene – gas oil maximization.

Modern hydrocrackers are units with two reaction stages: a hydrogenation step for desulfurization, denitrogenation, and deoxygenation using cobalt – molybdenum catalysts, and a hydrocracking step using nickel – tungsten catalysts. After fractionation of the liquid reactor product into gas, naphthas, and middle distillate, the bottom product of the fractionator is fed to the hydrocracking reactor. A typical example of a two-stage hydrocracker is shown in Fig. 2.20. Hydrocracking processes for treatment of residues ("hydroconversion" processes) are described in Section 2.3.6.3.2.

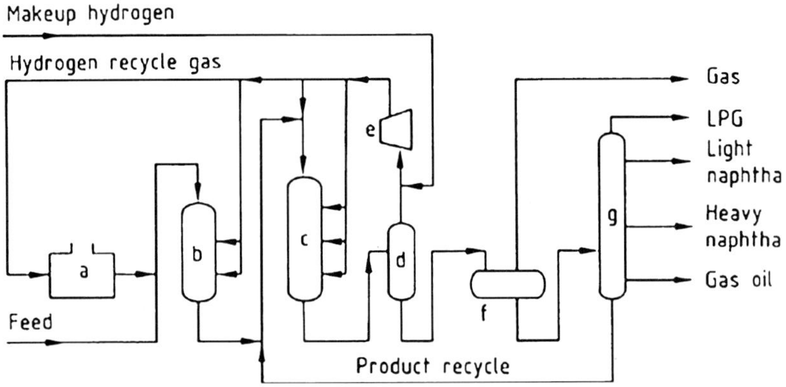

Fig. 2.20 Two-stage hydrocracking [26], [27] a) Hydrogen heater; b) First-stage reactor (hydrotreating); c) Second-stage reactor (hydrocracking); d) High-pressure separator; e) Hydrogen compressor; f) Low-pressure separator; g) Fractionator.

2.3.6 Residue Conversion Processes

2.3.6.1 Introduction

In the previous sections, the standard processes of refinery technology for refining of distillates were described. These are used to improve the product yields recovered from the crude oil as well as the quality of the individual products.

The standard processes include catalytic cracking and distillate hydroprocessing, i.e., processes with the main purpose of increasing the yield of light products from heavier crude fractions. These processes were developed and applied mainly in the United States, where the aim had been to achieve a high yield of motor gasoline as the main product of crude oil processing. In Europe and most other consumer areas, however, the objective was a high yield of diesel fuels, and light and heavy fuel oils.

In addition to these distillate conversion processes, the conversion of residues and heavy crudes in the modern refinery is of growing importance as a comprehensive process step for several reasons [28], [29]:

1. Since the drastic oil price increase following the first oil crisis in 1973 (from ca. \$2/bbl to more than \$ 70/bbl in 2006 with fluctuating but in the long term rising trends) there has been an increasing economic need to recover more light, high-value products from the crude oil barrel at the expense of the residue.
2. In many countries, heavy fuel oil has been substituted by other fuels, above all natural gas, because of the high sulfur and heavy metal content of residual oils.
3. The relative amount of residues also increases when the crude slate becomes heavier, i.e., richer in residue over the medium and long term (African and North Sea oils contain ca. 10 wt % vacuum residue, oils from the Persian Gulf ca. 30 wt %, and Venezuelan heavy oils still higher proportions of vacuum residues), see Table 2.5.
4. Finally, even the heavy crudes can no longer be processed in a normal refinery without additional treatment because of the high proportions of impurities, see also Table 2.5.

Important criteria to be considered when selecting suitable upgrading technologies for treating heavy oil are the range of crudes (i.e., the amount and quality of the residue to be processed), the product requirements (motor gasoline, diesel fuel, heating oil), and the flexibility of the refinery (existing structure, changing feed supply, and product demand).

The feedstocks for residue conversion plants are atmospheric and vacuum residues, heavy oils, or bitumens. In the future tar sands, and shale oils will play an increasing role in energy generation. The objectives and the products of the conversion vary according to the starting material and the method of conversion.

Table 2.5 Residue contents of typical crudes (in wt %) and concentration of impurities in vacuum residues

	Atmospheric residue	Vacuum residue	Sulfur, mass %	Vanadium, ppm
Nigeria	34.0	8.2	0.3	10
Algeria	30.9	8.3	0.2	5
North Sea	39.7	13.1	1.0	35
Mexican Isthmus	47.6	19.2	4	175
Arabian light	47.6	19.2	4	80
Russia	40.9	17.5	3	200
Kuwait	53.5	27.2	5	115
Arabian heavy	56.6	31.9	5	210
Venezuelan heavy	75.6	43.0	3	600

They range from processing a low-sulfur heavy fuel oil where only low yields of lighter products are aimed at, via maximization of the distillate yield to complete conversion of the residue. In the longer term the production of "synthetic crudes" (syncrudes) from tar sands and shale oils will gain in importance (Fig. 2.21). The technology is being developed in Canada and Australia where tar sands and shale oils are abundant.

2.3.6.2 Fundamentals of Residue Conversion and Process Options [30], [31]

The C:H ratio in the crude oil fractions increases with increasing molecular mass, i.e., the hydrogen deficiency of the residue (in comparison with lighter hydrocarbons)

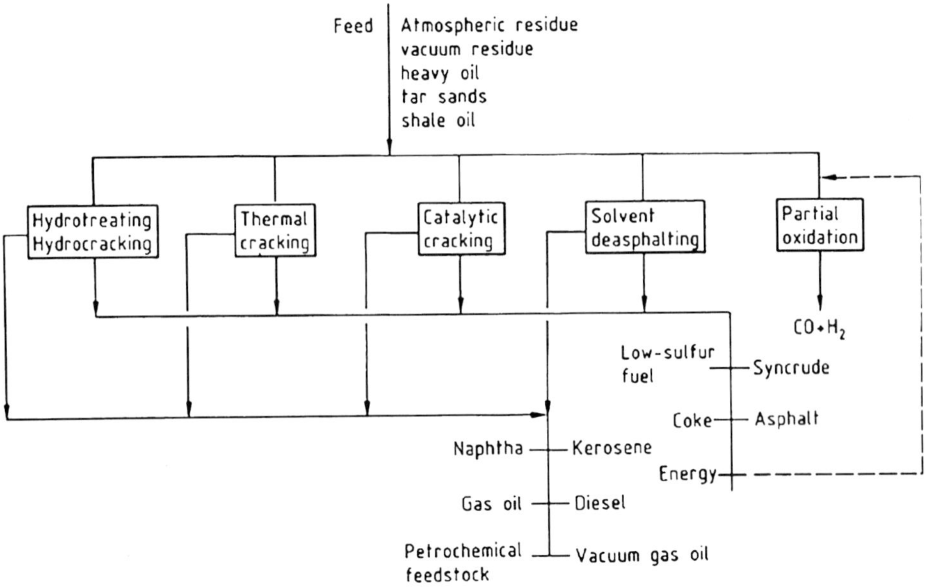

Fig. 2.21 Residue conversion processes.

increases in the sequence atmospheric residue < vacuum residue < visbreaker residue < coker residue < coke.

To obtain lighter products that are richer in hydrogen, two options exist for further processing which both lead to a change in molecular structure: (1) hydrogen addition ("H-in") processes (Section 2.3.6.3.); (2) carbon rejection ("C-out") processes (Section 2.3.6.4).

Additional processes for particular cases are the extraction of asphaltenes and the oxidation to synthesis gas or hydrogen (see Section 2.3.6.5.).

A combination of different process steps is often necessary because of special conditions in the plant and varying process solutions (see Section 2.3.6.6.).

Approximately 570×10^6 t/a of throughput capacity is installed in residue conversion plants worldwide (status 1997/1998). Nearly 60 % of this capacity consist of thermal processes, the most important being delayed coking and visbreaking. The thermal processes still predominate but the proportion of hydrogenation processes (hydrorefining and hydrocracking) is increasing, particularly the processes with high conversion rates, whereas the share of catalytic cracking capacities for treating of residual oils remains relatively constant.

Table 2.6 shows the approximate residue conversion capacities of the various world regions (1997/1998). The resulting worldwide capacity for conversion of residual oils amounted to 14–15 % of total refinery capacity. However, the regional ratios vary widely between 4% (Former Soviet Union, Africa, and Australia/Oceania) and 22 % (North America). In view of the heavier crudes to be processed and consequently higher residue yields anticipated for the near future, a large expansion of heavy oil upgrading technologies will be necessary.

Table 2.6 Estimated capacities (in 10^6 t/a) of residue conversion in 1997/1998 (derived from [2])

	North America	Latin America	Europe (excl. FSU[f])	Former Soviet Union	Asia (excl. FSU[f])	Africa, AUS/Oc[g]	Total
Refinery capacity	888	396	809	509	1126	190	3918
Thermal cracking[a]	105	49	92	20	64	5	335
Catalytic cracking[b]	36	7	18	3	20	2	86
Hydrorefining[c]	45	1	26		58	1	131
Hydrocracking[d]	9	2	1		9		21
Total[e]	195	59	137	23	151	8	573
% Residue conversion capacity of refinery capacity	21.9	14.9	16.9	4.5	13.4	4.2	14.6

[a] Thermal cracking of residues (visbreaking, coking).
[b] Catalytic cracking of residues (the capacity range for coprocessing of residual oils is estimated at 10–15 % of total cat. cracking capacity).
[c] Catalytic hydrofining of residues and heavy cat. cracking feedstocks.
[d] Catalytic hydrocracking of residual oils.
[e] Total residue conversion capacity.
[f] FSU = Former Soviet Union.
[g] AUS/Oc = Australia/Oceania.

Residue conversion continues to be in a state of rapid development, thus the present discussion deals mainly with the problems of residue conversion and the currently best-known processes. The details of process engineering will be discussed only insofar as they differ from those of established processes. The best-known commercial processes may be summarized as follows:

Hydrorefining	Resid/HDS (Chevron)
	Resid/HDS (Gulf)
	Residfining (Exxon)
	Unicracking (Union Oil)
	HDS (Shell)
	RCD Unibon (UOP)
Hydrocracking	H-Oil (HRI)
	LC-fining (Lummus)
	BOC Unibon (UOP)
	Combi-Cracking (VEBA)
	Canmet (Petro-Canada)
	Hycon (Shell)
Thermal cracking	Visbreaking
	Delayed coking
	Fluid coking (Exxon)
	Flexicoking (Exxon)
Catalytic cracking	Heavy oil cracking (Kellogg)
	Resid oil cracking (Total)
	Asphalt residue treatment (Kellogg)
	Reduced crude conversion (UOP)
Solvent deasphalting	ROSE (Kerr McGee)
	DEMEX (UOP)
Partial Oxidation	Texaco
	Shell

2.3.6.3 Hydrogen Addition ("H-in") Processes

As for hydrotreating of distillates, the term *hydrotreating* is also used for the catalytic hydrogen processing of residues if the main effect is an extensive removal of sulfur and nitrogen compounds (> 90 %) as well as of heavy metals, at a relatively low conversion rate (5–35 %). *Hydrocracking* means the operation under more severe conditions with the objective of high conversion rates of ca. 60–90 %.

The hydrogenation processes use high pressures (5–20 MPa) and are expensive; the principal cost factors are the technical equipment of the plants, the expensive hydrogen supply, and the catalysts. Estimated investment costs for hydrotreating are ca. $\$60 - 90\times10^6$ per 1×10^6 t residue per year; for hydrocracking ca. $\$150\times10^6$ per 1×10^6 t residue per year.

An advantage of H-in processes – especially of hydrocracking – is the recovery of pure, high-grade products in high yields.

2.3.6.3.1 Residue Hydrotreating (Demetallization, Desulfurization, Denitrification)

Residue hydrotreating processes have been developed by many oil companies and by well-known engineering firms on the basis of experience in the desulfurization of gasoline and middle distillates. They are also operated predominantly in fixed-bed re-

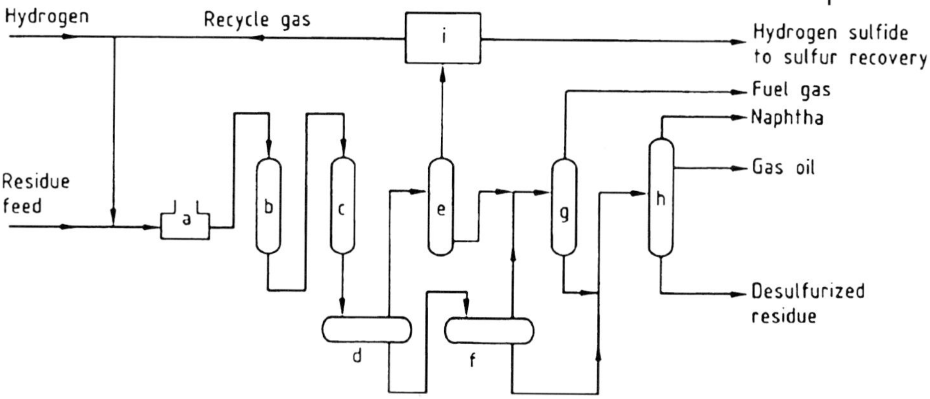

Fig. 2.22 Residue hydrodesulfurizer a) Process heater; b) Demetallization reactor; c) Desulfurization reactor; d) Hot high-pressure separator; e) Cold high-pressure separator; f) Hot low-pressure separator; g) Cold low-pressure separator; h) Product splitter (fractionator); i) Recycle gas treatment section.

actors, but with lower specific catalyst loadings (0.5–2 m^3 feed material per cubic meter of catalyst). So-called guard chambers are frequently installed upstream as traps for metal impurities and solids to extend the on-stream-time of the actual HDS and HDN reactors. Depending on the residue quality the on-stream times would be only up to 6–12 months without preliminary purification. The individual processes differ mainly in the type and number of reactors (1–4 in series) and in catalyst types required for the special purpose. In most cases Co-, Ni-, and Mo-containing catalysts are applied, similar to those used for distillate hydrogenation. A typical residue HDS process is shown in Fig. 2.22.

Some processes, such as the Shell HDS process, operate with a fluidized bed ("moving bed" or "bunker flow" reactors) in order to achieve continuous catalyst exchange and avoid fairly frequent plant shutdowns.

The objective of "mild" hydrotreating processes is to obtain low-sulfur residues which can be further processed catalytically (by FCC) or used as nonpolluting fuel oils. The main market for these processes has been in Japan, where "low-sulfur fuels" are mandatory.

2.3.6.3.2 **Residue Hydrocracking (Hydroconversion)**

Despite the advantage of a high yield of clean light distillates, residue hydrocracking is used worldwide in only a few plants, because of the high plant cost and the high hydrogen consumption of 300–400 m^3 per cubic meter of residue.

Fluidized-bed processes are especially suitable for hydrocracking because of the prevailing severe operating conditions in combination with rapid catalyst deactivation. An additional advantage of these processes is the low pressure drop in the reactor. The catalysts used are either expensive metal catalysts, which are continuously withdrawn and regenerated, or cheap, so-called "throwaway" catalysts, which are removed from the system in a once-through operation.

Processes Using Metal Catalysts The current commercial processes used are the *H-Oil process* developed by Hydrocarbon Research and Texaco and the Lummus *LC-fining process*. In these processes ca. 0.5–1 kg Co – Mo or Ni – Mo catalyst are used per ton of raw material in fluidized- or ebullated-bed reactors. The two processes are very similar in technology and process outline. The fuel oil – gas mixture passes upwards through the fluidized- or ebullated-bed reactors, of which up to 4 are connected in series according to the desired cracking severity. Suspended catalyst is added at the top of each reactor to maintain the desired catalyst activity and an equivalent amount of catalyst is withdrawn at the bottom. To maintain a flow rate that holds the catalyst in suspension in the fluidized or ebullated bed and which ensures uniform (isothermal) temperature control, so-called "ebullating pumps" circulate the reactor contents from top to bottom. In the first plants of the H-Oil process these pumps were arranged outside the reactor, where they could easily be cut off and were more easily accessible for repairs. In LC-fining, more effective

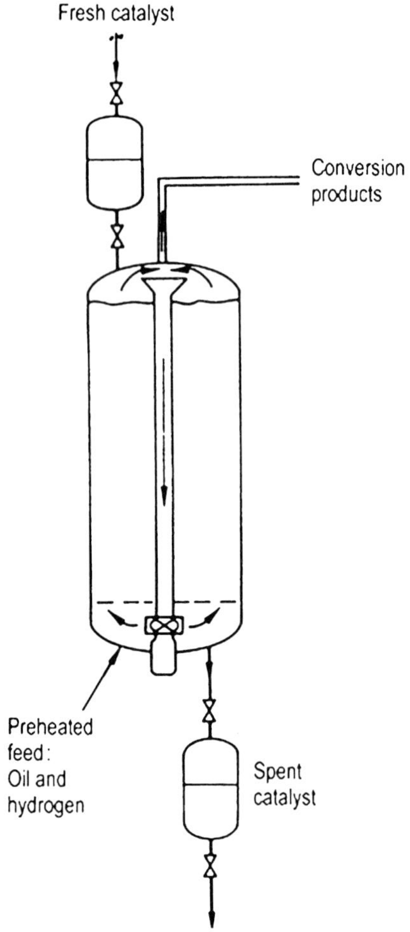

Fig. 2.23 Schematic representation of an LC-fining reactor.

Table 2.7 Product yields from hydrocracking at two conversion severities for an Arabian blend vacuum residue (process: LC-fining)

Product yield	Conversion, vol %	
	65	90
Gas	4.5	10.0
Gasoline	10.4	23.1
Gas oil	16.1	35.0
Vacuum gas oil	31.7	19.2
Residue	34.9	10.9
% HDS	83	89
% HDN	47	53
% Demetallization	93	96

internal pumps are used to achieve an internal recycle. A schematic representation of an LC-fining reactor is shown in Fig. 2.23.

Hydrogen is recirculated in the usual way, and the light conversion products escape rapidly overhead from the reactors, whereas the heavy material remains in the lower part so that the desired conversion can be obtained, if the residence time is sufficiently long.

Information on conversion, product yields, and degrees of desulfurization according to data from Lummus are given in Table 2.7.

A modified hydroconversion process, the *Shell Hycon process,* went on stream in 1989 in a large plant with a throughput of 4000 t/d. This process involves the combination of two already proven process stages. In the first step, the metal content of the feed is reduced by 80–90 % with an SiO_2 catalyst in three bunker-flow reactors. In the second step, the desulfurization and conversion is carried out in several fixed-bed reactors operated in series with an Ni – Mo catalyst on an Al_2O_3 support. The operating conditions (15–20 MPa and 400 °C) are similar to those for other hydrocracking processes. As the demetallization catalyst in the front-end reactors is deactivated within a short period of time, it is continuously replaced. The catalyst mass flow through the reactor is adjusted such that a shutdown cycle of ca. 2 years is obtained, which is normal for the tail-end fixed-bed reactors. The conversion is reported to be ca. 65–70 %, depending on residue quality. With improved catalysts at elevated temperature and with further developments in the reactor area (e.g., in material and catalyst replacement), higher conversion rates up to 90 % are possible even for extremely heavy residues [31].

BOC (Black Oil Conversion) is a fixed-bed process, licensed by UOP. The conversion is 60–70 % and desulfurization reaches 70–80 %. The reactor contains two catalyst beds separated by a specially developed “liquid – gas distributor”. In the first bed the catalyst poisons and solids are removed; in the second bed the actual conversion step occurs.

Processes with Coal-Based Catalysts A group of processes exist in which the residue is converted with pulverized coal as additive or doped with metals (preferably iron

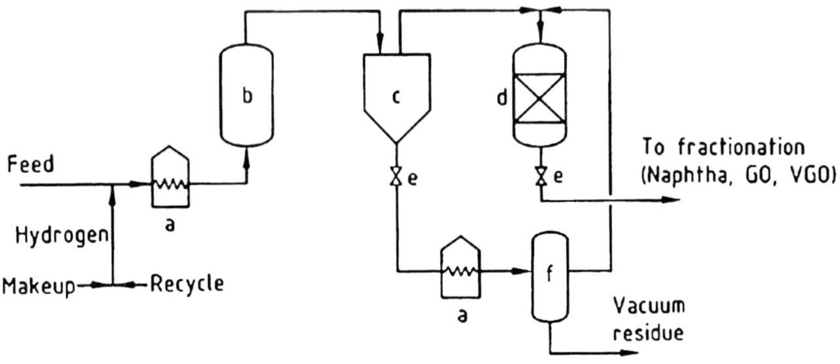

Fig. 2.24 Schematic representation of Veba-Combi-Cracking
a) Heater; b) Liquid-phase reactor; c) Hot separator; d) Gas-phase reactor; e) Depressurizer; f) Vacuum tower.

compounds) as catalyst. These processes are a further development of the coal liquefaction processes used before and during World War II, mainly in Germany. The amount of additive is between 0.5 and 5 wt %, according to the nature of the residue treated.

In *Veba-Combi-Cracking* (*VCC*), developed by Veba-Oel, liquid-phase cracking is combined with gas-phase hydrogenation of the conversion products formed. Both processes are performed at the same high-pressure level. In the first (liquid) stage (3–4 reactors), thermal cracking and hydrocracking reactions occur simultaneously in the fluidized bed at ca. 15–25 MPa and 430–470 °C. Oil residue, hydrogen, and catalyst are introduced at the bottom of the reactor, and a fluidized bed forms, the height of which depends on flow rate and desired conversion. The reactors have no internals and the temperature is controlled by quenching with recirculated gas. The second (gas) stage consists of a fixed-bed reactor of conventional construction with commercial HDS catalysts (Fig. 2.24).

The advantage of the direct combination of the two reaction stages in series is that the freshly cracked products are immediately hydrogenated before polymerization can occur to a significant extent. Furthermore, the specific energy consumptions are lower because of the omission of an intermediate condensation and depressurization step followed in turn by energy-consuming pressurization.

The *Canmet process* of Petro-Canada and Partec Levalin is another process based on a coal catalyst. The process has been tested for the heavy Canadian oils and residues in several pilot and demonstration plants. Plant sizes are such that they can perform up to ca. 1×10^6 t/a throughput in a single reactor. The hydrocracking reactions occur again in an up-flow reactor. A certain proportion of solid is withdrawn continuously from the bottom of the reactor. In contrast to the processes in which several reactors are connected in series, in this process the removal of sulfur- and nitrogen-containing compounds is less complete for a given conversion rate, so that larger reprocessing units are required [33].

2.3.6.4 Carbon Rejection ("C-out") Processes

The carbon rejection processes are designed as purely thermal or – increasingly – as catalytic processes. In these "C-out" processes, excess carbon is deposited as coke and the distillates, which still contain unsaturated hydrocarbons, must undergo further processing.

Disadvantages of these coking processes include the unwanted amount of low-priced coke formed and the high gasification rate.

2.3.6.4.1 Thermal Processes (Visbreaking, Coking)

Visbreaking and coking (with delayed and fluid coking and Flexicoking as variants) are process types which have been known and proven in crude oil processing for many decades. The pioneering work in fluid and Flexicoking was performed by Exxon.

Because of the relatively low investment and operating costs of these processes, the oil companies are working on their further development. A particular aim is the minimization and quality improvement of the coke produced by (1) suitable pretreatment of the raw materials (optimization of the vacuum distillation, desulfurization), (2) minimization of recycle, and (3) optimization of operating parameters (pressure, temperature).

Visbreakers are available from, e.g., Foster Wheeler, IFP, Kellogg, and Lummus Crest [34].

A typical visbreaker is shown in Fig. 2.25. The vacuum residue feedstock is heated and cracked in a tube furnace (a) at ca. 460–480°C. The furnace effluent is quenched with a cold oil stream, e.g., gas oil, which stops cracking reactions before the product enters the fractionator (b). The cracked product streams from the fractionator are routed to further treatment (e.g., gasoline hydrotreating, gas oil HDS). The visbroken residue usually has a lower viscosity and lower pour point than the feed stream.

Two-stage visbreaking is shown in Fig. 2.26. This process is applied for thermal cracking of atmospheric residues. The residue feed is cracked in the first stage residue furnace (a). The combined flashed distillates from the fractionator (b) and vacuum flasher (c) are subjected to a second cracking step in the distillate furnace (d). The temperature in this stage may be as high as 500 °C. This process maximizes the middle distillate (gas oil) yields, but a considerable portion of the gas oil is used as a blendstock to improve the viscosity of the cracked residue to match fuel oil specifications.

In the *coking* processes, the cracked residue is converted further to coke by allowing complete cracking within sufficient residence time [35]. Comprehensive

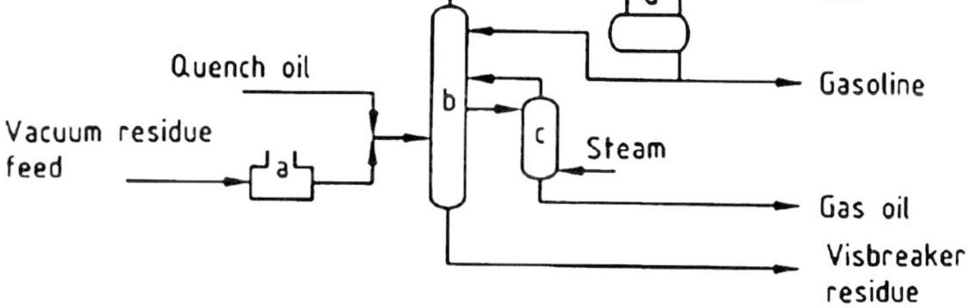

Fig. 2.25 Visbreaker [34–36] a) Cracking furnace; b) Fractionator; c) Gas oil stripper; d) Overhead accumulator.

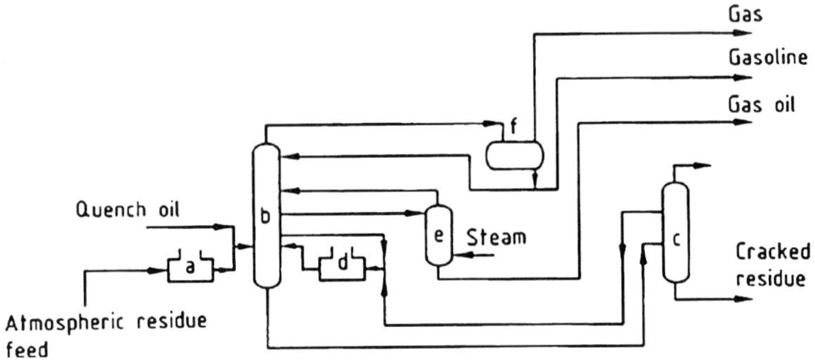

Fig. 2.26 Two-stage visbreaking a) Residue cracking furnace; b) Fractionator; c) Vacuum flasher; d) Distillate cracking furnace; e) Gas oil stripper; f) Overhead accumulator.

topical information on petroleum coke production development and trends is given in [35a].

The principle of the *delayed coking process* is shown in Fig. 2.27. The atmospheric residue feed is introduced to the fractionator (a) where it condenses some of the cracked vapors. The fractionator bottom product is heated in a tube furnace (b) to ca. 490 °C, and the cracked furnace effluent flows through one of the coke drums (c) in which coke is being formed and deposited. The cracked vapors from the coke drum are separated further in the fractionator. In a 24-h cycle, one of the coke drums is in use while the other is emptied by means of a hydraulic coke removal procedure.

The introduction of the *fluid coking process* brought the advantage of continuous operation, thus avoiding alternate use of the coke drums [37]. The cracking reactions occur at 500–550°C in the reactor in a fluid bed of coke particles into which the residue feed is injected. Coke fines are removed from the cracked vapors in cyclone separators before fractionation. The coke formed in the reactor flows continuously to the heater, where it is heated up to 600–650 °C by partial combustion in a fluid bed, from where the net coke production is withdrawn. Another portion of the heated coke particles is returned to the reactor.

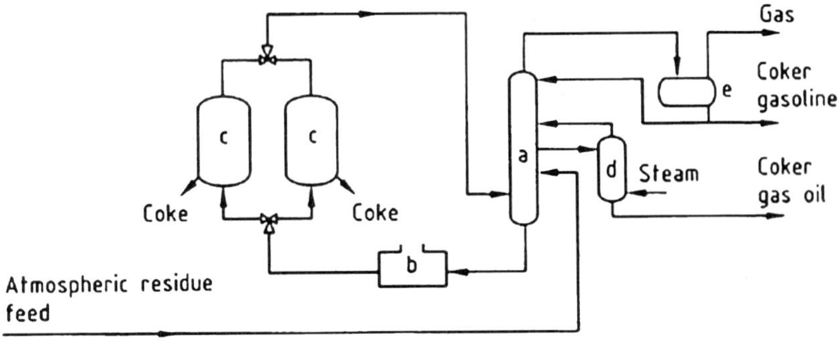

Fig. 2.27 Delayed coking a) Fractionator; b) Furnace; c) Coke drums; d) Gas oil stripper; e) Overhead accumulator.

Table 2.8 Product yields of delayed and fluid coking

Product %	Delayed coking, wt %	Fluid coking, wt
Gases (C_4 and lighter)	7.5	6.5
Coker gasoline	17	10
Coker gas oil	53.5	72
Coke, total	22	11.5

Table 2.8 shows a yield comparison of delayed versus fluid coking; delayed coking has higher gasoline and coke yields, whereas fluid coking maximizes coker gas oil. Delayed coking is the preferred process for production of high-quality petroleum coke grades.

The *Flexicoking process,* licensed by Exxon, combines fluid coking with gasification [38]. This process offers virtually full conversion of the residues to be cracked. A flow scheme is shown in Fig. 2.28. The coke from the heater (c) of the fluid coking section is fed to the gasifier (d) and converted with steam and air to coke gas. The hot coke gas is recycled to the heater (c) and its heat is transferred to the fluid coke bed. The gas mixture leaving the heater is then cooled and cleaned for use as a fuel gas.

2.3.6.4.2 **Catalytic Processes (Residue Cat Cracking)**

The residue cat cracking processes represent a further development of the well-known FCC technology which has been widely used for the cracking of vacuum gas oil (see Section 2.3.2.). The primary objective is maximum gasoline production; the C_3 – C_4 gases which are formed in increased yield are also converted into high-grade alkylate gasoline. For processing residues in fluid cat crackers it is essential that the contents of asphaltenes and metals, which deactivate the catalyst, are kept below a certain level. The metals content in the feed should not exceed 10–30 ppm, otherwise pretreatment is necessary. As the coke deposits on the catalyst in residue

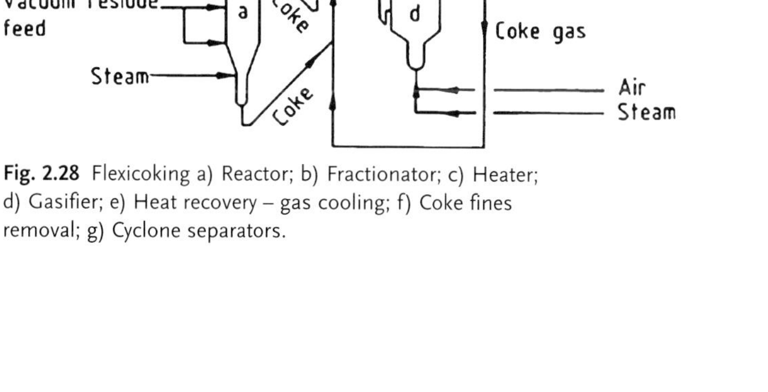

Fig. 2.28 Flexicoking a) Reactor; b) Fractionator; c) Heater; d) Gasifier; e) Heat recovery – gas cooling; f) Coke fines removal; g) Cyclone separators.

crackers are considerably greater than in vacuum gas oil crackers, research efforts are being undertaken mainly to develop new catalysts (use of zeolites, regeneration techniques) and to achieve a reduction of the residence time (riser know-how). At present only feedstocks with Conradson Carbon Residue < 15 wt % can be processed, which means that only 40–50 % of the available crudes can be used [39].

Most large-scale plants process mainly atmospheric residues with pretreatment of the feedstock. The best-known commercially operated processes are (1) Heavy Oil Cracking (HOC, Kellogg, Phillips Petroleum); (2) Reduced Crude Conversion (RCC, UOP) with regeneration in two stages; (3) Asphalt Resid Treatment (ART, Kellogg); and (4) Residual Oil Cracking (ROC, Total). In accordance with the development of the oil market, the trend in all these processes is towards conversion of heavier (vacuum) residues [40].

2.3.6.5 Other Processes

2.3.6.5.1 Extraction of Asphaltenes

Deasphalting is a process that was developed in the 1930s for the recovery of high-viscosity lubricating oil components. In this process the residue is separated with the aid of light hydrocarbons into two fractions: a deasphalted oil (DAO) and an asphalt-rich phase. In the original process, propane was mainly used as extractant and a DAO yield of 25–40 wt % was obtained. The process was not operated supercritically, i.e., the solvent and extract had to be separated by evaporation and condensation of the solvent with high energy consumption. The development of the supercritical extraction process began in the 1960s with the objectives of improving energy conservation, to achieve an increased DAO yield and to simultaneously reduce the amount of residue by employing a wider range of solvents.

The deasphalting processes are characterized by low investment cost, simple technical operation, and high flexibility with regard to extract yield and quality by choice of solvent.

Interest in deasphalting processes as an alternative to residue processing is increasing because the DAO can be used as a hydrocracker or FCC feed and because a large reduction in the unwanted residual products is achieved, with DAO yields of 75–80 % currently feasible (see Table 2.9).

Table 2.9 Deasphalted oil yields obtained with various extractants

		Extractant			
DAO Propane	Feed	Pentane	Butane	Isobutane	
Yield, wt %		80.0	65.0	50.0	35.0
Sulfur content, wt %	5.4	4.7	4.2	3.7	3.2
Nitrogen content, wt %	0.41	0.3	0.2	0.2	0.1
Conradson carbon residue, wt %	25	17.0	9.7	5.5	3.1
V+Ni content, ppm	250	60	25	11	5

The extract yield that can be obtained is, however, limited by its metal content, which rises disproportionately in DAO yields of > ca. 50 %. Depending on refinery type and product requirement, the optimum must be sought between DAO yield and acceptable metal content or the costs of a demetallization stage.

Commercial plants are operated mainly in the United States and Central America for processing the heavy crude oils rich in metals and asphalt that are produced there. The best-known industrial processes are the Demex process of UOP and the ROSE (Residuum Oil Supercritical Extraction) process of Kerr – McGee.

2.3.6.5.2 **Partial Oxidation**

The partial oxidation of hydrocarbons to a mixture of $CO+H_2$ (synthesis gas) is a well-known process (see Section 4.4.) and can be applied in residue processing within very wide limits. The noncatalytic reaction between hydrocarbons and oxygen in the presence of steam takes place at 1300–1500 °C in empty reactors lined with high-temperature-resistant brickwork or corundum refractory. Virtually all residues that are pumpable at the appropriate temperatures can be processed, and almost complete decomposition to $CO + H_2$ with a small (< 5 vol %) CO_2 fraction is achieved. The unburnt soot and ash fraction (≤ 5 wt %) is entrained in the synthesis gas and washed out with water in a downstream scrubber. Various possibilities exist for reprocessing this residual carbon, including complete soot recycling to the gasification reactor (Fig. 2.29).

The disadvantages of this universally applicable process are the relatively high oxygen costs and the production of a gas mixture of low calorific value that cannot be utilized in a standard refinery. If in the future methanol or higher alcohols become important as nonpolluting motor fuel components, the appropriate syntheses (from $CO+H_2$) could offer an interesting option. Similarly, the growing hydrogen requirement of the refineries can be covered by this route even in the medium term, because pure H_2 can be produced by a simple CO-shift conversion step.

The partial oxidation process can be used in particular as a final treatment step in combination with the aforementioned processes with incomplete conversion. This also applies to the coking processes, as even the gasification of solid substances is technically possible after appropriate slurrying with oil or water.

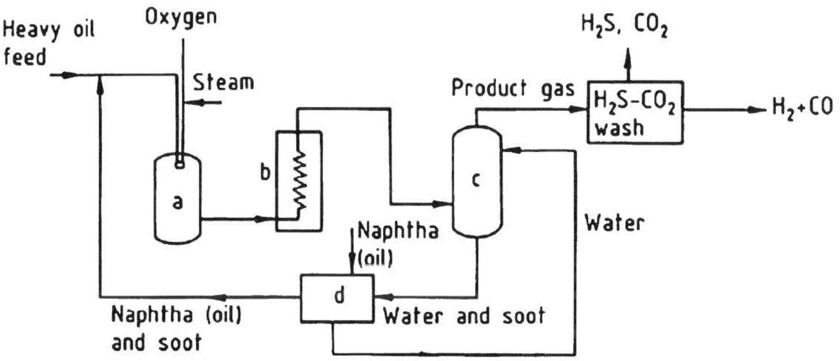

Fig. 2.29 Schematic representation of partial oxidation a) Reactor; b) Waste-heat boiler; c) Scrubber; d) Soot extraction.

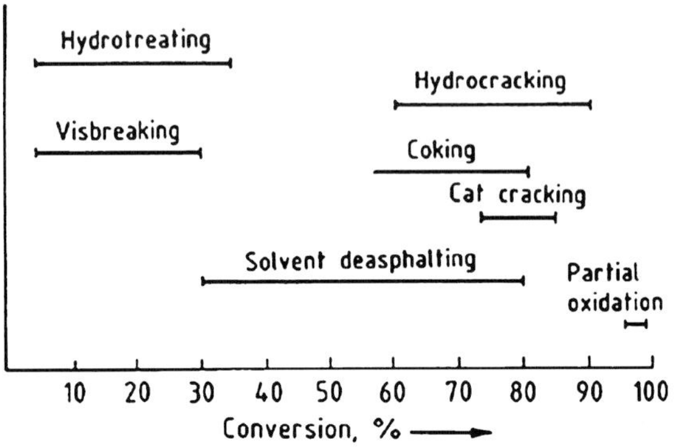

Fig. 2.30 Conversion rates of various residue conversion processes.

Commercial partial oxidation processes are offered by Texaco and Shell. On a large scale, besides atmospheric and vacuum residues, also heavy thermal cracking residues (with high density and viscosity and C:H ratios of $>$ 9.5) can be processed.

2.3.6.6 **Process Combinations**

Of the different process types of hydrogen addition and carbon rejection, which have been described in Sections 2.3.6.3. and 2.3.6.4., each has its advantages and disadvantages when used alone. A comparison of the residue conversion rates that can be achieved with the different processes is given in Fig. 2.30.

The "C-out" processes of thermal and catalytic cracking are proven processes with relatively low investment and operating costs, but they give a smaller distillate yield of poorer quality and a considerable amount of unwanted byproducts.

The "H-in" processes give much higher distillate yields, but they are more expensive. As H-in processes require catalysts, additional costs occur with increasing concentrations of catalyst poisons (reduced on-stream times, higher catalyst consumption).

Therefore, it is usually necessary to combine various conversion processing steps. A great variety of possible combinations exist with regard to conversion, amount, and quality of the desired products, as well as to the existing refinery infrastructure. They include both combinations of different residue conversion processes and further conversion units for the higher-boiling vacuum gas oils fraction, which is formed in considerable amount during the residue conversion.

The medium- and long-term trend towards the use of heavy crudes that are rich in sulfur, nitrogen, and metals, in combination with environmental problems makes it increasingly necessary to add upstream hydroprocessing stages (principally hydrotreating, and in special cases hydrocracking), followed by deeper conversion [41].

The following combined steps in the residue line are used or discussed for the planning of conversion refineries:

hydroprocessing + coking

hydroprocessing + catalytic cracking

hydroprocessing + solvent deasphalting
and as a more far-reaching step:
coking
hydroprocessing + deasphalting + partial oxidation
hydrocracking

In most cases the hydrotreater is installed in the atmospheric residue line, so that the overhead and the bottom products from the down-stream vacuum tower can be fed to the subsequent conversion units in sufficiently good quality.

In delayed coking, up to now the most frequently used residue conversion process, an up-stream hydrotreater, results in a lower yield of coke with improved quality and increased yields of lighter products (Table 2.10).

Hydrotreating of the vacuum residue alone can be considered for crudes that give a vacuum gas oil of sufficient quality to be used without pretreatment in an FCC plant or a hydrocracking process. In this case, the hydrotreater can be smaller and the yields of liquid products from the downstream coker can be increased further.

The hydrotreating – deasphalting combination, likewise, leads to an increased yield of liquid products, because the proportion of a DAO with reduced sulfur and metal contents, which is suitable for further processing, can be increased by up to 30 % after pretreatment of the residue in the downstream extraction step.

Partial oxidation as the possible final step in a complete residue conversion is described in Section 2.3.6.5.2. (Fig. 2.29). An important advantage of gasification is that it solves the environmental problem of sulfur, which concentrates in all residues and in coke to such an extent that these residual materials are unsuitable for combustion. In partial oxidation the sulfur compounds in the feed materials are completely converted into H_2S, which can be processed to sulfur in a Claus plant.

The combinations of residue conversion processes given as examples here represent only some of the possible and useful variations. The anticipated increasing complexity of refinery processing and the more difficult technical problems to be solved, will result in a steep rise in investment and operating costs. Increasing energy prices, an increasing shortage of crude oil and its poorer quality, in addition

Table 2.10 Delayed coking (feedstock: Arabian heavy vacuum residue), yield improvement by upstream hydrotreating

	Feed	
	Untreated	**Pretreated**
C_4, wt %	9.5	12.0
C_5<200°C[a], wt %	19.0	22.1
200–360°C [*], wt %	31.3	29.9
360–500°C [a], wt %	19.7	40.0
Coke, wt %	32.5 (fuel)	12.0 (anode)
S content, wt %	6.5	2.5
V content, ppm	300	75

[a]Boiling range of hydrocarbon fraction.

to growing environmental problems worldwide, mean that the residue problem will in the future be one of the most important tasks for crude oil processing.

2.3.7
Gasoline Upgrading Processes

2.3.7.1 Alkylation

The alkylation process is based on the reaction of isoparaffins (e.g., isobutane) with olefins (e.g., propene, butenes, pentenes) to form higher molecular mass branched paraffins in the presence of a strong acid, e.g., sulfuric acid. The alkylation product, or alkylate, is obtained as a mixture of isoparaffins with high research octane numbers, making the product an excellent blending component for motor gasolines.

A typical example of alkylation is the reaction of isobutane (1-methylpropane) with 1-butene to form isooctane (2,2,4-trimethylpentane):

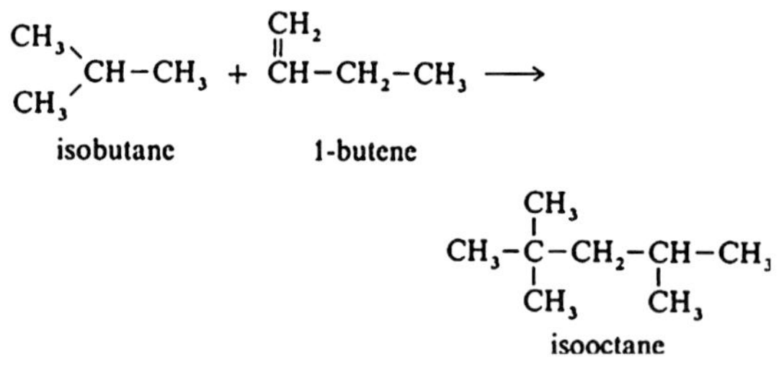

Alkylation reactions are catalyzed by strong acids. The first generation of commercial units applied sulfuric acid as the catalyst. During World War II alkylation plants with large capacities were installed to satisfy the drastic increase in the demand for high-octane gasoline. Alkylation units could easily be installed in refineries where olefins were readily available from catalytic cracking units. Isobutane was supplied by LPG streams, e.g., from reformers, and by C_4-isomerization. The newer generation of alkylation units used hydrogen fluoride (HF) as the catalyst, which has a better regeneration ability than sulfuric acid.

The most commonly applied HF alkylation processes were developed by Phillips Petroleum [42] and UOP [43], [63]. A scheme of the last process is shown in Fig. 2.31. The olefin feed and recycle isobutane are contacted with HF in the reactor (a). The reaction product is separated in the settler (b), from where HF is recycled to the reactor, while the alkylation product is fractionated in the isostripper (c) together with mixed butane feed. The isostripper overhead product is further separated in the depropanizer (d), from where the isobutane is recycled to the feed. The propane product is freed from entrained catalyst in the HF stripper (e). *n*-Butane is drawn off as a byproduct from the isostripper. The bottom product of the isostripper is alkylate, which is supplied for gasoline blending.

2.3.7.2 Polymerization

The polymerization process to produce additional gasoline is used mostly in refineries where excess amounts of olefins, e.g., propene and butenes from catalytic

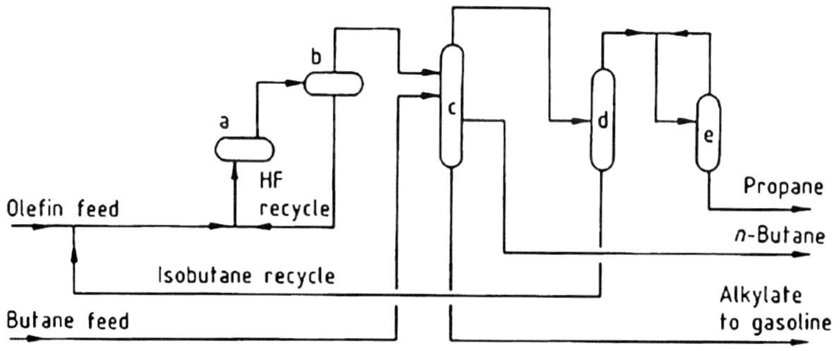

Fig. 2.31 HF alkylation a) Reactor; b) Settler; c) Isostripper; d) Depropanizer; e) HF stripper.

crackers, are available. The most widely applied process is the UOP polymerization, using phosphoric acid as catalyst. The C_3/C_4-olefin feed is contacted with the acid in a reactor. After separation and stabilization, the polymerate is obtained and used as a gasoline component. Another polymerization development is the IFP Selectopol process [44], [45].

The more recent IFP Dimersol process [46], [47] is a dimerization process for olefins. Three types of Dimersol processes are now available for producing gasoline components from olefins [64]:

Type G	for high-octane gasoline from propene
Type X	for octenes from butenes, or for hexenes – heptenes – octenes mixtures by co-dimerization of propene and butenes
Type E	for gasoline from lean ethylene streams

A simplified scheme of the G-Type Dimersol version is shown in Fig. 2.32. The feed is contacted with the liquid catalyst in the reactor. After separation from the catalyst, the reaction product is debutanized in a stabilizer column.

2.3.7.3 **Isomerization**

Isomerization of low molecular mass *n*-paraffins has been used for a long time, particularly for isomerizing *n*-butane to isobutane in order to produce makeup

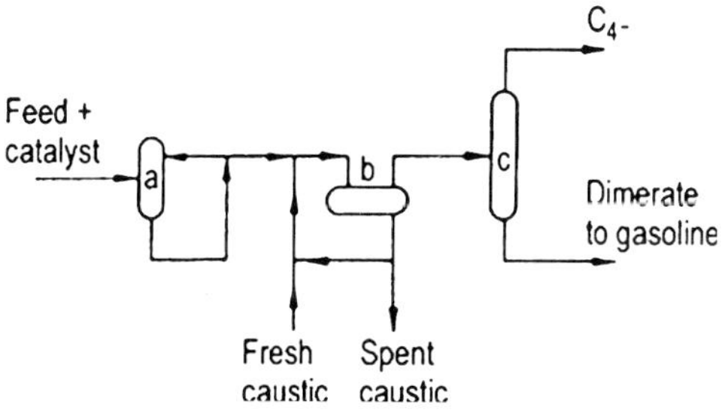

Fig. 2.32 Dimersol process a) Dimerization reactor; b) Catalyst recovery; c) Stabilizer column.

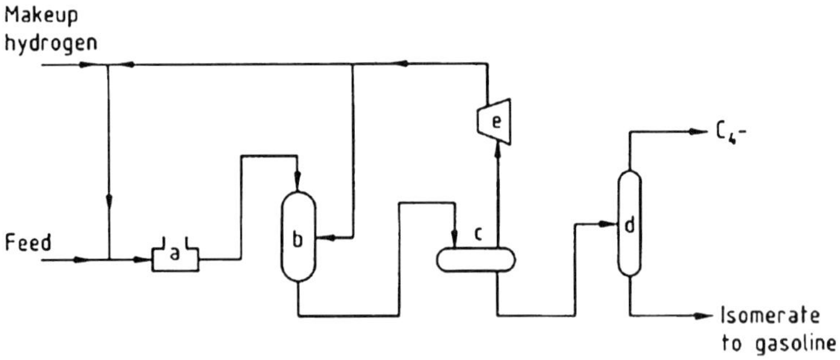

Fig. 2.33 Hysomer process a) Process heater; b) Isomerization reactor; c) Reactor product separator; d) Stabilizer column; e) Recycle gas compressor.

feedstock for alkylation units. Isomerization gained importance after the introduction of lead-free gasolines, to compensate for the reduction in octane number. This was achieved by isomerizing C_5 and C_6 *n*-paraffin fractions to the corresponding isoparaffins and led to the development of various catalytic isomerization processes that used hydrogen and operated at moderate reaction conditions.

A typical example of this type of hydroisomerization is the Shell Hysomer process [48], [49] shown in Fig. 2.33. The C_5/C_6-rich light gasoline feed is heated up together with hydrogen in the furnace (a) and isomerized in the reactor (b) containing a noble metal catalyst. The reaction product is separated, and the stabilized isomerate run down as a blending component for motor gasoline.

To maximize production of C_5/C_6 isomers, the total isomerization process (TIP) can be applied. This is a combination of the IsoSiv (Union Carbide) and the Hysomer processes, providing an iso/*n*-paraffin separation in a molecular sieve unit and subsequent isomerization of the fraction containing the *n*-paraffins.

2.3.7.4 **Production of Ethers (MTBE, ETBE)** [47a]

When high-octane unleaded motor gasoline was required as a prerequisite for the introduction of exhaust catalysts in the 1970s in the USA and in the 1980s in Europe, the use of methyl *tert*-butyl ether (MTBE, see Chap. 6, Section 6.1.) as a blending component became attractive. MTBE was first produced in 1976 as a byproduct at Chemische Werke Hüls, Germany. Its high octane rating of RON 117 and its rather low boiling point of 55 °C made it a valuable gasoline component. The ether is produced by reaction of isobutene with methanol at 40–70 °C on an acidic ion exchange resin followed by distillative separation. The presence of water in the reaction mixture leads to the formation of *tert*-butanol (TBA), another excellent gasoline component. In a similar way ethyl *tert*-butyl ether (ETBE) can be produced from isobutene with ethanol as reactant. As the utilization of ethanol from agricultural production is promoted because of political reasons, an MTBE plant can be modified to produce ETBE [47b]. In the USA, production of *tert*-amyl methyl ether (TAME) is favored, with methylbutenes as unsaturated hydrocarbon reactants. The characteristics of ethers as blending components for motor gasolines are listed in Table 2.11.

Table 2.11 Characteristics of ethers as components for motor gasolines

Property	MTBE	ETBE	TAME
RON	112–130	120	105–115
MON	97–115	102	95–105
Vapor pressure, bar	0.54	0.28	0.17
Boiling point, °C	55	72	88
Density, kg/L	0.742	0.743	0.788

While the blending properties of these ethers are extremely good, their chemical constitution is a disadvantage. All ethers contain a tertiary carbon atom, resulting in a very slow biological decomposition. Soil that has been contaminated accidentally can not be cleaned easily. For this reason the ethers have been banned from gasoline in sensitive areas of the USA (see Chap. 6, Section 6.1.13.).

2.3.8 Other Processes

2.3.8.1 Gas Treating

Most refinery gas and LPG streams contain poisonous and/or corrosive contaminants, e.g., H_2S, mercaptans, COS, and CO_2, which have to be removed by suitable physical or chemical methods. The gases should be moisture-free, as water in hydrocarbon mixtures may form hydrates which can plug lines and equipment.

Commonly used gas treatment processes for H_2S and other contaminants are based on chemical absorption, using regenerable solvents such as alkanolamines, e.g., monoethanolamine (MEA), diethanolamine (DEA) or diisopropanolamine (DIPA). The Sulfinol process, using a mixed sulfolane – alkanolamine solvent, is also widely applied.

As a typical example the purification of a gas stream in a DEA treater is shown in Fig. 2.34. The gas is contacted with the "lean" (pure) solvent in

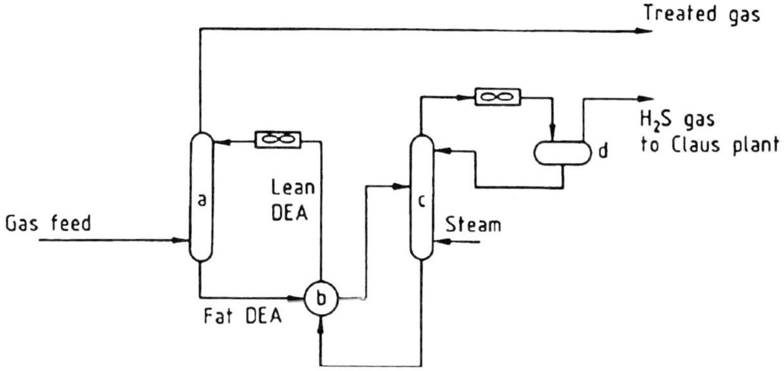

Fig. 2.34 Alkanolamine (DEA) gas treatment a) Absorber column; b) Lean – fat solvent heat exchanger; c) Regenerator; d) Overhead drum.

countercurrent flow in an absorber column (a), where H_2S and CO_2 are absorbed by the solvent.

The "fat" (loaded) solvent enters the regenerator (c), where the contaminants are stripped off by heating and steam injection. The H_2S-concentrated gas is then routed to the sulfur recovery plant.

Caustic treatment is the preferred process for the removal of mercaptans from LPG, using alkali hydroxide solutions (e.g., caustic soda) as solvent. Dryers containing caustic soda or other desiccants are applied to remove traces of water from the LPG.

2.3.8.2 Sulfur Recovery

The Claus process is the most widely used method for sulfur recovery, and is indispensable for most refineries with gas streams containing H_2S. The process is based on burning part of the H_2S to form SO_2, which subsequently reacts with H_2S to elemental sulfur:

$$H_2S + \frac{3}{2}O_2 \rightarrow SO_2 + H_2O \tag{6}$$

$$2H_2S + SO_2 \leftrightarrows 3S + 2H_2O \tag{7}$$

Owing to the reaction equilibrium, the recovery of sulfur is limited to 94–95 %.

The Claus off-gas (containing sulfur gases) is usually burnt in an incinerator to form SO_2. The sulfur recovery can be increased to above 99.5 % by applying a separate treatment process, e.g., the Shell Claus Off-gas Treating (SCOT), in which the SO_2 is catalytically converted into H_2S, and the H_2S is absorbed by means of solvent (DEA) circulation to and from the Claus unit.

2.3.9 Catalysts

With the exception of distillative and purely thermal processes as visbreaking and coking, processes based on catalysts are the backbone of modern refinery production. Starting with catalytic cracking in the 1930s and catalytic reforming in the early 1950s combined with catalytic hydrodesulfurization, the number of catalyst-based processes has steadily increased. In parallel, the types and quality of catalysts have changed. This has been demonstrated in Section 2.3.2. for cracking, in Section 2.3.3. for hydrodesulfurization and in Section 2.3.4. for reforming catalysts. Furthermore, alkylation, isomerization, hydrocracking, and other processes rely on catalysts, which are steadily being improved by their manufacturers. New developments encompass the carrier material in addition to the active elements and possible promoters. The main aim is to increase activity, stability, and selectivity. As the improvement in catalyst technology is essential for the competitiveness of modern refineries, the state-of-the-art is being presented in numerous articles in scientific journals and discussed at specific international seminars [49a][49b][49c].

2.4
Environmental Protection in Oil Refining

2.4.1
Introduction [50–52]

The natural product crude oil and the products obtained from it are generally easy to handle and with careful treatment – which is necessary because of the flammability – cause little pollution to the environment. (The CO_2 emissions associated with all fossil fuels and the corresponding effects on the earth's climate cannot be considered in this context.)

Environmental problems arise, however, in dealing with crude oil and its products because the petroleum industry – unlike virtually no other branch of industry – maintains exploration, transport, and refining installations for crude oil that are scattered over the entire globe. The distribution and consumption of heating and transportation fuels are even more scattered (Fig. 2.35). As the geographical location of crude oil fields does not coincide with that of the large consuming and refining regions, enormous transportation distances exist. The large amounts handled – the annual refining capacity in 2004 was about 4×10^9 t in more than 650 refineries worldwide – means that the industry needs to take precautions in all areas with regard to air emissions, water and soil pollution and, to a lesser degree, noise. The problems of introducing uniform standards are caused by legal standards which differ from country to country depending on the local concentrations of industrial areas with their corresponding environmental pollution and population density.

Because of the high cost of technical measures for environmental conservation, harmonization of legislation is absolutely necessary. As these costs amount to more than 25 % of the total processing costs in the refineries, equal competition between different countries must be guaranteed (see Section 2.4.4.).

In the following sections, the environmental problems of crude oil refining, storage and loading, and in the application of the products are discussed separately, because the problems in the various fields differ.

2.4.2
Manufacturing Emissions

Although closed, gas-tight systems are generally used in refinery units, emissions into air and water cannot be completely avoided even with careful handling during refining and storage of the crude oil and its products. This is due to the management and control of the process and the properties of the products concerned.

Hydrocarbons are discharged into the air because of their high vapor pressure and they appear in refinery wastewater effluents because of their water solubility, which is, however, small. Aromatic hydrocarbons, especially benzene, are considered particularly dangerous due to carcinogenic effects.

Further attention must be paid to the sulfur and nitrogen compounds originating from the heteroatomic compounds in the crude oil, both because of their smell

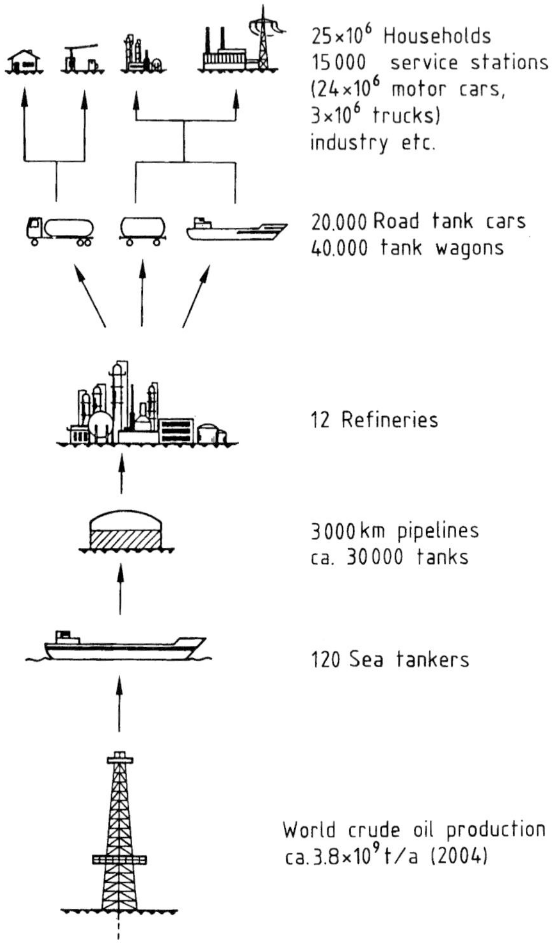

Fig. 2.35 Petroleum's route from the well to the consumer shown for Germany as an example (data from 1998 or 2004).

and toxicity and because of the air pollution which arises in the form of SO_2 and NO_x emissions during combustion in process plant heaters.

2.4.2.1 **Hydrocarbons** [53]

2.4.2.1.1 **Hydrocarbons in Air**

Hydrocarbon emissions can arise in *production plants* during normal operation: (1) from leaking flanges in the pipework system; (2) at the seals of valves, pumps, and compressors; and (3) in the course of sampling. In the case of an accident or a malfunction of a unit, gases are led in closed systems to gas recovery systems, where they are compressed and returned to the process. The remainder of the gases is burnt in elevated or ground flares at efficiencies which can exceed 99 %. Liquid products from leakages of fittings or seals are collected in closed loop systems that are equipped with pressure reservoirs and tanks and later returned to the production circuit.

In new plants measures to reduce emissions are taken during their construction, whereas continuous retrofitting is necessary in existing plants. Examples are flangeless piping, low-emission stuffing boxes, and seals, such as duplicated slide ring packings. For intensely odorous, poisonous, and carcinogenic substances more far-reaching measures are necessary (canned-motor pumps, special extraction devices, etc.).

Most hydrocarbon emissions in processing occur in the *storage areas,* i.e., tank farms for crude, feedstocks, intermediate, and final products. Pressure – vacuum relief valves and floating cover tanks are generally used to reduce emissions. More recent developments are emission-free tank farms, where several fixed roof tanks fitted with internal floating roofs breathe into a closed system at one common gas holder, which normally absorbs all changes in the tank level. In the rare event of unusually large changes in the system (large amounts flowing into or out of storage, solar irradiation, heavy rain), the surplus quantity is burnt in an associated flare or, in the case of a pressure drop, the system is topped up with inert gas.

Considerable amounts of hydrocarbons are also emitted in the *loading facilities,* especially when loading gasolines. Here low-emission or emission-free loading for transport by ship, rail, and road has become largely accepted. Various methods are used (Fig. 2.36):

1. Vapor recovery, where in a closed system the displaced gasoline vapors are either returned to the product tank or collected in a gas holder and used for the firing of process plants.
2. Regenerative adsorption of the vapors on suitable adsorbents.
3. Recovery of the products in liquid form after cooling or washing out the vapors.

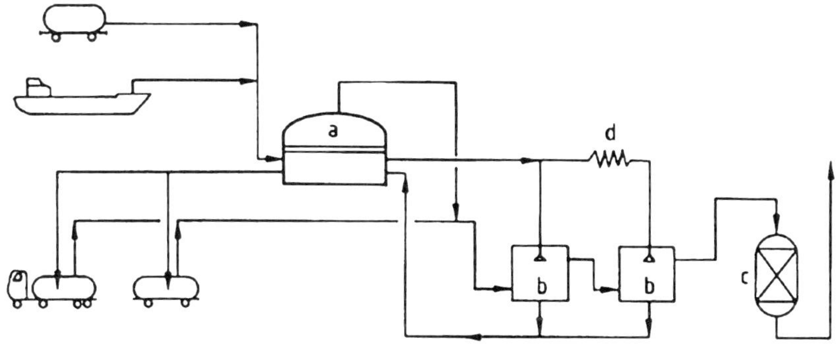

Fig. 2.36 Emission-free loading and unloading of gasoline hydrocarbons with off-gas purification (refinery, intermediate storage) a) Storage tank with floating roof; b) Exhaust gas washes (gasoline); c) Fine purification (adsorption); d) Low-temperature cooling (to – 40° C).

Care must be taken to prevent the formation of explosive gasoline – air mixtures which can occur in the road tanker to be loaded and in the adjacent piping and equipment.

This can be achieved by: (1) keeping the concentrations outside the explosive range, (2) short transportation paths and exclusion of ignition sources, and (3) extremely strict control of the oxygen contents.

Dispensing of gasoline from the road tanker to the service station is also increasingly carried out with vapor balancing between the road tanker and the installed storage tank of the service station. The possibilities of reducing emissions in the final step (filling the customer's motor vehicle) are described in Section 2.4.3.1., Transportation Fuels.

The measures described have reduced the atmospheric emissions from processing, storage, and distribution of refinery products in the mid-1980s in Western Europe to < 8 % of the total man-made HC emissions. The portion from the refineries themselves was in turn only a quarter of this value. More recent investigations by Concawe show further improvements. Thus, relative to the crude oil throughput, refinery emission losses of ca. 0.05 wt % are expected, depending on the structure and size of the plant. For inland refineries in Germany, for example, values of 0.02–0.03 wt % have been published.

2.4.2.1.2 **Hydrocarbons in Wastewater**

Hydrocarbon-containing wastewater is generated at various points in the refinery because of the water content of the crude oil itself and because steam is employed in various processing steps. The total amount of hydrocarbon-containing wastewater in a normal refinery is of the order of 60–100 m^3/h.

This wastewater must be removed from the process units after separation from the oil phase, and led via a closed system to a wastewater treatment plant. Rainwater from exposed plant areas and from tank yards, and possibly contaminated cooling water from leaks or accidents must be treated in the same way. Considerable buffer volumes must be made available for the latter amounts of water that are formed discontinuously and sometimes in large quantities. However, some refineries are using rainwater as an additional salt-free freshwater source.

Treatment in the wastewater purification system is carried out stepwise by:

1. mechanical separation (sieves, filters, oil – water separators)
2. physicochemical purification (stripping, flocculation, flotation)
3. biological treatment

Biological treatment of hydrocarbons in refinery wastewater is normally possible without any problems. However, the incoming streams and the corresponding buffer volume must be continuously monitored to detect pollution by sulfides or nitrogen compounds and by oxygen-containing components such as phenols.

In many countries there is an increasing legal requirement for covered water treatment plants to avoid odor and for total nitrogen removal to improve protection of surface waters (rivers, lakes, etc.). The latter usually requires an additional purification stage with increased residence time.

After the biological stage, the water is clean and can be discharged into the receiving water. The average analyses of the wastewater from a modern refinery (in mg/L) are:

COD	60–100
BOD	5–15
Oil	0.5–2
Settleable solids	0.1–0.3
Phenols	0.1–0.2

As a result of increased wastewater purification, the amount of hydrocarbons discharged by the refineries in Western Europe was reduced by about 95 % in the period 1970–2000, for about the same amount of crude oil processed (Table 2.12) [54].

2.4.2.1.3 Hydrocarbons in Soil and Groundwater

As hydrocarbons are water-soluble (even if only to a small degree), their penetration into the soil with possible contamination of the groundwater must be carefully avoided, whenever crude oil or any of its products are handled.

Transport from the oil terminal to the refinery is carried out almost exclusively in underground pipelines, which are also the safest means of transport. The choice of high-grade steels as construction material, good insulation, cathodic corrosion protection, and continuous monitoring for leaks, including visual monitoring from aircraft or ground inspection, ensure a high level of safety. On difficult terrain and in areas of extreme temperature fluctuations, additional measures must be taken (pipeline compensation, elevated piles, intermediate tanks, etc.).

The location of the refinery must be carefully selected with regard to possible dangers to drinking water. According to new legislation, all HC-handling units must be erected so as to prevent the discharge of spilled products underground, to

Table 2.12 Oil discharged from European refineries between 1969 and 2000

Year	Number of refineries reporting	Total oil discharged, t/a	Oil discharged, g per tonne capacity	Reported capacity, 10^6 t/a
1969	73	44 000	127	(400)
1974	101	30 700	45	(730)
1978	109	12 000	16	754
1981	105	10 600	15	710
1984	85	5090	8	607
1987	89	4640	7.9	587
1990	95	3340	5.8	570
1993	95	2020	3.3	618
1997	105	1170	1.74	670
2000	84	750	1.32	566

adjacent streets, or canals; the storage tanks are placed in collection spaces which are made impermeable to oil using clay layers, plastic tilts, or concrete lining; in case of a leak they must be capable of receiving the entire tank contents.

If hydrocarbon contamination occurs in the soil, the affected portion of soil must be removed to prevent subsequent pollution of groundwater. If small amounts have escaped, the contaminated soil is usually combusted in incinerating plants. With larger amounts – and particularly if large areas are polluted with chemical residues and dangerous refuse – the damage must be treated *in situ*. Depending on the nature of the soil and the corresponding migration of the oil, this can be done by pumping off and purifying the contaminated water, possibly by additional injection of fresh water into adjacent wells.

Degradation of the oil by microorganisms is becoming increasingly important. This can also be done *in situ* or by excavating the soil and treating it externally [55].

Loading and storage of the products outside the refinery are subject to similar regulations. However, these are still very different in individual countries. As a result of the large number of external distribution depots for transport fuels and heating oils and of service stations, and because of the enormous number of oil-heated households, special care must be taken against overfilling and escape of products due to leaks. Corrosion-resistant and nonageing steels, plastic-lined steel tanks, and novel, glass-fiber-reinforced, plastic tanks are widely used. These tanks must also be constructed in such a way that the whole volume can be collected if leakage occurs (double-wall tanks).

In some countries with a high population density and dense housing and industrial areas, the regulations which apply to pure hydrocarbons (i.e., products) are also applied to the handling of process water with much lower HC-concentrations.

2.4.2.2 **Sulfur and Nitrogen Compounds**

As a natural product, crude oil also contains heteroatomic compounds containing sulfur, nitrogen, and oxygen in addition to hydrocarbons. Whereas the nitrogen and oxygen contents are in the ppm range and play only a secondary role in atmospheric emissions, the sulfur content of the crudes can be as high as several percent. The distribution over the individual refinery fractions varies, but the content increases with increasing molecular size.

2.4.2.2.1 **Sulfur Compounds**

Sulfur and its compounds are catalyst poisons and adversely affect atmospheric emissions. Hydrogen sulfide, mercaptans, and disulfides are odor nuisances, and sulfur dioxide is formed during the combustion of crude oil products. Therefore, sulfur and its compounds must be removed or their contents reduced.

The light refinery products, liquefied petroleum gas and gasoline, must be almost completely sulfur-free (Chap. 3). For diesel fuels and for light heating oils a substantial sulfur reduction to 0.1–0.5 % is required by recent legislation.

Serious problems exist with heavy fuel oil, which is used almost exclusively as fuel in large industrial furnaces and power stations and leads to considerable SO_2 emissions. Many countries have established a maximum sulfur content in fuels of 1–2 wt %. This value can be reached without additional treatment only with a few, low-sulfur crude oils, whose supplies are limited.

Sulfur is generally removed from distillates by hydrodesulfurization, whereby the chemically bound sulfur is converted into hydrogen sulfide. The H_2S is removed from the reaction gas stream in a gas scrubber, converted into elemental sulfur in the downstream Claus process, and supplied to the chemical industry as a raw material.

The intense smell of the H_2S-containing gases and the high toxicity of H_2S, even at high dilution, means that stringent precautions must be taken. All sulfur-processing plants must be completely gastight: in the hazard zones, instruments and alarm devices are installed which automatically shut down the plants in case of danger.

Sulfur recovery in refineries has increased greatly as the result of the lower permitted sulfur content of the products. In conventional Claus plants with two reactors connected in series, the conversion of the H_2S feed is ca. 95 %, a third reactor increases the conversion to 96 %. For a further reduction of the H_2S content an aftertreatment step (Scot process, Sulfren) is required. Almost complete removal, which is required in various countries (e.g., Germany 10 ppm H_2S in off-gas), can only be achieved with an additional high-temperature combustion step. Figure 2.37 compares the sulfur removal from the processed crude oil with the crude oil capacity in West German refineries for the period 1970–1988. During the subsequent decade the amount of recovered sulfur has increased to more than 3.7 t per 1000 tonnes of crude throughput. The increase in sulfur removal resulted in a dramatic reduction of overall SO_2 emissions, as most of the desulfurized hydrocarbons are used as fuels.

Sulfur dioxide emissions from the refinery stacks are relatively small. Depending on the refinery complexity, only 4–6 % of the crude oil input is required for fueling in the process plants. About half of this amount is covered by

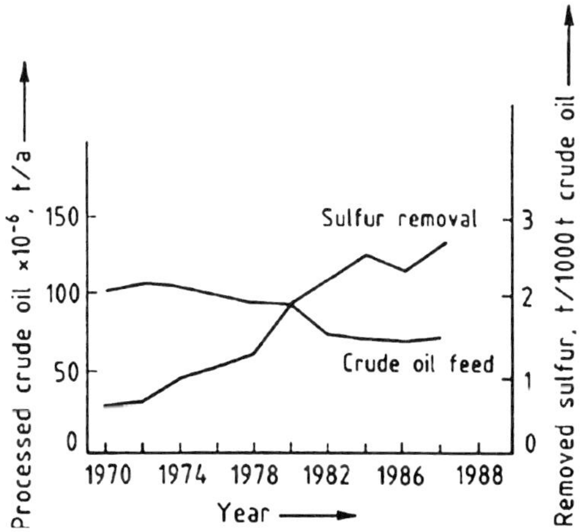

Fig. 2.37 Sulfur removal during crude oil processing.

low-sulfur or sulfur-free gases produced during processing, and the remainder is heavy fuel oil. With more severe requirements to lower SO_2 emissions, the refiner might be forced to use all of the LPG that is produced in the refinery as fuel.

Legislation differs greatly in individual countries: the stipulated limit values can be related to a single process unit, a stack, or the refinery as a whole and can refer to the total sulfur discharged or the sulfur concentration in the flue gas. Combinations of several measures are also possible, and so-called "bubble concepts" are used, particularly in the United States. (Bubble concept means that for a given amount of emission from the total complex, different limit values or amounts discharged are allowed for each individual plant.)

The expensive process of flue gas desulfurization is normally uneconomical for firing installations of the size of those employed in a normal refinery. Therefore it is used only in special cases, e.g., if the refinery is connected with a large power station or if the sulfur content of the fuel is extremely high.

2.4.2.2.2 Nitrogen Compounds

The nitrogen content of most crude oils is relatively low, and in the distillates it is reduced during hydroprocessing to a few ppm. The residual nitrogen content causes no further difficulty in the use of the products.

Ammonia, which is formed in the hydrogenation steps and added in many refineries for process control in various process stages (pH adjustment), enters the wastewater and can be removed in the biological stage of wastewater purification (see Section 2.4.2.1.2.).

2.4.2.3 Noise

In the past, noise pollution in the neighborhood of petroleum refineries has played only a secondary role. Worldwide, pumps and compressors were installed not in closed buildings but in the open air, because of the flammability of petroleum and its products. However, because the plants were smaller at that time, the noise emission of the units was relatively small. In addition, the sound radiation from furnaces and their burners produced relatively low sound levels in the refinery surroundings, because of the compact construction of the main process plants that were simultaneously screened by low-noise auxiliary installations (tank farms) at the periphery.

Problems arose, though, in densely populated areas with insufficient distances between industrial and residential areas. Most noise problems were caused by flare noise in the event of process disturbances.

Owing to growing environmental awareness in the public and a correspondingly greater strictness of legislation, further measures will have to be taken in the future to reduce sound emissions.

Important measures for noise reduction could cover the following items:

1. Low-noise burners and additional noise insulation on process heaters and piping.
2. Sound hoods on drive motors and turbines of pumps and compressors. A complete "in-housing," however, may also give rise to safety problems, because fire fighting is more difficult and the danger of explosion is greater in closed rooms.

3. Sound insulation on control valves.
4. Low-noise flare stack tips for elevated flares or additional ground flares.

A noise reduction by 10 dB(A), i.e., a reduction of the observable noise level by about one-half, may be possible depending on the specific refinery and the noise abatement already installed, and according to the state-of-the-art, though only with considerable expenditure.

2.4.3 Consumer Related Emissions

The emissions of SO_2 and NO_x from the combustion of refinery products outside the refinery are much greater than the emissions during refining itself. In addition, the HC emissions in the transportation field, both in refueling and in running motor vehicles, are also considerably greater than in the steps from production to the filling station (see Table 2.13). A newer emission assessment for 1998 based on German figures [52] indicates that although the relationship between sources has remained fairly constant, the total amount has decreased to about 2500 t. Significantly lower figures were estimated for distribution and refueling.

2.4.3.1 Transportation Fuels

When gasoline is burned in a spark ignition (Otto) engine, varying amounts of CO, NO_x, and unburnt hydrocarbons are emitted with the exhaust gas in addition to the combustion products water and CO_2. The amounts differ according to the driving style, type of fuel, and engine construction. Besides these substances SO_2 and soot particles are emitted from diesel engines (Section 3.3.2.9.).

In addition to exhaust emissions, the emission of hydrocarbon vapor during refueling and running of motor vehicles is an important environmental pollution factor.

Table 2.13 Hydrocarbon emissions in the petroleum field (Western Europe 1986)

Source of emission	Proportion	
	10^3 t	%
Crude oil production	20	0.4
Refinery, storage	170	4.1
Distribution	310	7.4
Vehicles		
Refueling	180	4.3
Exhaust gas	2500	59.7
Evaporation	1010	24.1
Total	4190	100.0

2.4.3.1.1 **Motor Gasoline** [56]

Different factors cause emissions of hydrocarbons from motor vehicles: firstly, the unburnt hydrocarbons in the motor vehicle exhaust, and secondly, the hydrocarbons emitted during refueling and running, because of their high vapor pressure. In Western Europe in 1986, the emissions (in 10^3 t/a) were distributed as follows:

Exhaust emissions	2500	≙	68 %
Evaporative emissions	1010	≙	27 %
Refueling losses	180	≙	5 %

Specific measures are provided to reduce these different types of emission either in the motor vehicle or at the filling station.

To effectively reduce exhaust emissions, the use of catalysts in the exhaust gas line has become state-of-the-art (Chap. 7).

Efforts to reduce evaporative emissions and refueling losses are less clear-cut. Their proportion in the emissions from the motor vehicle sector is ca. one-third of the total output, corresponding to ca. 12 % of the total man-made emissions of organic substances.

Two solutions are under discussion (Fig. 2.38) [57]:

1. Vapor recovery at the service station, possibly supplemented by a smaller activated carbon canister in the motor vehicle.
2. A larger activated carbon canister fitted into the car, whose charge can adsorb all of the displaced hydrocarbons.

Vapor recovery at the service station is only a partial solution, because most of the emission is produced when the car is running and is not influenced by this measure. The large activated carbon canister ("LCC, on-board system") is considered to be the better solution in the long term because of its greater effectiveness and lower cost. Technical trials had been carried out with various types of vehicles. Installation in new cars presents no problems; while retrofitting to old vehicles will only be possible in a few cases. However, when the choice was made, policy makers in Germany were of the opinion that an efficient system at service stations could be installed within a fairly short period of time, while it would take longer to equip a high portion of the automobile population with the LCC. So vapor recovery at service stations was made mandatory. This system has been adopted by several other EU member states. Others, however, are rather reluctant, so only between 5 and 10 % of the UK filling stations have been equipped with vapor recovery equipment. In the United States, however, the large carbon canister became mandatory by an EPA final rule of January 24, 1994. A yearly increasing percentage of new cars, starting in 1998 with 40 % and ending in 2000 with 100 %, had to be equipped with LCC, so eventually the larger part of the car population will possess this canister. Equipping of light trucks followed in 2003.

Owing to the wide use of motor fuels, particular attention must be paid to benzene emissions. The limit for the benzene content of gasoline (1 % vol) presented a

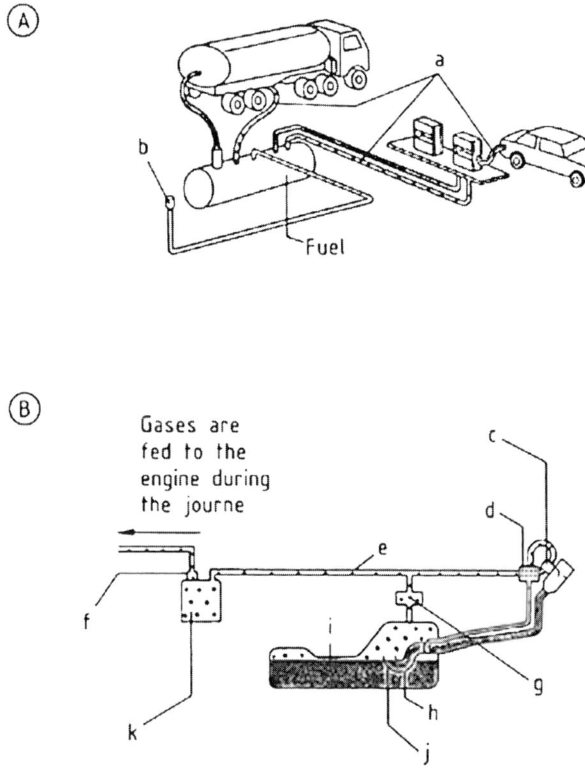

Fig. 2.38 Reduction of hydrocarbon emissions A) Vapor recovery at the service station; B) Large carbon filter in the motor vehicle a) Gas displacement pipe; b) Vent; c) Gas venting valve actuated by filling nozzle; d) Gas – liquid separator; e) Gas line; f) Magnetic valve and regeneration control orifice; g) Standard gas vent and overturn protection; h) Outlet; i) Fuel tank; j) Liquid seal in filling tube (reduces escape of gases); k) Activated carbon filter with 4.5 L capacity (traps gases).

great challenge for the oil industry, because the benzene content of various blending stocks can be considerably higher, depending on the quality or origin of the crude, up to 8 vol % for high severity reformate and 18 –40 vol % for pyrolysis gasoline.

2.4.3.1.2 **Diesel Fuel**

The main exhaust gas problems of diesel vehicles result from the emissions of soot particles and SO_2. The NO_x and CO emissions are less important than in the spark ignition (Otto) engine exhaust. Demands are increasingly being made for a reduction of these emissions, because of the smell of the exhaust gases, and because the soot particles may have a carcinogenic effect. Emission reduction of diesel fuels is discussed in more detail in Chap. 3, Section 3.3.2.9).

2.4.3.1.3 **Marine Fuels** (see Chap. 9)

For a long time no specific attention had been given to exhaust emissions from sea-going ships, as most of these emissions are generated on the open seas. When it was realized that in ports, rivers leading to them, on coastal waterways and even in special sea areas as the North Sea or the Baltic, SO_2 and NO_x emissions from ships using heavy marine fuels contributed markedly to inland pollution, the IMO, the International Maritime Organisation, became engaged. After intense international negotiations a schedule to reduce the sulfur content of marine fuels (which had had an upper limit of 5 wt %) was agreed on. Marketing of high-sulfur residues as marine fuels, a welcome outlet for the sulfur-stressed refinery, will therefore become more and more restricted.

2.4.3.2 Fuels for Heat Generation

The largest proportion of refinery products worldwide is used as fuels for domestic or industrial heating as well as for electricity generation in power stations. These are LPG, the middle distillates, and heavy fuel oil. Emissions of SO_2, NO_x, and CO that are associated with combustion, as well as emission of ash and unburnt carbon (in the form of soot) can be considerably influenced by selection and pretreatment of the products used and by the burner design. This is particularly true for the large heating installations operated with sulfur-rich heavy fuel oils.

Liquefied Petroleum Gas The liquefied petroleum gases used as fuels are almost sulfur-free as a result of pretreatment in the refineries, and they burn without formation of soot. Their use presents no problem for the user.

Light Heating Oil (US No. 2 Fuel) Middle distillates contain up to 1.5 wt % sulfur, depending on the origin of the crude. Most of the sulfur can be removed in the refinery by hydrodesulfurization (see Section 2.4.2.2.).

The permissible sulfur limits for light heating oil in most countries are <0.5 wt %, because of its use in domestic heating. In the EU, the sulfur content of domestic heating oil is now limited to 0.2 wt %. If lower sulfur contents are to be achieved, desulfurization costs will rise steeply with the current range of crudes, according to studies by Concawe [50] (Fig. 2.39).

Heavy Fuel Oil (US No. 6 Fuel) The residues from crude oil refining are used as heavy fuel oil for industrial heating and energy production. Sulfur dioxide, nitrogen oxides, CO, and particulate emissions must be considered for heavy fuel oils.

2.4.4
Cost of Environmental Conservation

Because the understanding of environmental problems, technical solutions, and legislation are still in rapid development, a view of the expenditure on environmental conservation can only be estimated. Such an assessment is necessary for an overall view of the problem, because the costs of environmental conservation have reached such a high level that they markedly influence the total processing

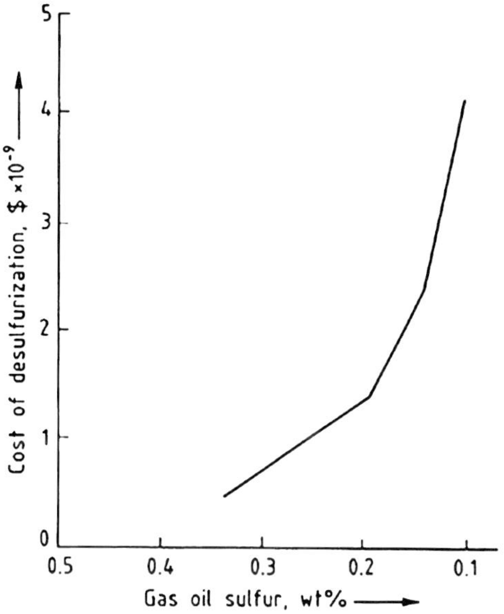

Fig. 2.39 Investment to reduce gas oil sulfur in Western Europe.

costs. The great variety of possible solutions worldwide and the different legislation in the various countries and regions also have a corresponding impact on costs and can throw the market off-balance in specific areas, e.g., in Europe.

Besides the investment costs required for new plants (15–20 % of the total cost is used for environmental conservation) and the conversion of old plants, there are continuous operating costs for energy, maintenance, personnel, etc. It is difficult to give current, generally applicable figures due to the increasing debate on the environment and the rapid development in this sector. Table 2.14 shows the growth of environmental conservation costs in Germany from 1977 to 1991.

A study was set up in 1986 by the Commission of the European Communities to determine the environmental conservation costs for refineries in the EC member countries [59]. This considered the current (1985) and future (1993) operation of typical refineries and determined the costs of keeping to environmental regulations – at the refinery sites and with regard to product specifications. The current standards

Table 2.14 Investment and operating costs for environmental conservation in the Federal Republic of Germany in comparison with crude oil processing

	1977	1979	1981	1983	1985	1987	1989	1991
Crude oil processed, 10^6 t/a	99.5	109.6	98.9	90.2	87.3	81.4	83.5	97.9
Investment costs, 10^6 DM	294	127	130	305	103	121	197	310
Operating costs, 10^6 DM	826	894	1129	1212	1097	900	921	1036

and regulations and those expected in future were taken into account. The most important measures in the processing field are:

1. Gas washing systems and Claus plants for sulfur recovery from H_2S followed by fine purification.
2. Closed systems for the discharge of gaseous (via a flare) and liquid hydrocarbons.
3. Floating roofs and floating covers in storage tanks for crude oil and products.
4. Hydrocarbon vapor recovery systems for the storage and loading of volatile products.
5. Collection spaces for escaping hydrocarbons and precipitation water in production plants, tank farms and loading plants. Drainage to wastewater treatment plants.
6. Reduction of emissions (SO_2, NO_x) from refinery furnaces.
7. Closed sampling systems, laboratory analysis, and on-line instruments for pollutant measurement in air and wastewater.

The following measures are for the production of less polluting products:

1. Hydrodesulfurization of gasoline and middle distillates.
2. Reforming and isomerization for the production of high-octane components as a basis for unleaded gasoline.
3. Synthesis of suitable components for unleaded gasoline (MTBE).
4. Conversion of heavy residues into light, clean products.
5. Separate storage and loading of unleaded types of transportation fuels.

In determining the measures required and the resulting costs, the study considered the environmental conservation laws applicable in each EC member country at the time concerned or scheduled up to 1993. The results are given in Table 2.15. They show that: (1) the environmental costs rose in all EC countries; (2) there are large differences in cost between the individual countries; (3) for comparable costs arising from the EC directives, the additional national legislations lead to different

Table 2.15 Environmental costs of petroleum processing in the EC member states (in ECU/t)

		1993	1993
Country	**1985**	**EC directives only**	**EC+national regulations**
Belgium	1.01	6.25	8.57
France	1.05	3.11	4.54
Germany	3.38	5.87	19.28
Italy	0.34	2.15	2.52
Netherlands	1.01	6.12	17.07
Spain	0.31	3.21	3.58
United Kingdom	0.28	6.64	7.01

costs; (4) the burden of environmental costs is in some countries of the same order of magnitude as that of the other processing costs.

The figures given are more likely to be near the lower limit: the study did not, e.g., consider noise protection measures, soil clean-up or special regional regulations, because these can differ greatly according to the location of the refinery. Moreover, the conservation of groundwater and consideration of climatic changes are becoming increasingly important, and are continually leading to a tightening of the limits of the (current) standards.

The investigations carried out for the EC can be applied to all other regions of the world. Taking into account the high expenditure for minimizing potential dangers (flammability, explosion risk, transport risks) and for protection of the plant workers, crude oil refining is burdened with extremely high additional costs related to product value and technical efforts.

Although this study has not been validated for the present year (2006), the unceasing political demands to reduce, for environmental reasons, plant emissions on the one hand and the emissions caused by the consumer by improving product quality on the other, as discussed in Sections 2.2.3. and 2.4.3. have increased the 1986 figures considerably. This can easily be derived from Table 2.14, where data for Germany are presented at least up to 1991. Figure 2.39 demonstrates the same tendency, as sulfur levels for gasoline and diesel fuel have actually dropped from 1000 mg/kg in 1985 to 10 mg/kg in 2006 and from 0.2 % in 1988 to 0.005 % in 2006, respectively.

2.5 Integrated Refinery Models

2.5.1 Trends of Refinery Structures

The drastic increase in crude oil prices in between the early 1970s and the first decade of the 21st century, in addition to changes in the product demand pattern in major consumer regions, i.e., North America, Europe, and Japan, have led to new considerations in integrated refinery processing. The objectives are to achieve better utilization of residual materials, a maximization of distillate production, an improvement in process economics, and for current environmental standards to be met.

These trends resulted in a worldwide restructuring of oil refining with the aim of increasing the yields of the high-value products – chiefly transportation fuels, light heating oil, and industrial feedstocks – at the expense of low-value heavy fuel oils. The simple "hydroskimming" type refining scheme has meanwhile been abandoned and great effort has been taken to clear the way for more complex refinery structures with integrated residue conversion processes. A comparison of three conversion concepts versus hydroskimming is given by the following schemes, which are based on Arabian light crude processing [60]. Examples of developments in thermal and catalytic cracking processes for heavy fuel oils and of developments in residue hydroprocessing, both in integrated refinery structures, are given in [61] and [62], respectively.

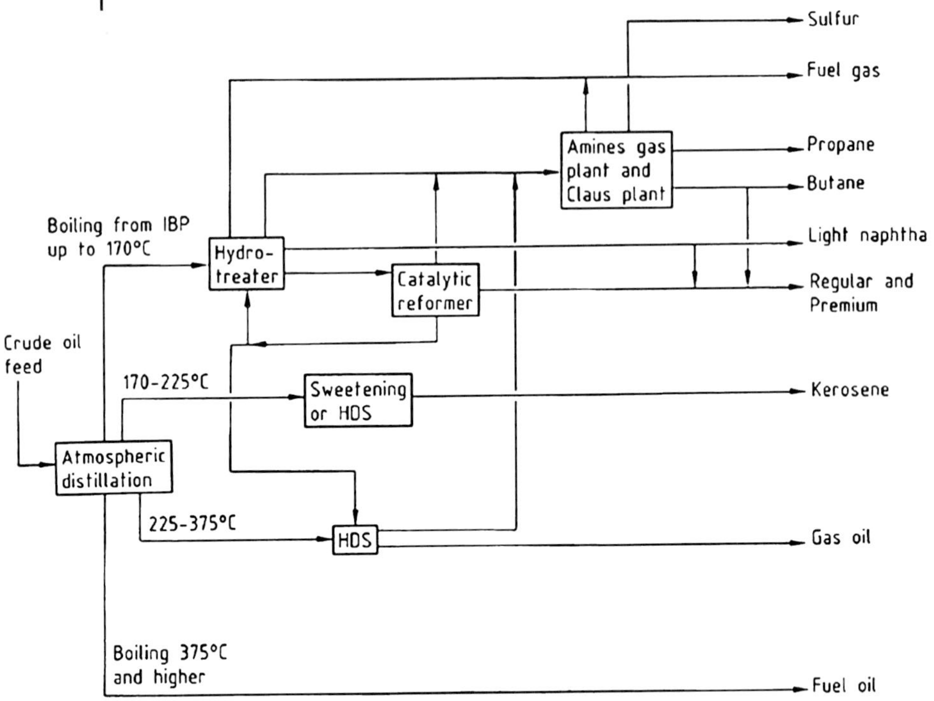

Fig. 2.40 Hydroskimming.

2.5.2
Hydroskimming Refinery

The now obsolete process configuration, known as hydroskimming or "fuel oil refinery," is shown in Fig. 2.40 as a basis for comparison. The plant layout consists of a crude distillation unit, hydrotreaters for gasoline and gas oil, a catalytic reformer, and a gas and sulfur workup section. It does not include a residue conversion unit. Atmospheric residue with a yield of 43.5 % of the crude represents by far the largest product portion, which has to be marketed as heavy fuel oil.

The naphtha yield is 17.9 % of the crude (15.9 % gasoline and 2 % chemical feedstock), and the middle distillates amount to 34.9 % (9.6 % kerosene, 25.3 % gas oil). A comparison of the product yields of the hydroskimming refinery with the yield structures attainable with different conversion concepts is given in Table 2.16.

2.5.3
Conversion Concepts

2.5.3.1 Cat Cracking – Visbreaking Refinery

A typical refinery for maximum gasoline production using an FCC unit is shown in Fig. 2.41. The FCC feedstock, vacuum gas oil, is produced by vacuum distillation and

Table 2.16 Yield structures of refinery conversion schemes for Arabian light crude processing [60]

Product	Hydroskimming [a]	FCC –VB[b]	HC – SDA –FCC [c]	HC –coking [d]
Intake, wt %				
Crude oil	100.0	100.0	100.0	100.0
Hydrogen		0.4	0.5	
Output, wt %				
Fuel gas	1.8	3.4	3.1	4.7
LPG	1.6	3.0	2.9	2.9
Chemical feed	2.0		1.7	
Gasoline	15.9	33.0	32.9	24.4
Kerosene	9.6	9.6	9.6	9.6
Gas oil	25.3	26.3	40.2	50.9
Fuel oil	43.5	22.7	9.2	
Coke		1.0 *f*	1.1 *f*	5.0 *f*
Sulfur	0.3	1.0	1.4	1.3

[a] Hydroskimming = refinery configuration without residue conversion processes.
[b] FCC – VB = conversion scheme with cat cracker and visbreaker.
[c] HC – SDA – FCC = conversion scheme with hydrocracker, solvent deasphalting, and cat cracker.
[d] HC – coking = conversion scheme with hydrocracker and coker.
[e] Coke burnt in FCC unit.
[f] Coke produced in coker unit.

pretreated in a hydrodesulfurizer for sulfur removal and FCC yield optimization. In a visbreaker unit, the vacuum residue is thermally cracked to produce additional distillates. As can be seen from Table 2.16, the residue output from this refinery type is 22.7 % of the crude, i.e., only about half that of the hydroskimming type. The gasoline yield amounts to 33 % and middle distillates make up 35.9 % of the crude.

2.5.3.2 Hydrocracking – Cat Cracking Refinery

A high-conversion refinery which combines fluid catalytic cracking and hydrocracking is shown in Fig. 42. The HC feed is vacuum gas oil, whereas the FCC feed (deasphalted oil) is produced in a solvent deasphalting unit (SDA) and desulfurized in an HDS unit. The asphalt product of the SDA unit can be used as heavy fuel oil; however, it must be blended with a "fuel thinner" (middle distillate) to meet fuel specifications. The hydrogen produced in the catalytic reformer is not sufficient for supplying the HC unit, therefore makeup hydrogen is produced in a

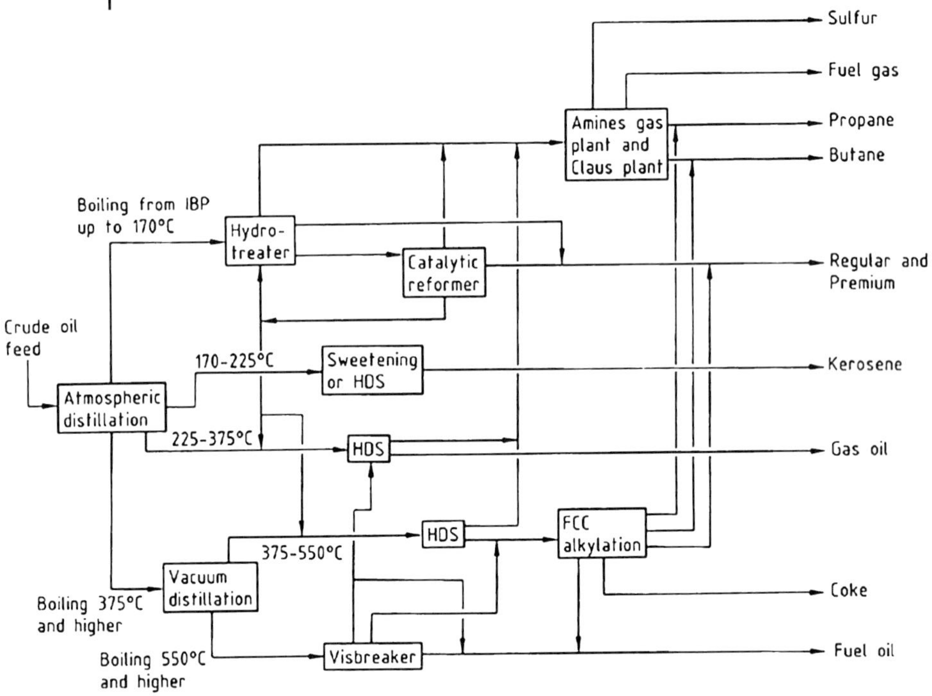

Fig. 2.41 Catalytic cracking – visbreaking.

steam reformer. As shown in Table 2.16, the net fuel oil portion is only 9.2 % of the crude. Gasoline and middle distillate yields are 32.9 % and 49.8 % respectively, i.e., an improvement in comparison with the two schemes previously described.

2.5.3.3 Hydrocracking – Coking Refinery

The hydrocracking – coking refinery is illustrated in Fig. 2.43. In this configuration the production of middle distillates is maximized by using a hydrocracker and a coker. Vacuum residue is used as coker feed while the HC feed is a combination of vacuum gas oils and heavy coker distillates. As in the previous scheme, extra hydrogen has to be supplied from a steam reformer.

The only residual product is 5 % coke (Table 2.16), which can be sold to the chemical industry or for electrode manufacturing. As no heavy fuel oil is produced, this process configuration is often referred to as a "zero-fuel" refinery. The yield of middle distillates (kerosene and gas oil) is >60 % and the naphtha yield (gasoline and chemical feedstock) is 26 %.

2.5.3.4 Integration of Existing Refineries

Instead of rebuilding or retrofitting an existing refinery to cope with the new challenges, petroleum companies have also chosen another route. This is the integration of two or more neighboring refineries, irrespective of the original ownership. Thus the main strength of each of the refineries can be used, supplementing the

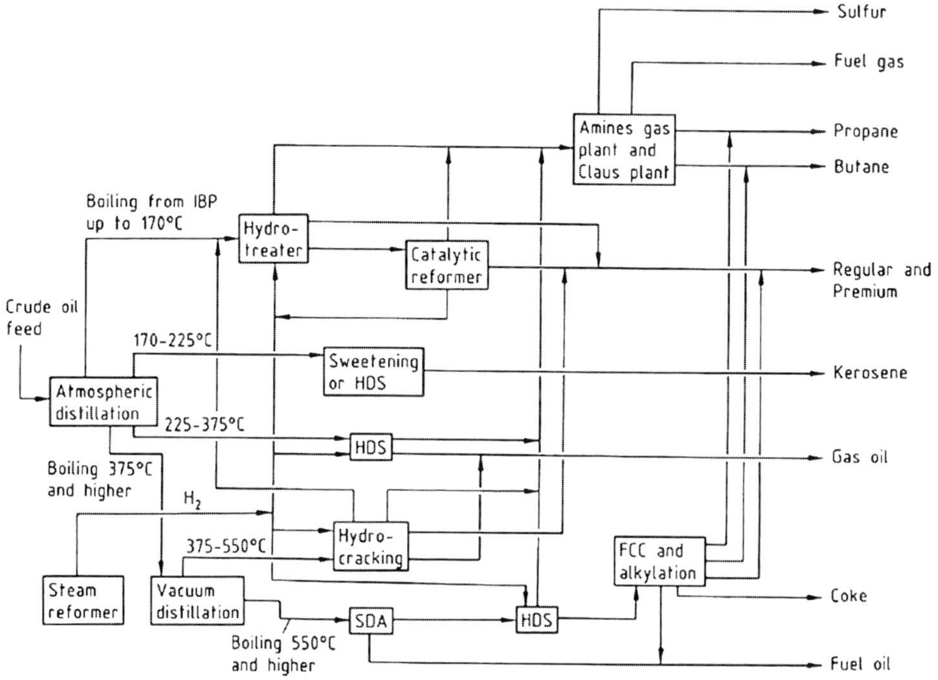

Fig. 2.42 Hydrocracking – catalytic cracking.

partners to increase flexibility and to reduce costs. An example is Bayern Oil Refinery in Southern Germany, where initially two, and several years later three, refineries were made into one by installing a network of pipelines and a common production control system. Surplus plants in one or the other location were shut down, others were enhanced. The system would not work without the highly developed process control systems. Other examples for this integration route are Mineralölraffinerie Oberrhein MIRO in Karlsruhe or Shell's Rheinland Raffinerie in Cologne [62a].

2.6 Corrosion and Materials

Prevention of corrosion in refining equipment is determined by material selection in plant construction and day-to-day operational control. Only the main refining-related corrosion problems and the selection of corrosion-resistant materials are summarized as follows:

Corrosion problems are mainly caused by organic sulfur compounds and inorganic chlorides in crude oils. The danger of steel embrittlement by hydrogen in all processes involving hydrogen must also be taken into account.

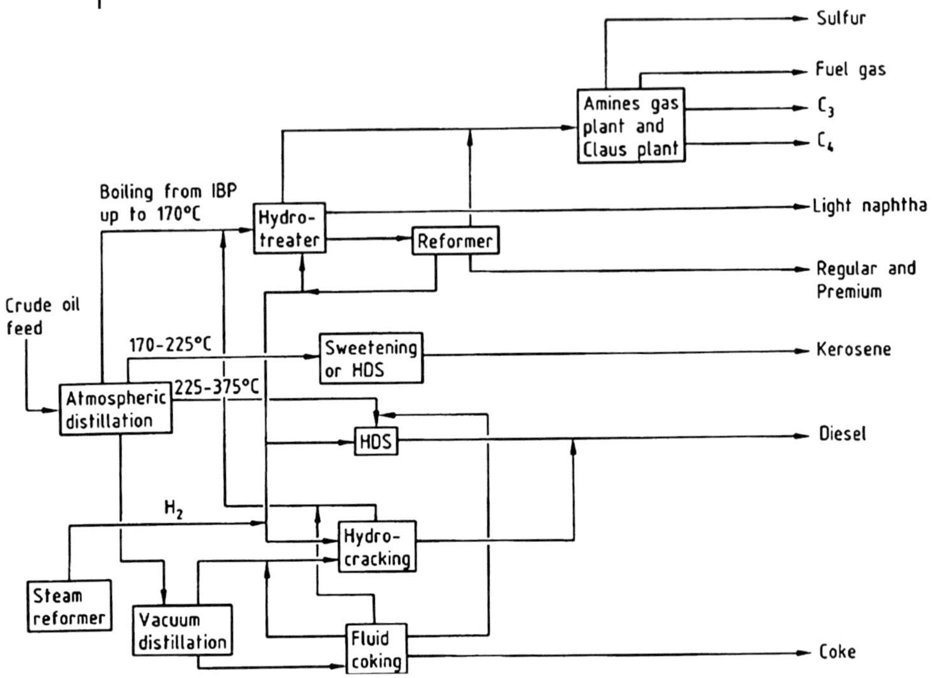

Fig. 2.43 Hydrocracking – coking.

Corrosion in distillation processes is caused by the following:

1. Formation of iron sulfide and subsequent corrosion by H_2S and mercaptans at elevated temperatures. Suitable materials for corrosion prevention are Cr and austenitic Cr – Ni alloys, particularly for transfer lines and reboiler tubes, and Cr – Mo alloys for furnace tubes. The lower parts of the crude and vacuum columns should be lined with Cr steels.
2. Attack by naphthenic acid to form iron naphthenates and subsequent corrosion. Austenitic Cr – Ni – Mo alloys can be used and naphthenic acid can be removed upstream of the columns.
3. Hydrogen chloride corrosion in overhead condensation systems. Monel or other Ni – Cu alloys can be used. Measures that can be taken to prevent corrosion include desalting of crude oils, neutralization of overhead vapors with ammonia, and pH-control, and use of corrosion inhibitors in endangered equipment sections.

Corrosion in cracking processes is caused by the following:

1. *High-temperature sulfur corrosion.* Cr – Mo alloys in the furnace and heat-exchanger tubes and Cr steel lining in the columns are suitable materials for corrosion prevention.

2. *Erosion by catalysts.* Ceramic and concrete linings in vessels and cyclones and regenerator internals made of Cr – Mo and austenitic Cr – Ni alloys are suitable materials.

Corrosion in hydrogen-carrying processes is caused by hydrogen attack at high temperatures (blistering and embrittlement of steels). Cr or Mo steels, at higher temperatures austenitic Cr – Ni alloys, can be used to prevent corrosion. In the presence of hydrogen and H_2S, i.e., in hydrotreating and hydrocracking processes, steels with high chrome contents must be used.

2.7 Testing and Analysis

The chemical and physical properties of crude oils and their fractions depend on the types and concentrations of their constituents. The typical data for crude oils and the properties of their products must be determined by analytical methods, to meet requirements for production, transport, storage, handling and processing of crude oils, and to predict and define criteria for processing in the refinery.

This quality testing is performed by regular laboratory analysis or by automatic instruments.

2.7.1 Crude Oil and Product Properties

Density and API Gravity. The density ranges of various crudes are as follows:

Paraffin-based crudes	815–835 kg/m^3
Mixed-based crudes	835–860 kg/m^3
Naphthalene-based crudes	860–955 kg/m^3

The API gravity (introduced by the American Petroleum Institute) is the internationally used density measure for crude oils. The relation between degrees API and density at 15 °C is:

$$°\mathrm{API} = \frac{141.5}{\mathrm{Density}\ [\mathrm{kg/L}]} - 131.5$$

Distillation The *boiling range* of crude oils is one of the most important quality criteria for determining the yield distribution of the various product fractions. Boiling analyses under atmospheric conditions can be supplemented by distillation under reduced pressure for oil fractions that boil above 360 °C.

More detailed and precise results are obtained by applying fractional or *true boiling point* (TBP) distillation.

Viscosity The viscosity of crude oils is determined by the ratio of low and high molecular mass constituents in the oil. The kinematic viscosity of crude oils (at 38 °C = 100 °F) can vary from 4 mm^2/s (0.04 St) for light (North Sea) crudes to 700 mm^2/s (7 St) for heavy (Venezuelan) crudes.

Flash point The flash point depends on the concentration of light (low-boiling) constituents. For crude oils, flash points between – 20 °C and + 80 °C are observed.

Pour Point The pour point is the temperature at which an oil or oil product ceases to flow. It is an important measure of its cold flow properties (Chap. 3, Section 3.5.2.3.).

Other properties that are analytically tested include: specific heat; calorific value; contents of sulfur, nitrogen, oxygen, and inorganic compounds; the C/H ratio; the hydrocarbon composition; and product-specific criteria (see Section 2.2.3.).

2.7.2 Testing Methods and Standards

In the oil industry a framework of standard analytical procedures has been established by national standardizing bodies, e.g., ASTM, DIN and the British Institute of Petroleum, (IP) standards. A large number of these former national standards have been transferred into the international standardizing bodies ISO or EN (European Standardizing Committee) with respect to the international trade of petroleum products. In addition, various in-house standards are still in use by most international oil companies.

A selection of former frequently applied DIN methods and their replacement by international standards or other fate are listed in Table 2.17.

Analytical testing by *automatic continuous analyzers* has come into use worldwide. The continuous analysis of process streams by automatic "onstream analyzers" has

Table 2.17 Comparison of DIN and EN standards

	Former standard (1995)	New standard (2006)
Density	DIN 51757	EN ISO 3675
Distillation (boiling range)	DIN 51751	EN ISO 3405
Vacuum distillation	DIN 51356	abandoned, obsolete
Fractional distillation	DIN 51567	abandoned, obsolete
Flash point	DIN 51755,	valid, altern. EN ISO 2719
Viscosity	DIN 51 562-1	EN ISO 3104
Cold properties (cloud point, pour point, freezing point)	DIN 51583, 51597, 51421	DIN EN 23015 DIN ISO 3016 DIN ISO 3013
Water content	DIN 51582	DIN 51 777 or DIN EN ISO 12937
Water and sediment content	DIN 51793	
Salt content	DIN 51576	valid
Sulfur content	DIN 51400, 51768, 51409, 51418, 51450	valid/abandoned/ replaced/ valid/abandoned

proven to be an indispensable tool for the quality control of products in integrated refining systems.

2.8 Storage and Transport

Surface Storage *Crude oils* are stored in cylindrical fixed-roof or floating roof tanks; *LPG* is stored in spherical or cylindrical pressure vessels (Chap. 3, Section 3.8.1.5.). Suitable storage tanks for *gasolines, kerosenes,* and *gas oils* are cylindrical floating-roof or fixed-roof tanks (optionally with vapor balancing) (Section 3.7.). *Heavy fuel oils* and *bitumen* are stored in cylindrical fixed-roof tanks (optionally with gas blanketing). Heating by steam is required.

Underground Storage Crude oil and liquid (e.g., light heating oil), and gaseous products (LNG) can be stored in underground caverns (Chap. 3, Section 3.8.1.5.). The underground storage of crudes and liquid products in caverns is usually intended as long term (emergency) storage; caverns for gas are used as buffer storage.

Subsea Storage Crude oils (from offshore production) and hydrocarbon condensates (from offshore separation) are stored in undersea tanks.

Transport of Crude Oils and Products Crude oils and bulk (mass) products are normally transported by large tankers and pipelines. Finished products are usually transported by smaller tankers and river barges, product pipelines, rail tank cars, and road tank trucks.

2.9 References

1 H. S. Bell: American Petroleum Refining, D. van Nostrand, Princeton, NJ 1959.
2 Oil Gas J. **95** (1997) Dec. 22, "OGC" Special," 41–44 and Oil Gas J. (2005) Dec. 19.
3 Hydrocarb. Process. **48** (1969) no. 2, 77–88.
4 Oil Gas J. **71** (1973) no. 45, 88–92.
5 Hydrocarb. Process. **51** (1972) no. 5, 85–88.
6 Hydrocarb. Process **51** (1972) no. 5, 92–97.
7 Hydrocarb. Process. **65** (1986) no. 9, 93, 95.
8 In [7], p. 94.
9 Hydrocarb. Process. **63** (1984), no. 9, 106.
10 1984 Refining Process Handbook, Hydrocarb. Process. **63** (1984) no. 9, 70–84, 89–90.
11 1986 Refining Process Handbook, Hydrocarb. Process. **65** (1986) no. 9, 84–87.
12 Hydrocarb. Process. **52** (1973) no. 9, 131–133.
13 Petrol. Eng. Manage. **28** (1956) no. 3, C 37–C 44.
14 Hydrocarb. Process. **58** (1979) no. 9, 136–142.
15 Hydrocarb. Process. **58** (1979) no. 5, 103–106.
16 Hydrocarb. Process. **64** (1985) no. 3, 51–59.
17 Oil Gas J., **81** (1983) no. 3, 103–113.

17a Oil Gas J. 2001 Nov. 19, 72–76.
18 Hydrocarb. Process. **63** (1984) no. 9, 70.
19 Hydrocarb. Process. **65** (1986) no. 9, 98–100.
20 Hydrocarb. Process. **63** (1984) no. 9, 115–119.
21 In [20] , p. 118.
22 In [20] , p. 115.
23 Oil Gas J., **74** (1976) no. 3, 48–52.
24 Oil Gas J., **70** (1972) no. 21, 52.
25 Hydrocarb. Process. **65** (1986) no. 9, 99.
26 Oil Gas J., **79** (1981) no. 22, 102–105.
27 Hydrocarb. Process. **63** (1984) no. 9, 104.
28 E. Naber et al.: "Refining in the 1990's," Erdoel Erdgas **107** (1991) no. 3, 124–132.
29 A. K. Logwinuk et al.: Factors Affecting Choice of Heavy Hydrocarbon Processing, NPRA Ann. Meet., San Antonio 1981.
30 H. N. Olson: "Cost and Cost Aspects in Residue Conversion," Proc. World Pet. Congr. 11th, 1983, 1–16.
31 H. R. Siewert et al.: "Optimize Design for Heavy Crude," Hydrocarb. Process. **64** (1985) no. 3, 61–66.
32 Shell Internationale Petroleum Maatschappij, EP 264 158, 1987 (R. H. van Dongen et al.).
33 M. A. Menzies, T. F. Scot, J. Denis: "The Canmet Residuum Hydrocracking Process," Proc. Intsoc. Energy Convers. Eng. Conf., 15th 1980, 26–32.
34 Hydrocarb. Process. **65** (1986) no. 9, 97, 98.
35 Hydrocarb. Process. **65** (1986) no. 9, 95, 96.
35a Erdöl Erdgas Kohle **121** (2005) 10, 348–352
36 V. D. Singh, Erdoel Kohle **39** (1986) no. 1, 19.
37 S. W. Martin, L. E. Wills: "Coking Petroleum Residues," in K. A. Kobe, J. J. McKetta (eds): Advances in Petroleum Chemistry and Refining, vol. 2, Interscience Publishers, New York–London 1959.
38 Oil Gas J. **80** (1982) no. 20, 93–102.
39 H. Predel: Delayed Coking, Erdoel Kohle Erdgas Petrochem. **43** (1990) no. 4, 145–146.
40 M. J. Humbach et al.: "Residue Conversion, RCC in the 1990's," Erdoel Kohle Erdgas Petrochem. **43** (1990) no. 12, 477–482.
41 A. Rhoe et al.: Meeting the Refiners Upgrading Needs, NPRA Annu. Meet., San Francisco 1983.
42 Hydrocarb. Process. **63** (1984) no. 9, 123.
43 Hydrocarb. Process. **64** (1986) no. 9, 124.
44 Hydrocarb. Process. **64** (1986) no. 9, 126.
45 Hydrocarb. Process. **60** (1981) no. 9, 134–138.
46 Hydrocarb. Process. **61** (1982) no. 5, 110–112.
47 Hydrocarb. Process. **63** (1984) no. 9, 125.
47a Winnacker-Küchler, 5th Ed. Vol. 4 "Die Verarbeitung von Erdöl" Section 6.2.4.
47b Erdöl Erdgas Kohle 120 (2004) 10, 317-322.
48 Hydrocarb. Process. **63** (1984) no. 9, 121.
49 Hydrocarb. Process. **59** (1980) no. 5, 110–114.
49a OG Oil Gas-Euro. Mag. 1/2003, 44–46
49b OG Oil Gas-Euro. Mag. 2/2004, 95–96
49b Erdöl Erdgas Kohle 122(2006) 2, 83–84
50 The Oil Refining Industry and the Environment, CONCAWE, The Hague, Sept. 1987.
51 O. Neuwirth: "Umweltfragen bei der Nutzung von Mineralöl," Erlanger Forsch. Reihe B; Naturwiss. Med. **14** (1984) 119–134.
52 Mineralöl und Umweltschutz, Publ. Mineralölwirtsch. Verband, Hamburg 1999.
53 A. H. Edwards et al.: "Volatile Organic Compound Emissions in Western Europe," CONCAWE Rep. 1986, no. 2.
54 D. A. J. Dando et al.: "Trends in Oil Discharged with Aqueous Effluents from Oil Refineries in Western Europe," CONCAWE Rep. 2004, no. 4/04.

55 C. L. F. Webb: "Biodegradation of Contaminated Soil," Erdoel Erdgas **106** (1990) no. 11, 453–455.
56 J. S. McArragher et al.: "Motor Vehicle Emission Regulations and Fuel Specifications, CONCAWE Rep. **1990**, no. 2.
57 International Workshop Berlin: "Emissionsarme Betankung von Kraftfahrzeugen," Erdoel Kohle Erdgas Petrochem. **42** (1989) 468.
58 R. J. Ellis: "Cost Estimating Aspects of Environmental Control Technology — HDS as a Case Example," CONCAWE Rep. 1988, no. 51.
59 Commission of the European Communities: Study on the Costs Borne by the Refining Industry in the Member States in Order to Comply with Environmental Legislation, prepared by Chem. Systems Int., London 1986.
60 G. Heinrich, Y. Jacquin, A. Vidal, Proc. World Pet. Congr., 10th, 1979, vol. 4, paper PD 18–1, Heyden & Sons, London 1979.
61 F. G. Turpin, C. R. Lipuma, A. M. Edelman, in [60] , paper PD 18–2.
62 C. T. Douwes, J. van Ken, J. B. Wyffels, W. C. van Zijll Langhout, in [60] , paper 18–3.
62a Erdöl Erdgas Kohle **116** (2000) no. 5; Erdöl Erdgas Kohle **122** (2006) no 2.
63 Erdoel Erdgas Kohle **105** (1989) no. 9, 369–372.
64 Erdoel Erdgas Kohle **106** (1990) no. 7/8, 309–315.

Abbreviations

BTX: benzene – toluene – xylene
CC: catalytic cracker
CR: circulating reflux
DAO: deasphalted oil
DEA: diethanolamine
DIPA: diisopropyl amine
ETBE: ethyl *tert*-butyl ether
FCC: fluid-bed catalytic cracking
GO: gas oil
IBP: initial boiling point
ICR: intermediate circulating reflux
HC: hydrocracking, hydrocarbons
HDN: hydrodenitrification
HDS: hydrodesulfurization
HGO: heavy gas oil
HT: hydrotreating
LCR: lower circulating reflux
LGO: light gas oil
LHSV: liquid hourly space velocity
LPG: liquefied petroleum gas
MEA: monoethanolamine
MON: motor octane number
MTBE: methyl *tert*-butyl ether
OPEC: Organization of the Petroleum Exporting Countries
RON: research octane number
SDA: solvent deasphalting
TAME: *tert*-amyl methyl ether
TBP: true boiling point
TIP: total isomerization process
UCR: upper circulating reflux
VB: visbreaker
VGO: vacuum gas oil

3
Automotive Fuels

Werner Dabelstein, Arno Reglitzky, Andrea Schütze, Klaus Reders

3.1
History

3.1.1
The Spark Ignition (Otto) Engine and Its Fuel

In 1876 NIKOLAUS AUGUST OTTO developed a stationary, single-cylinder, four-stroke engine, that ran on coal gas. This invention, which was later named after him, set a development in motion that substantially shaped the industrial age and has not yet come to a halt [1, 2]. About ten years later the four-stroke engine was first used to drive vehicles (Daimler and Maybach automobiles, Benz patent car). The first vehicle engines ran on so-called *light gasoline* which had a low density (< 700 kg/m^3) and a very low final boiling point (ca. 85 °C). Light gasoline was available as a cleaning agent at pharmacies and could be evaporated in the first surface carburetors. The development of spray nozzle carburetors finally enabled the use of *straight-run gasoline* with a broader boiling range and higher final boiling point [3]. A higher proportion of crude petroleum could thus be used as a fuel, and the rising demand for gasoline could be satisfied.

After the development of a reliable mixture preparation system, attention focused on the further improvement of the engine power by increasing the compression ratio (from ca. 1918 on). The increase in compression ratio was shown to be limited by abnormal combustion, so-called engine knock, which can damage the engine. Since it was observed that knocking combustion can be related to gasoline formulation, systematic research began to improve the knock resistance of fuels.

Discovery of the strong antiknock effect of tetraethyllead by MIDGLEY and BOYD in 1922 was a milestone in the development of gasolines. A little later, refinery conversion processes were established that allowed the demand for gasoline to be met by improving the yield, and which also led to a further improvement in antiknock performance, so-called octane quality (→ Oil Refining). The further advances in the development of engine technology and in the quality of gasoline were mutually

Handbook of Fuels: Energy Sources for Transportation. Edited by Barbara Elvers

ISBN: 978-3-527-30740-1

dependent. From 1950 on, increasing attention was given to the further improvement of gasoline by the addition of small quantities of highly active additives (see Section Gasoline Additives).

Since 1970, measures for improving the environmental compatibility of automotive use have become increasingly important worldwide. The first step was the successive reduction in the lead additive content of gasolines. Later, completely unleaded gasolines were developed, which were a prerequisite for the introduction of catalytic converters.

The reduction of sulfur content in gasoline has occurred more recently and it is still ongoing. This measure improves the efficiency and durability of modern exhaust gas aftertreatment devices and reduces the emission of sulfur compounds. In this context "sulfur free" gasolines (less than 10 ppm sulfur) have been made available on a wider scale. In countries with tax incentives for "sulfur free" fuels like Germany this fuel quality has a high market share. The limitation of benzene, other aromatics, and olefins concentration in gasoline is intended to reduce the resulting-post catalyst exhaust emissions. The worldwide rising demand for oil products has led to noticeable price increases and because of this development alternative fuels or fuel components (e.g., alcohols from biomass) have again achieved a certain significance.

3.1.2
The Diesel Engine and Its Fuel

The diesel engine, patented in 1892 by RUDOLF DIESEL, is characterized by self-ignition of the fuel [4]. Self-ignition is achieved by compressing the incoming air to such an extent that the resulting temperature increase is sufficient to ignite the fuel.

In the development of his engine, DIESEL'S aim was an engine operation as close as possible to the Carnot cycle — which gives the highest theoretical efficiency — with the result that the first diesel engine built in 1893 at the Maschinenfabrik Augsburg Nürnberg AG (MAN) operated at a hitherto unachieved efficiency of 26.2 %. The power of this engine was 17.8 metric horsepower (13.1 kW) at 194 rpm, and it ran on a mixture of gasoline and lamp oil. After 1897, experiments with other fuels started, which had little success at first.

The injection of small quantities of fuel at high pressure caused the greatest problem in the initial phase of the development. Blowing the fuel in with compressed air, which was tried at first, did not solve the problem satisfactorily. High-pressure liquid injection was developed in 1910; and in 1927, large-scale production of fuel injection systems started and the way was opened for broad expansion of the diesel engine. MAN and Daimler–Benz introduced the diesel engine to commercial vehicles in 1923–1924. About 12 years later, Daimler–Benz introduced the first passenger car diesel engine (type 260 D, 2.6 L, 45 metric horsepower at 3000 rpm).

Even the first diesel engines were often run on the so-called gas-oil fraction of crude petroleum (boiling range 200–350 °C). This type of fuel, which is provided in sufficient quantity mainly by crude oil distillation at moderate production cost is basically still used today, because of its suitable ignition quality (see Section Ignition Quality).

In addition to high ignition quality (cetane number; see Section Ignition Quality), cold flow properties and low tendency toward coking and sooting became important at an early stage.

During World War II, large quantities of diesel fuel were produced from coal in Germany. Mixtures of components of low ignition quality (tar oil) and very high ignition quality (from the Fischer–Tropsch synthesis) were used [3].

Because of the increasing demand for light oil products that followed World War II, gas-oil fractions from conversion processes have also been used in diesel fuel. However, these components are generally less suitable because of their lower ignition quality. For environmental protection and improved engine performance a number of measures were taken to raise the diesel fuel quality. For example, the maximum permitted sulfur content of diesel fuel was lowered stepwise from 1.0 (1955 in Germany) to 0.035 mass % (2000) and further to 0.005 mass % (2005) in Europe. Also a fuel quality with a sulfur content less than 0.001 mass % (sulfur free) must be available. The maximum permitted sulfur content in different areas is given below:

United States	0.001 (max average of refinery production) mass %
EU	0.005 and 0.001 mass %
Japan	0.0015 mass %

The reduction of sulfur in the diesel fuel helped also in the introduction of particulate traps in the exhaust systems.

Synthetic components with ultra-high ignition quality available from, e.g., natural gas synthesis (basicly Fischer–Tropsch synthesis with a newly developed catalyst) are used in some markets to formulate a diesel fuel with superior quality. Also high-quality components produced from biomass will shortly be available in limited quantities for commercial use.

Reliable engine operation and a reduction in exhaust gas emissions is also the primary aim in the treatment of diesel fuels with multifunctional additives, which has come into increased use from 1985 onwards. In addition, acceptability of diesel fuels could be improved (foaming, odor; see Section Additives for Diesel Fuel).

3.2 Engine Technology

3.2.1 Otto Engines

The basic principle of the spark ignition (Otto) engine—a combustion engine with timed ignition from an external source — can be realized in a number of variants with different types of fuel metering and mixture preparations [5].

The most frequently used type is the *four-stroke* reciprocating Otto engine in which gas exchange processes during these four strokes are controlled by valves. The sequence of strokes is

1. intake of the fuel–air mixture,
2. compression of the mixture and timed ignition,
3. combustion and expansion (working stroke), and
4. exhaust of combustion gases.

To carry out one working stroke the crankshaft turns twice. The valves are operated by a camshaft connected to the crankshaft.

The *rotary engine* (frequently called the *Wankel* engine after its inventor) is a variant of the four-stroke engine. It differs from the reciprocating piston engine in its rotary piston and in the use of ports for control of the inlet and exhaust gases [6, 7]. The rotary piston engine has not been very successful up to the present, compared to the reciprocating engine because of its relatively high fuel consumption and exhaust gas emissions.

In the standard *two-stroke* Otto engine the gas exchanges are also controlled by ports (inlet, transfer, and exhaust ports) in the cylinder. A working cycle is achieved with each rotation of the crankshaft, and high power output at low swept volume is achieved. However, in the two-stroke process, scavenging losses must be accepted that lead to relatively high fuel consumption and hydrocarbon-exhaust emission. The two-stroke engine is used only in small motorcycles, outboard engines and light-weight machines [8]. Technically advanced models controlled by valves can overcome many of the disadvantages normally associated with these engines to a certain degree. However, more stringent exhaust gas requirements will reduce the use of two-stroke engines in the future.

The Otto engine runs with throttled intake air (quantity governing) and an almost homogeneous, approximately stoichiometric fuel–air mixture. Traditionally, the composition of the mixture depended on the running condition of the engine: i.e., at full load and directly after cold start the engines operated with an excess of fuel (rich mixture) whereas under part load nearer to the stoichiometric mixture. For the preparation of the fuel-air mixture previously carburetors [9] were used.

Modern electronic fuel injection systems, are superior to carburetors with respect to performance, fuel consumption, and accuracy of mixture strength control necessary for the efficient use of three-way catalytic converters [10]. Carburetors were therefore gradually replaced by fuel injection systems. Latest development are the so-called direct injected engines, which offer optimum fuel efficiency [11]. The main advantage of this technology is the avoidance of throttling losses at part load conditions. Whereas conventional Otto engines run at fixed air-fuel ratio and adjust power at part load by reducing the cylinder charge by the throttle valve, direct injected engines run in this respect rather like a diesel with almost unthrottled cylinder charge and varying amounts of fuel injected. To ensure proper ignitability of very lean mixtures at low load a locally gasoline enriched layer of air–fuel mixture ("stratified charge") must be created.

In the Otto engine the combustion process is started by an electric spark. Ignition occurs shortly before the piston has reached the top dead center and is

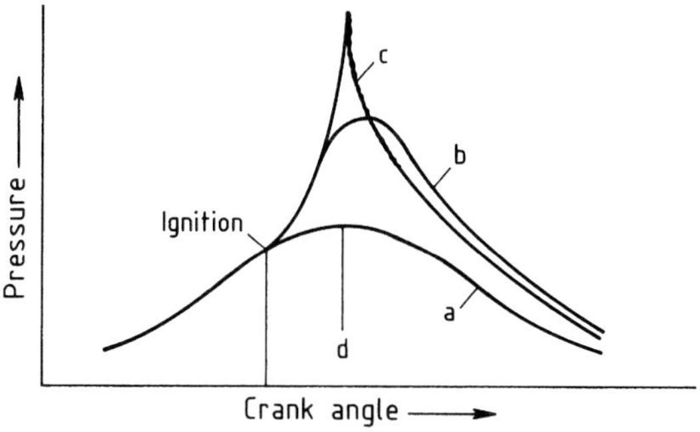

Fig. 3.1 Combustion in the Otto engine a) Without ignition; b) Normal combustion; c) Knocking combustion; d) Top dead center.

advanced with increasing speed to allow a timed combustion beginning shortly before the top dead center [12].

To achieve the highest efficiency in the Otto engine, combustion induced by the ignition spark must develop in a controlled way under all operating conditions. If the resistance of the gasoline against autoignition, i.e., the octane quality of the fuel does not meet the octane number requirement of the engine, after compression by the piston either before or after ignition by the spark plug, autoignition of the yet unburned portion of the mixture occurs. This autoignition (engine knock) leads to a considerably higher combustion rate with a very sharp increase in pressure and temperature (Fig. 3.1). An anomalous combustion of this type manifests itself by a knocking or pinking noise. Whereas occasional knock that occurs on acceleration usually does not damage the engine, permanently knocking combustion at high speeds and loads over a longer period increases the risk of engine damage. Modern engine knock sensors allow safe engine operation even if gasolines of varying octane qualities are used, because they adjust the spark timing according to octane quality. However, gasolines with low octane quality require retarded ignition which leads to efficiency losses and, e.g., higher exhaust gas temperatures.

Apart from design parameters the octane number requirement of an Otto engine varies with the engine speed and the load. Furthermore increasing deposit formation in the combustion chamber with increasing milage can raise the octane number requirement [13]. In designing an engine on the basis of a particular type of fuel, a safety margin must therefore be provided, to avoid knocking combustion even under unfavorable conditions.

3.2.2
Diesel Engines

The diesel engine differs from the Otto engine in the heterogeneous composition of its mixture and the self-ignition of the fuel. The temperature necessary for self-ignition is produced in the compression stroke, in which compression of air to ca. 3.0–5.5 MPa leads to temperatures of 700–900 °C. Fuel is injected into the heated air of the combustion chamber shortly before the end of the compression stroke, where it self-ignites. In principle the diesel engine can be either of the four- or two-stroke type. In practice, four-stroke engines are found in the vast majority of vehicles, whereas a valve-controlled two-stroke process with precompressed air is used in large ship engines.

In the diesel engine, air is generally taken in unthrottled and the fuel–air ratio varies with the quantity of fuel injected (quality governing) thus leading to a very lean fuel–air mixture ($\lambda > 1.3$, for definition of λ, see → Automobile Exhaust Control).

Two types of combustion systems have been developed for the diesel engine that differ in the preparation of the combustible gases [14]. *Direct injection* is characterized by a one-piece combustion chamber, into which fuel is injected without further preparation. Indirect injection is characterized by a two piece combustion chamber and was a mechanical solution to control the rate of combustion.

One of the main problems of the early diesel engines was the control of the combustion rate. Depending on the fuel and engine parameters ignition was often delayed and the subsequent rate of combustion more rapid than desired for best efficiency, low noise and emissions, etc. In those days before the development of modern electronic control units, various mechanical solutions were available, such as divided combustion chambers or specific mechanically controlled injection spray patterns to adjust the rate of combustion. Very common, especially for passenger car diesel engines, were concepts with divided combustion chambers, so called indirect injection engines (IDI engines). These *precombustion chamber* processes, such as the *swirl chamber* and *prechamber* processes, have a two-piece combustion chamber. Fuel is injected into the precombustion chamber, partially homogenized with air, and self-ignited. Partially oxidized combustible gases and evaporated fuel reach the main combustion chamber via connecting channels and burn there after further mixing with air. Indirect injection engines were used predominantly in passenger cars. Their engine noise is less harsh, but their fuel consumption is higher than that of the direct injection diesel engine. The split combustion chamber gives rise to pumping losses which impair the efficiency of precombustion chamber engines. This is also brought about by their higher compression ratio leading to higher friction losses.

In contrast to the divided combustion chamber engine concepts with a one-piece combustion chamber, i.e., direct injection engines (DI engines), were originally used mainly for heavy-duty transport and machinery. Their undivided combustion chamber allows higher engine efficiency. In modern direct injection engines sequentially electronically controlled high-pressure fuel injection systems allow optimum combustion and pressure rise thus avoiding previous problems, e.g., noise or limited speed range. The so-called *common rail technology* allowing high injection pressures is now well implemented for passenger car diesel engines.

The compression ratio of direct injection engines (ca. 15–17: 1) lies near that optimum for most favorable total efficiency, which occurs at a compression ratio of ca. 15: 1.

As mentioned before the diesel fuel is injected into the combustion chamber near the end of the compression stroke shortly before the piston has reached the

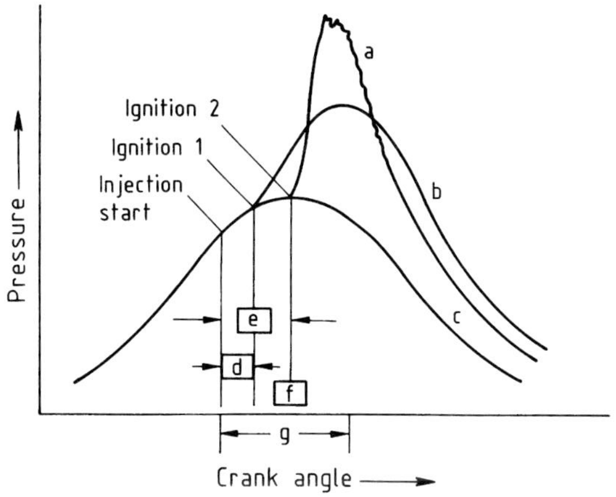

Fig. 3.2 Combustion in the diesel engine a) Noisy combustion; b) Normal combustion; c) Without combustion; d) Ignition delay 1; e) Ignition delay 2; f) Top dead center; g) Injection period.

top dead center. The period between the beginning of the injection mode and the start of the combustion is called ignition delay. The length of ignition delay depends on ambient conditions, e.g., temperature, design and construction of the engine and the ignition quality of the diesel fuel. Optimum engine operation, necessary for high efficiency, low noise, and exhaust emissions, requires short ignition delay and therefore high diesel fuel ignition quality, i.e., a high cetane number. Prolonged ignition delay causes an unacceptable steep pressure rise causing, e.g., high noise and nitrogen oxide emissions (see Fig. 3.2) [15, 16].

3.3 Fuel Composition and Engine Efficiency

Because of the different modes of operation of Otto and diesel engines (see Chap. Engine Technology), completely different requirements exist for gasoline and diesel fuels and their composition. The effects of important fuel properties on the performance of engines are discussed below. The standardization of minimum requirements derived from them is described in Chapter Fuel Standardization and Testing.

3.3.1 Quality Aspects of Gasoline

The most important quality criteria for gasoline are its resistance against autoignition (engine knock), i.e., octane quality, the evaporation properties to provide an ignitable air–fuel vapor in the combustion chamber, environmental acceptability and low toxicity, and also cleanliness and stability. While all of these aspects are influenced by the main gasoline components and thus by refinery technologies, some can also be improved by additives, i.e., small proportions of organic substances, added to the gasoline mainly at the end of the production process and prior to delivery to the service stations.

3.3.1.1 **Octane Quality**

High octane quality is required to allow knock-free combustion in high compression engines to enable optimum engine efficiency. While the engine efficiency can be raised generally by increasing the compression ratio—leading to a higher octane requirement of the engine—the production of matching high octane fuels is connected with a higher energy demand in the refinery. The overall optimum efficiency (taking into account engine and refinery) is obtained with the most widely used premium and super gasolines with research octane numbers of 95 to 100 and engines designed for that octane quality.

The octane level of a gasoline depends on that of the individual refinery streams and on that of other, e.g., oxygen containing components such as ethers or alcohols. The octane quality of gasoline is described by the *octane number.* The octane number of a gasoline is determined by comparative measurements of its octane quality and that of binary mixtures having variable concentrations of *n*-heptane (low octane quality) and 2,2,4-trimethylpentane (isooctane; high octane quality). By definition the octane number of *n*-heptane is 0 and that of isooctane is 100. The octane numbers of mixtures are given by their percentage by volume of isooctane; octane numbers >100 can be determined with lead-containing isooctane or toluene-containing mixtures.

Octane number determination (e.g., according to EN 25164 or EN 25163) is carried out in single-cylinder, four-stroke test engines specially developed for the purpose. The CFR (Cooperative Fuel Research) engine is most widely used worldwide. The test engines have a mechanically adjustable compression ratio. This ratio is raised during octane number measurement of a fuel until a certain standardized level of knocking, shown by a so-called knock meter, is obtained. The gasoline's octane number is determined by finding the *n*-heptane–isooctane mixture that gives the same knock level. Determination of the octane number is carried out under two different running conditions:

1. **Determination of the research octane number (RON):** according to EN 25164, ASTM D 2699, in which comparatively mild conditions are used (600 rpm, no prewarming of the mixture, constant ignition timing).
2. **Determination of the motor octane number (MON):** according to the EN 25163, ASTM D 2700, in which a higher mechanical and thermal load is used (900 rpm, prewarming of the mixture to 150 °C, and variable ignition timing).

Because of the more severe test conditions the MON is usually lower than the RON; the difference between both is called *sensitivity* [17]. The RON describes the knocking performance of a gasoline at low and medium engine speeds ("acceleration knocking") whereas the MON defines the knock behavior under high speed and load ("high speed knocking"). The knock rating of a fuel is generally sufficiently characterized by its RON and MON. Historically it was necessary to use an additional characteristic, the *front octane number* (FON) describing the RON knock rating of the low boiling fuel components. An insufficient FON leads to knocking during acceleration, particularly in carburetor vehicles. This problem does no longer exist in fuel-injected vehicles.

For research and development purposes it is also possible to measure the octane quality in vehicle Otto engines under normal road use. The so-called *road octane number* gives information on the octane quality in vehicle Otto engines under normal road use. The road octane number is specifically related to the vehicle type and is also affected by driving conditions and weather. Usually the road octane number is given by an expression of the type a RON + b MON + c, where a, b, and c are constants that depend on the vehicle. The road octane number of a gasoline lies between the RON and MON, nearer the RON under acceleration conditions and approaching the MON under constant speed.

3.3.1.2 **Volatility**

Fuel volatility, expressed by distillation behavior (EN ISO 3405, ASTM D 86) and vapor pressure, controls the formation of an ignitable air – fuel mixture and the ability of the engine to start and run under cold and hot conditions. For safe operation the fuel must have a balanced distillation performance [13]. A certain proportion of volatile fuel components are needed for trouble-free performance at low temperature. An excess of these components leads, however, to hot fuel handling problems with a warm engine or high ambient temperature (see Fig. 3.3). To ensure trouble-free engine running, the boiling characteristics are usually matched to

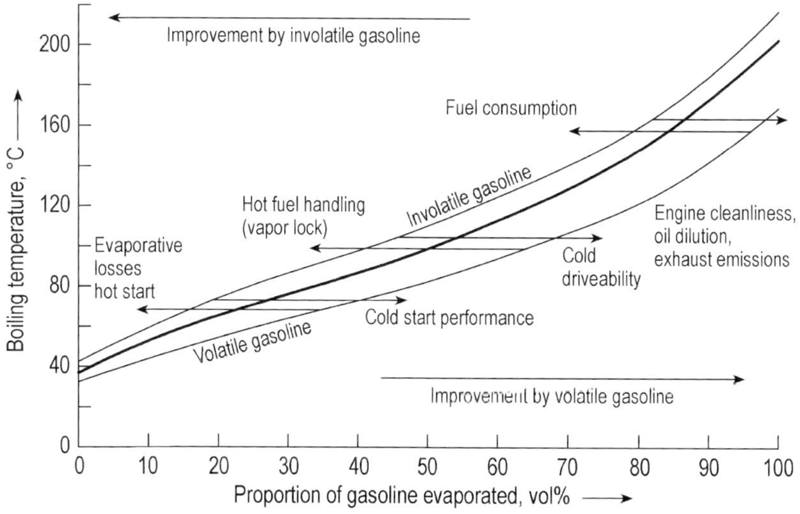

Fig. 3.3 Significance of gasoline distillation properties for engine performance and emissions [19].

climatic conditions and change during the year. An excessively high final boiling point can cause deposit formation and lubricant dilution, and consequently engine wear.

Gasoline volatility also controls exhaust gas emissions and evaporative emissions. Vehicle design elements such as the activated carbon canister have of course a larger effect on these emissions. However, in a vehicle of given design the volatility of the fuel can contribute to guarantee low emissions. For example, a sufficiently high vapor pressure and high volatility ensures low exhaust gas emissions at low ambient and engine temperatures and sufficiently low vapor pressure reduces evaporative losses at high ambient and fuel system temperatures.

3.3.1.3 Fuel Composition to Reduce Toxicity and Exhaust Emissions

Some of the gasoline constituents, especially benzene and previously used lead antiknock additives, are toxic and other components can form toxic materials during combustion. Both benzene and alkylleads were formerly used to improve the octane quality of gasolines. Their maximum allowed concentration was legally reduced stepwise. While lead has completely disappeared from gasoline in most countries also because of its poisoning effects on exhaust gas catalysts, benzene content has been restricted in Europe initially to maximum 5 vol % and since the year 2000 to 1 vol %. The major German oil companies produced their top quality Super Plus since 1995 already with benzene concentrations below 1 vol % on a voluntary basis. Benzene has not been used as an active blending component for a long time. It is extracted from the relevant gasoline components (e.g., reformate) mainly by distillation and also its formation in the catalytic reformer is minimized by extraction of "benzene precursors" in the reformer feed material.

To keep the octane quality of gasolines without lead and benzene above the standardized level, contents of other nontoxic materials were increased, namely of isomerates, higher aromatics, and to some extent ethers (→ Methyl *Tert*-Butyl Ether) and alcohols.

Ideally the gasoline is combusted completely to water and carbon dioxide, however, in practice other unwanted components are emitted because a small proportion of the gasoline (usually less than 1 %) leaves the combustion chamber unburned, e.g., because of quenching at the combustion chamber surfaces, or partly oxidized due to the high pressures and temperatures. Apart from carbon monoxide, unburned hydrocarbons, and nitrogen oxides, other toxic substances such as aldehydes, benzene, and 1,3-butadiene are emitted at very low concentrations. All of these are oxidized or reduced by the exhaust gas catalysts by more than 95 %. Their residual concentration can be influenced by gasoline composition and hence by the gasoline formulation in the refinery. Gasolines forming less toxic and environmentally harmful material have been called "reformulated gasolines".

The effects of gasoline properties and composition on exhaust gas and evaporative emissions have been studied in a wide range of modern vehicles, partly in joint motor–oil industry projects, (e.g., auto–oil programs) in the US and in Europe. The results of the European program are summarized in Tables 3.1 and 3.2 [20].

Table 3.1 Summary of effects of gasoline properties on emissions in noncatalyst cars (source: EPEFE program)

	Change	Lead	CO	HC exhaust	HC evaporation	NO_x	Benzene	Butadiene	Aldehydes
Add oxygenate	0.0 to 2.7 mass % O_2	0	▼▼	▼	0–▼	±0	0	0	▲▲
Reduce aromatics	40 to 25 vol %	0	▼	▼	0	▼	▼▼	0	▲
Reduce benzene	3 to 2 vol %	0	0	0	– 0	0	▼▼	0	0
Reduce olefins	10 to 5 vol %	0	±0	▲	– 0	▼	0	▼▼	0
Reduce sulfur	300 to 100 ppm	0	0	0	0	0	0	0	0
Reduce Reid vapor pressure	70 to 60 kPa	0	0	±0	▼▼▼	0	0	0	0
Increase E 100*	50 to 60 %	0	+ 0?	▼?	±0	0	0	0	0
Increase E 150**	85 to 90 %	0	0	▼▼	0	▲?	0	▲?	▲?

0 = no effect. ± 0 = −2 to +2 %. ▼ or ▲ = 2–10 % effect. ▼▼ or ▲▲ = 10–20% effect. ▼▼▼ or ▲▲▲ = > 20 %. ?=insufficient information.
* E 100 = % evaporated at 100 °C.
** E 150 = % evaporated at 150 °C.

Table 3.2 Summary of effects of gasoline properties on emissions in catalyst cars (source: EPEFE program)

	Change	Lead	CO	HC exhaust	HC evaporation	NO_x	Benzene	Butadiene	Aldehydes
Add oxygenate	0.0 to 2.7 mass % O_2	0	▼▼	▼	0 – ▲	+ 0	0	0	▲▲
Reduce aromatics	40 to 25 vol %	0	▼	▼?	0	▲?	▼▼	+ 0	▲?
Reduce benzene	3 to 2 vol %	0	0	0	– 0	0	▼▼	0	0
Reduce olefins	10 to 5 vol %	0	0	+ 0	– 0	– 0	0	▼▼	0
Reduce sulfur	300 to 100 ppm	0	▼?	▼?	0	▼?	▼?	▼?	▼?
Reduce Reid vapor pressure	70 to 60 kPa	0	0	– 0	▼▼	0	0	0	0
Increase E 100*	50 to 60 %	0	+ 0?	▼?	0	0	0	0	0
Increase E 150**	85 to 90 %	0	0	▼▼	0	▲?	0	▼?	▼?

0 = no effect. ± 0 = −2 to +2 %. ▼ or ▲ = 2–10 % effect. ▼▼ or ▲▲ = 10–20 % effect. ▼▼▼ or ▲▲▲ = > 20 %. ?=insufficient information.
* % evaporated at 100 °C.
** % evaporated at 150 °C.

The reduction of sulfur and benzene has a clear-cut positive effect. The other measures, e.g., the reduction of total aromatics, the increase of mid-range volatility and the use of methyl *tert*-butyl ether have positive effects on some of the emission components but negative on others. This applies also to the reduction of vapor pressure (reduction of butane), which reduces the evaporative losses but increases emission of CO_2 and unburned hydrocarbons under cold start conditions.

A further objective is to reduce the formation of ozone and smog that occurs as a result of photochemical reactions between hydrocarbons and nitrogen oxides. The ozone forming potential of hydrocarbons differs depending on their structure and size. Thus, the composition of the unburned hydrocarbons can be optimized by the composition of the gasoline to reduce the ozone forming potential of evaporative losses and exhaust emissions. As an example the reduction of the highly reactive olefins in the fuel is a measure to reduce the ozone forming potential, however olefins are desired fast burning gasoline components which are also readily converted in the exhaust gas catalysts.

The evaluation of the results of the above mentioned programs have led to the production of specific "reformulated gasolines" in some areas, e.g., California and to the stepwise introduction of more stringent gasoline specifications, e.g., in Europe, as listed below.

	2000	2005
Aromatics	42 vol %	35 vol %
Benzene	1.0 vol %	1.0 vol %
Olefins	18 vol % *	18 vol %
Sulfur	150 ppm	50 ppm and 10 ppm

* max. olefin content regular grade 21 vol %

In Germany some mineral oil companies had voluntarily introduced sulfur free gasoline qualities, e.g., Shell Optimax. Since 2005 at least one sulfur free gasoline must be available at the gasoline stations. Currently a tax incentive for sulfur free gasoline ensures that the majority of gasoline sold in Germany is sulfur free. Low and ultra-low sulfur levels can also be found in other countries, e.g., in Japan and in the US. Sulfur free gasolines (sulfur content less than 10 ppm) are required for new highly efficient exhaust gas aftertreatment systems especially for lean burn direct injection engines. The new specifications require changes to the refinery units including substantial investment.

3.3.1.4 **Stability, Cleanliness, etc.**

Problem-free gasoline storage, distribution, and vehicle/engine operation requires a practical freedom from water, acids, and foreign matter particles. The gasoline must have long term and high temperature stability in the presence of air, (oxygen) and must be neutral to metals and elastomers. A number of laboratory tests with specified maximum tolerable limits are used to define these properties (see Chap. Fuel Standardization and Testing). Some of these properties can be influenced by additives (see Chap. Fuel Additives), e.g., water shedding, anticorrosion properties, oxidation stability. The long-term effect of fuel on elastomers in the fuel system is controlled by the selection of suitable materials by fuel system manufacturers. These materials have been tested in standardized and critical reference fluids.

3.3.1.5 Performance Additives

To ensure safe operation of the engine, high-quality gasolines contain special additives which ensure optimum inlet system cleanliness (see Section Detergents) and prevent corrosion (corrosion inhibitors, see Section Corrosion Inhibitors).

3.3.2 Quality Aspects of Diesel Fuels

The diesel combustion process requires a fuel, which in contrast to gasoline ignites easily. Ignition quality is therefore the main quality criterion of diesel fuel. Density is important as an indicator for fuel consumption. Cold flow performance, fuel lubricity, viscosity, stability, and cleanliness control mainly reliability of engine operation. All the above mentioned aspects together with sulfur content have effects on emissions. Fuel volatility is far less important than that of gasoline.

3.3.2.1 Ignition Quality

Since the fuel is not ignited by an external source in the diesel engine, good ignition performance of diesel fuel is an important criterion for its quality. Ignition quality is determined by characteristics of the components used in production of the fuel (see Section Diesel Fuel Components) and can be improved by additives (ignition improvers, cetane improvers, see Section Ignition Improvers (Cetane Improvers)).

The *cetane number* is used to describe the ignition quality of diesel fuels. Like the octane number, the cetane number is determined by comparative measurements in single-cylinder test engines (CFR swirl chamber engine (EN ISO 5165, ASTM D 613), or BASF prechamber engine) [21]. Mixtures of α-methylnaphthalene with very low ignition quality (cetane number 0) and cetane (*n*-hexadecane) with very high ignition quality (cetane number 100) are used as references. The cetane number of a reference mixture is given by the volume percentage of cetane in α-methylnaphthalene. For the determination of ignition quality the ignition delay (time interval between start of injection and start of self-ignition) is kept constant by varying the compression ratio of the engine depending on the test fuel quality.

The ignition quality of diesel fuels can also be calculated using the *cetane index* method (EN ISO 4264, ASTM D 4737). However the cetane index gives only an indication of the ignition quality of a diesel fuel without cetane improvers. This index is calculated by means of a complex formula based on density and distillation range. The formula is updated in regular intervals depending on changes in diesel fuel quality and production technology.

A high cetane number is advantageous for the ignition and starting behavior, the reduction of exhaust gas and noise emission ("clatter") [22, 23].

3.3.2.2 **Density**

The density of a diesel fuel (EN ISO 12185, ASTM D 4052) has a considerable effect on the engine performance. Because the quantity of fuel injected into an engine is metered by volume, the mass of fuel introduced into the engine increases with density. A higher fuel density leads to an enrichment of the fuel–air mixture which in principle yields a higher engine power output. On the other hand a high density increases also particulate emissions and black smoke [24]. The maximum density for diesel fuels was therefore reduced from 860 to 845 kg/m^3 in Europe.

3.3.2.3 **Sulfur Content**

Exhaust emissions, specifically particulates, are increased by the sulfur in the diesel fuel. In addition acidic combustion products arising from sulfur (sulfur oxides) can lead to engine corrosion. In Europe the maximum sulfur content had therefore been limited to 350 ppm from January 1, 2000 onward and was further reduced to 50 ppm by 2005. Furthermore, it was decided to make "sulfur free" diesel fuels (sulfur contents less than 10 ppm) generally available to allow the use of improved exhaust gas after treatment systems. A tax incentive in Germany for sulfur free diesel fuels led to a widespread availability of these fuels.

3.3.2.4 **Cold Flow Properties**

The composition of a diesel fuel affects its filterability at low temperature to a great degree. Particularly, *n*-paraffins with high ignition quality tend to form wax crystals at low temperature, which can lead to clogging of the fuel filter. For this reason during winter periods the proportion of low boiling components such as kerosene and others with reduced concentration of *n*-paraffins is often increased in relation to higher *n*-paraffin-containing gas-oil fractions. Diesel fuels with filterability at very low temperatures, such as arctic fuels, consequently may have lower densities, viscosities, and also ignition performance than normal diesel fuels.

The low-temperature filterability of diesel fuel has been substantially improved since the 1970s by use of highly active flow-improving additives at reasonable costs [25]. The cold filter plugging point (CFPP), a laboratory test EN 116, gives an indication of the low-temperature behavior of diesel fuels (see Section Cold Flow Additives). The cold flow performance of fuel components is also characterized by the cloud point, EN 23015, i.e. the temperature at which wax crystals begin to form. For fuels containing cold flow additives the cloud point has little relevance. However, it is used for the specification of arctic fuels as a safeguard with values 10 °C above the CFPP filterability.

3.3.2.5 **Lubricity**

The introduction of low and ultra-low sulfur fuels initially led to severe problems due to wear in fuel injection pumps, because the desulfurization process removes substantial amounts of molecules acting as natural lubricants. These have to be substituted by highly efficient lubricity additives (see Section Additives for Diesel Fuel). The lubricity performance of diesel fuels is defined by a laboratory rig test according to EN ISO 12156-1.

3.3.2.6 Viscosity

To ensure optimal engine performance the fuel's viscosity at 40 °C has to comply with narrow limits, e.g., in Europe it must be between 2.0 and 4.5 mm^2/s. (EN ISO 3104). Too low a viscosity can lead to wear in the injection pump; too high a viscosity deteriorates injection and mixture formation.

3.3.2.7 Volatility

The diesel engine is far less sensitive to changes in the fuel volatility than the Otto engine. However, the boiling range of a diesel fuel (EN ISO 3405) influences other properties such as viscosity, flash point, autoignition temperature, cetane number, density, and cold flow performance. Raising the back end temperature, for example, also raises the cloud point; lowering the temperature of the front end also lowers the flash point of the fuel. Generally speaking the variations in volatility of commercial diesel fuels are modest and usually far less significant than the influence of individual engine design features. To reduce exhaust emissions the amount of high boiling fuel components has been limited by lowering the specified 95 % point (temperature for 95 % evaporation) from previously 370 °C to 360 °C. Because of the intercorrelation with the other above mentioned physical properties it is not necessary to specify the light end of the diesel fuel by distillation properties.

3.3.2.8 Diesel Fuel Stability, Cleanliness, etc.

Safe storage and transportation of diesel fuels, their filterability, injector cleanliness, and compatibility with metals is achieved by meeting a number of specification criteria:

The minimum flash point (EN ISO 2719) of diesel fuel has been set to 55 °C in Europe by law. This limit has been adapted to the safety design elements of diesel fuel tanks which are less rigid than those for gasoline. Contamination of diesel fuel with small proportions of gasoline reduces the flash point below the legal limit and makes storage and transportation unsafe. However, it is legal to blend diesel fuel with gasoline in the vehicle tank to improve the cold flow performance at extremely low temperatures.

Total dirt (EN 12662) and water content (EN ISO 12937) are limited to guarantee acceptable filterability especially at low temperatures when filter elements may additionally be loaded with wax formed from paraffinic fuel components.

Carbon residue (EN ISO 10370) and oxidation stability (EN ISO 12205, ASTM D 2274) give an indication of possible deposit formation in the fuel system. Deposit formation is also controlled by addition of additives (see Section Detergent Additives).

The anticorrosion properties are specified by means of the copper corrosion test (EN ISO 2160). Again they can be improved by specific additives (see Section Additives for Increasing Storage Stability—Antioxidants).

3.3.2.9 Diesel Fuel Effects on Exhaust Emissions

The performance of a diesel engine is less sensitive to changes in fuel quality than that of the gasoline engine. However, in the diesel engine the fuel effects on emissions are more complex for several reasons: (1) The reduction of undesired exhaust gas constituents was so far achieved mainly during the combustion process and to a lesser extent by aftertreatment; (2) there is a wider range of engine technology in use (however there is now more conformity, i.e., direct injection engines with electronically controlled high pressure injection systems); (3) the diesel fuel composition is more complex than that of gasoline, i.e., because of the higher boiling range the individual molecules are less well defined (e.g., monoaromatics are mainly chemically bonded to long chain paraffins). Also properties and composition are often interlinked, such as aromatics concentration and density.

The effect of fuel properties and composition on exhaust emissions has been studied by oil companies and motor manufacturers—partly in joint programs, e.g., the European Program on Emissions, Fuels and Engine Technologies (EPEFE).

The effect of some fuel parameters, i.e., sulfur and monoaromatics content had been sufficiently documented previously. Sulfur in the fuel increases the emission of sulfur dioxide and of exhaust gas particulates, it has also negative effects on aftertreatment systems. Monoaromatics in the fuel have no measurable effect on exhaust emissions. The effect of other important fuel properties, i.e., density, polyaromatics content, cetane number, and final boiling point has been studied in the EPEFE program. The results of this investigation are summarized in Tables 3.3 and 3.4 [20].

For overall reduction of unwanted exhaust gas components it was decided to tighten the specification of diesel fuels in two steps in the years 2000 and 2005, (see also Chap. Fuel Standardization and Testing) by increasing the cetane number, reducing the maximum sulfur content, the final boiling point and the density and by introducing a maximum limit for polyaromatics.

3.3.2.10 **Performance Additives**

To ensure optimum operation of the car or truck, high-quality diesel fuels contain a variety of special additives which ensure maximum cleanliness of the engine (detergents), minimize the foaming of the diesel during fueling processes (antifoaming

Table 3.3 Summary of effects of diesel fuel properties on emissions in light duty vehicles[a] (Source: EPEFE program)

	Change	CO	HC	NO_x	PM[c]
Reduce sulfur	2000 to 500 ppm	0	0	0	▼
Reduce density[b]	850 to 820 Kg/m^3	▼	▼	0	▼▼▼
Reduce polyaromatics[b]	6 to 3 vol %				
Increase cetane number	50 to 55	▼▼	▼▼	± 0	▼
Reduce T 95[d]	370 to 330 °C	−0	−0	−0	−0

0 = no effect. ± 0 = −2 to +2 %. ▼ or ▲ = 2 – 10 % effect. ▼▼ or ▲▲ = 10 – 20 % effect. ▼▼▼ or ▲▲▲ = > 20 %. ?=insufficient information.

[a] Effects are not necessarily additive.

[b] Insufficient data are available to reliably separate the influence of these parameters.

[c] PM = particulate matter.

[d] T 95 = temperature for 95 % evaporation.

Table 3.4 Summary of effects of diesel fuel properties on emissions in heavy duty vehicles[a] (Source: EPEFE program)

	Change	CO	HC	NO_x	PM[c]
Reduce sulfur	2000 to 500 ppm	0	0	0	▼▼
Reduce density[b]	850 to 820 Kg/m^3	0	0	▼	▼
Reduce polyaromatics[b]	6 to 3 vol%				
Increase cetane number	50 to 55	▼	▼	▼	▼
Reduce T 95[d]	370 to 330 °C	−0	−0	−0	−0

0 = no effect. ± 0 = −2 to +2 %. ▼ or ▲ = 2–10 % effect. ▼▼ or ▲▲ = 10–20 % effect. ▼▼▲ or ▲▲▲ = > 20 %. ?=insufficient information.
[a] Effects are not necessarily additive.
[b] Insufficient data are available to reliably separate the influence of these parameters.
[c] PM = particulate matter.
[d] T 95 = temperature for 95% evaporation.

additives), prevent corrosion (corrosion inhibitors), improve lubricity (lubricity improver) and improve the cold flow performance (flow improver, wax antisettling additives). Also ignition improvers may be used (see Section Additives for Diesel Fuel).

3.4 Fuel Components

The minimum requirements for fuel quality are defined in standards or specifications (see Chap. Fuel Standardization and Testing) to guarantee trouble-free engine operation and acceptable exhaust emissions. These requirements must be fulfilled in refineries by best use of the available product streams [26]. The blending of gasoline and diesel fuels therefore requests the careful selection of suitable refinery components, with sometimes very different physical and chemical properties in different refineries (see also → Oil Refining) or as a result of crude oil properties. More stringent requirements for the improvement of exhaust emissions post catalyst bring with them more and more intense refining processes such as, e.g., desulfurization.

3.4.1 Gasoline Components

The most important criteria that determine the suitability of gasoline components are the octane numbers (RON and MON), volatility (boiling properties and vapor pressure) and more recently chemical composition (aromatics, absence of benzene, olefins and sulfur) (Table 3.5). Gasoline fuels consist of hundreds of individual chemical substances belonging mainly to one of the following three groups: aromatics, paraffins, and olefins (Table 3.6).

Table 3.5 Important properties of various gasoline components (approximate values) [3]

Component	ρ_{15}, g/mL	MON	RON	E 70,[a] vol %	E 100,[b] vol %
Straight-run gasoline	0.680	62	64	70	100
Butane	0.595	87–94[c]	92–99	100	100
Pyrolysis gasoline[d]	0.800	82	97	35	40
Light coker gasoline	0.670	69	81	70	100
Light catalytically cracked gasoline	0.685	80	92	60	90
Heavy catalytically cracked gasoline	0.800	77	86	0	5
Light hydrocracked gasoline	0.670	84	90	70	100
FR reformate (94)[e]	0.780	84	94	10	40
FR reformate (99)[f]	0.800	88	99	8	35
FR reformate (101)[g]	0.820	89	101	6	20
Isomerizate (isopentane)	0.625	87	92	100	100
Alkylate	0.700	90	92	15	45
Polymer gasoline	0.740	80	100	5	10
Methyl *tert*-butyl ether	0.745	98	114	100	100
Methanol–*tert*-butanol (1:1)	0.790	95	115	50	100

[a] E 70 = amount of fuel components with *bp* >70 °C.
[b] E 100 = amount of fuel components with *bp* >100 °C.
[c] Depending on the isobutane content.
[d] BTX aromatics extracted.
[e] Full range platformate with RON = 94.
[f] Full range platformate with RON = 99.
[g] Full range platformate with RON = 101.

Oil refineries normally have extremely interdependent structures (→ Oil Refining) and, accordingly, many different fuel components are produced. The properties of fuel components, however, depend not only on the type of conversion process and the feed used, but also substantially on the mode of operation of the process.

3.4.1.1 Straight-Run Gasoline

This gasoline component is obtained by distillation and is therefore easiest to produce and also longest in use. The composition of straight-run gasoline depends on the type of crude oil used (frequently high paraffin content). The octane quality is usually low.

In the past, when highly leaded fuels were used, straight-run gasoline was a primary component of fuel. Currently (ca. 2000), the proportion of straight-run gasoline in motor fuel is generally very low (0–20 %).

Table 3.6 Approximate composition of various fuel components (in volume percent)

Component	Paraffins	Olefins	Aromatics
Straight-run gasoline	94	1	5
Butane	100		
Pyrolysis gasoline*	ca. 20	ca. 10	ca. 70
Light coker gasoline	ca. 57	ca. 40	ca. 3
Light catalytically cracked gasoline	61	26	13
Heavy catalytically cracked gasoline	29	19	52
Light hydrocracked gasoline	100	0	0
FR reformate (94)**	45		55
FR reformate (99)**	38		62
FR reformate (101)**	29	1	70
Isomerizate	98		2
Alkylate	100		
Polymer gasoline	5	90	5

* Reflux from petrochemical plants; after partial hydrogenation.
** For abbreviations, see Table 3.5.

Butane also belongs to the straight-run products but it has a high octane level. It is added to gasoline, besides being used as liquefied petroleum gas (→ Liquefied Petroleum Gas). Because of its high vapor pressure, butane is good for engine cold-start. On the other hand, too high a vapor pressure can lead to hot-fuel handling problems (warm start problems, stumbling engine, vapor lock) and increased evaporative hydrocarbon emissions [27]. Thus, addition of butane is limited by the vapor pressure specifications.

The yield of straight-run products supplied naturally by crude petroleum cannot meet the high demand for gasoline either qualitatively or quantitatively. Thus, high-boiling components of crude petroleum are cracked either thermally or catalytically into lower boiling products that can be used in gasoline production.

3.4.1.2 Thermally Cracked Gasoline

Thermally cracked gasoline is often used in the form of *pyrolysis gasoline*. It is obtained in particular as a byproduct from steam crackers (→ Ethylene) in petrochemical plants and has a high content of olefins and aromatics (with good octane quality). BTX (benzene, toluene, xylene) aromatics are frequently extracted from pyrolysis gasoline as valuable raw materials for chemical processes, so only the residues from petrochemical plants are sometimes used in gasoline production [28].

Variants of thermal cracking are *visbreaking* and *coking* (→ Thermal Processes (Visbreaking, Coking)). The relatively small quantities of gasoline obtained (visbreaker gasoline and coker gasoline) are olefin-rich and have a low octane quality.

3.4.1.3 **Catalytically Cracked Gasoline**

Catalytically cracked gasoline is more important for gasoline production. It has a high concentration of olefins and has to be desulfurized to be suitable for modern gasolines. Catalytically cracked gasoline is usually separated into high-boiling and low-boiling fractions, which have different applications. The octane quality (especially the MON) of catalytically cracked gasoline is usually sufficient only for regular gasoline. In order to meet the legally defined olefin specification of gasolines the amount which can be used in gasoline production is limited. The high olefin content of these fuel components also requires special precautions to be taken against polymerization and oxidation during their use (see Section Antioxidants).

Another variant of catalytic cracking is *hydrocracking*, a process in which hydrogen is added. The hydrocracked gasoline produced by this process contains practically no unsaturated components and consists principally of iso- and cycloparaffins. Hydrocracked gasoline has a high MON, a moderate RON, relatively low density, and high volatility [29].

3.4.1.4 **Catalytic Reformate (Platformate) (→ Catalytic Reforming)**

One of the most important motor gasoline component is catalytic reformate, sometimes referred to as platformate. Reformate has an exceptionally high aromatic content and a very high octane quality. It is obtained by reforming (isomerization and dehydrogenation reactions) from paraffin-rich straight-run gasoline with a low octane quality [30]. The properties of the reformate essentially depend on the mode of operation in the reforming process [31]. When very high octane numbers are achieved, the aromatics content and the density increase, but at the same time the yield and volatility decrease. Because of its high octane quality, reformate is the backbone for the production of lead-free premium fuels. The restrictions of benzene and aromatic content in gasoline fuels limit the use of these components and also requires specific treatment of the reformer feed and/or the reformed product to remove the benzene from this stream (mainly distillation and hydrogenation techniques).

3.4.1.5 **Isomerate (→ Isomerization)**

The demand for volatile gasoline components with sufficient octane quality cannot be met by, e.g., unrefined straight-run gasoline [32]. For this reason, from the 1970s onward the low-octane straight-run gasoline was increasingly converted into the higher octane isomerate (mainly isoparaffins). Because it is free from aromatics isomerate is becoming increasingly attractive for environmental reasons.

3.4.1.6 **Alkylate (→ Alkylation)**

Another attractive fuel component consisting mainly of isoparaffins is alkylate, which is moderately high in octane quality (especially with regard to the MON), however, less suitable to formulate higher octane premium fuels like Super Plus (minimum RON 98). Alkylate is obtained by the reaction of isobutane with olefins and can only be produced in limited capacity because of the low availability of isobutane.

Since the introduction of mandatory maximum aromatic levels in gasolines alkylate has become one of the most important components for the production of premium fuels. Alkylate is the main component in aviation gasoline (Avgas).

3.4.1.7 Polymer Gasoline (→ Polymerization)

A less common gasoline component is polymer gasoline, obtained by oligomerization of propene or butenes produced in cracking plants. It consists primarily of $C_6 - C_{12}$ olefins.

3.4.1.8 Oxygenates

High octane quality of unleaded gasoline can be obtained more easily by the addition of oxygen-containing fuel components [33, 34]. In particular, the addition of methyl *tert*-butyl ether (MTBE) (→ Methyl *Tert*-Butyl Ether) has proved to be very useful for maintaining the octane quality of unleaded high octane fuels, even more since the introduction of maximum aromatic levels of 42 vol % (2000) and 35 vol % (2005). Besides its high octane quality MTBE is a very clean burning gasoline component.

Furthermore, methanol, ethanol produced from biomass, or higher alcohols can be used in low concentrations.

3.4.2 Diesel Fuel Components

The self-igniting diesel combustion process requires molecules with different structures than used in gasoline, i.e., in contrast to gasoline diesel fuel components must decompose easily at high temperature and pressure. Normal *n*-paraffins are for that reason ideal diesel fuel constituents, in contrast to olefins and aromatics. However, because of the higher boiling range of diesel fuel, the hydrocarbon molecules often consist of a combination of different structural moieties (e.g., an aromatic ring with a long paraffinic or olefinic side chain). The main characteristics determined by the composition of the diesel fuel are its ignition performance, cold flow properties, volumetric calorific value, and density. The latter can affect the air–fuel ratio and therefore influences particulate emissions (smoking tendency) under maximum torque conditions.

None of the classes of substances present in diesel fuel fulfills all the criteria equally well; for example, *n*-paraffins, which have a very good ignition performance and low smoking tendency, show poor low-temperature behavior and have a low calorific value (Table 3.7). However the combination of, e.g., paraffinic and

Table 3.7 Properties of various hydrocarbon groups with regard to their suitability as diesel fuels

	Ignition quality	Cold flow properties	Volumetric calorific value	Density	Smoking tendency
n-Paraffins	good	poor	low	low	low
Isoparaffins	low	good	low	low	low
Olefins	low	good	low	low	moderate
Naphthenes	moderate	good	moderate	moderate	moderate
Aromatics	poor	moderate	high	high	high

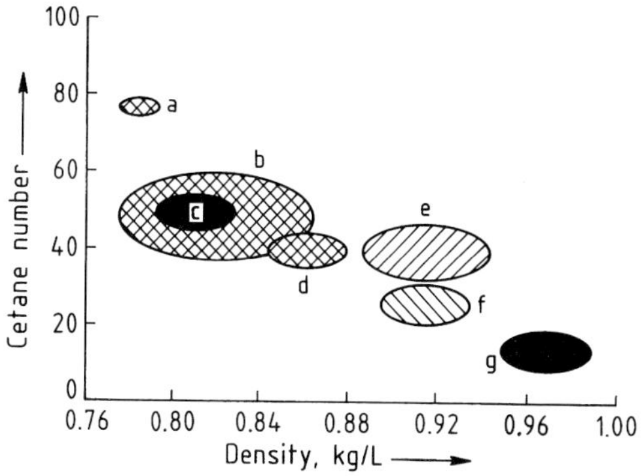

Fig. 3.4 Typical properties of gas-oil components a) Synthetic middle distillate; b) Straight-run gas oil; c) Hydrocracked gas oil; d) Thermally cracked gas oil; e) Syncrude gas oil; f) Coker gas oil; g) Light catalytically cracked cycle oil.

aromatic substructures in the molecules can reduce cold flow problems without significantly impairing the combustion properties.

The diesel engine can also theoretically be run on lower and higher boiling hydrocarbons. Marine diesel engines for example use higher boiling residual fuels (→ Residual Fuel Oil). The particular value of the automotive diesel engine lies, however, in its ability to utilize economically the middle distillates in the distillation range 150–360 °C. Accordingly, the diesel fuel components produced in refineries must correspond to the aforementioned requirements as far as possible.

3.4.2.1 Straight-Run Middle Distillate

Straight-run middle distillate (distillate gas oil) represents a valuable diesel fuel component because of its usually high cetane number (see Fig. 3.4). The quality of straight-run middle distillate nevertheless depends strongly on the quality of crude petroleum used. In many cases, distillate gas oil has a relatively high paraffin and low aromatic content.

Although middle distillates can be used for the diesel engine without further processing (apart from desulfurization), other diesel fuel components are also employed. The following reasons are decisive: (1) the natural middle distillate fraction in crude petroleum is generally not sufficient to meet the demand for middle distillate; (2) unwanted heavy components in crudes are present in excess and are, therefore, converted into lighter products; (3) the refinery structures also provide a certain "compulsory fraction" of middle distillates from conversion processes; and (4) particular quality requirements can also make the production of special diesel fuel components necessary.

3.4.2.2 Thermally Cracked Gas Oil

Thermally cracked gas oil originates from either the visbreaking or the coking process and tends to have a lower cetane number and higher density than distillate gas oil

Table 3.8 Properties of diesel fuel components

		Composition	
Process	**Component**	**Before treatment**	**After treatment**
Distillation	distillate gas oil (straight run)		content of aromatics: medium; olefins: very low; paraffins: high
Thermal cracking	thermocracked gas oil	content of aromatics: low; olefins: medium; paraffins: high	content of aromatics: low; olefins: none; paraffins: high
Catalytic cracking	catalytically cracked gas oil	content of aromatics: high; olefins: low; paraffins: low	content of aromatics: high; olefins: none; paraffins: low
Hydrocracking	hydrocracked gas oil		content of aromatics: very low; olefins: none; paraffins: very high
SMDS synthesis	synthetic gas oil		content of aromatics: none; olefins: none; paraffins: very high
Distillation or hydrocracking	kerosene		content of aromatics: low; olefins: none; paraffins: very high

(Table 3.8). Thermally cracked gas oil usually goes through a subsequent hydrogenation step because of the formation of unstable olefinic components during the cracking process. After hydrogenation, it represents a usable diesel fuel component.

3.4.2.3 Catalytically Cracked Gas Oil

Catalytically cracked gas oil is generally characterized by a very low cetane number and high density. Because of its instability, catalytically cracked gas oil must normally be hydrogenated [35]. Because of the poor ignition performance resulting from the high aromatic content and the high smoking tendency, it can be added to diesel fuel only in limited quantities.

3.4.2.4 Hydrocracked Gas Oil

Hydrocracked gas oil is a very valuable diesel fuel component. Because of its very high paraffins content, the ignition performance is good and the smoking tendency low.

3.4.2.5 Kerosene

Improved cold flow properties of diesel fuel can be achieved by the addition of kerosene. Because kerosene is predominantly used as jet fuel and, therefore, expensive and in short supply, its addition is limited mainly to meet the requirements of low temperature climatic conditions. The density, calorific value, and viscosity of kerosene are in most cases, however, much lower than those of other diesel fuel components and therefore its use is limited anyway. The low density and low calorific value would lead to decreased engine power and increased volumetric fuel consumption. The low viscosity would require additives to avoid wear in the injection pump because of poor lubricity.

3.4.2.6 Synthetic Diesel Fuel

Diesel fuel can also be obtained from fossil energy sources other than crude petroleum. During World War II diesel fuel was produced from coal by the Fischer–Tropsch process in Germany. This synthetic fuel (*kogasin*) had a range of outstanding properties: cetane number ca. 100, almost sulfur-free composition, low density, and high gravimetric calorific value. This synthetic diesel fuel component was used principally for enhancing low-value coal tar oil. Synthetic diesel fuel is produced in small quantities by the Shell Middle Distillate Synthesis (SMDS) process in Malaysia from natural gas [36]. The process includes the following steps:

1. Steam reforming of natural gas (conversion to synthesis gas),
2. Buildup of long-chain *n*-paraffins (Fischer–Tropsch synthesis),
3. Isomerization to isoparaffins
4. Distillative separation into kerosene and diesel fraction

The product was initially used mainly for improvement of the poor quality of the straight-run middle distillate otherwise available in Malaysia. The middle distillate produced by the SMDS process results in lower exhaust emission in diesel engines because it has a high cetane number (>60). To meet increasing requirements for diesel fuel quality of modern low emission engines synthetic middle distillates are meanwhile used as components in top-quality diesel fuels.

A process is currently under development is to convert biomass to gas on a commercial scale with subsequent use of the SMDS process.

3.5 Fuel Additives

By the generally accepted definition only agents that are added to fuels in a concentration of <1 % should be described as "additives." For substances present in higher concentration, the term "fuel components" is more appropriate.

Fuels are treated with additives for various reasons: in the past the addition of octane enhancers to gasolines enabled the production of high octane grades. Nowadays the use of cetane improvers offers the possibility of upgrading a base fuel very economically compared with refinery measures.

Performance additives, on the other hand, which are added to modern high quality fuels, improve the behavior of a base fuel in operation, and offer technical advantages that often cannot be achieved by measures taken in the refinery [37]. In many cases, these types of additives offer the only possibility of guaranteeing trouble-free engine performance over a longer running period. Finally, treatment of fuels with additives is a major route to achieve product differentiation and trademarked quality.

Thus it is not surprising that in Western Europe further increase in the use of additives is generally anticipated [38] with the exception of the alkyllead compounds which due to their incompatibility with exhaust catalysts and their toxicity have been phased out as of January 1, 2000 in the EU. A derogation of two years for the lead phase out was granted to Spain, Italy, and Greece by the EU commission. Lead phase out is a general trend worldwide, except for developing countries with low car population.

3.5.1
Gasoline Additives

The treatment of gasoline with additives is almost as old as the fuel itself [39]. At first (around 1900), the search for so-called brisance-increasing additives was predominant (Fig. 3.5). In 1918, treatment with certain additives was recognized to allow an increase of compression ratio without dangerous knocking. An increase

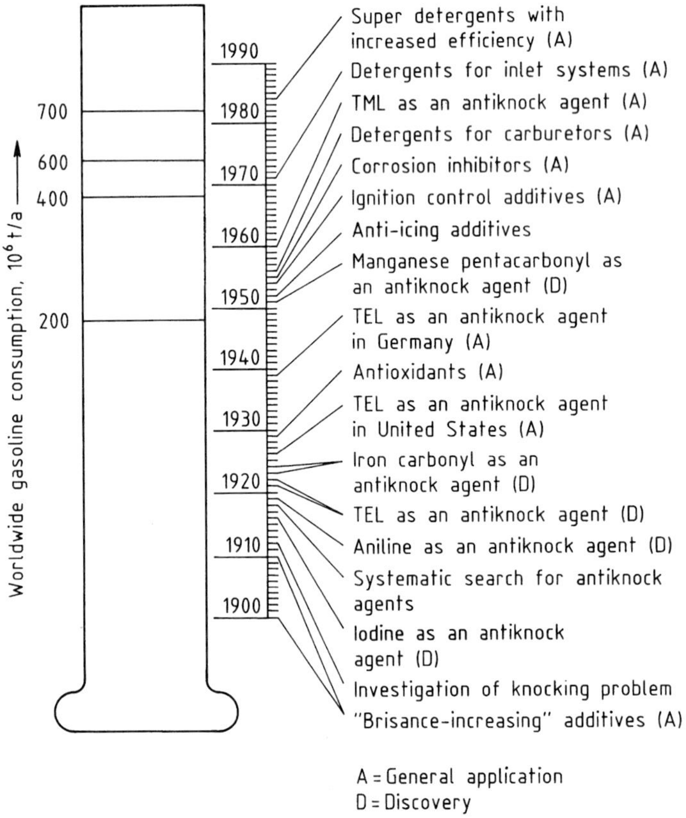

Fig. 3.5 Historical development of gasoline additives.

in the octane number became the major aim for decades in additive technology. Systematic research into additives that could positively influence many aspects of engine behavior began around 1950 on a wider scale. Since ca. 1970, conventional, ash-forming antiknock additives became increasingly less important for environmental reasons. Nowadays additive development is driven by environmental specifications as well as demanding new technologies like direct injected gasoline cars.

3.5.1.1 Corrosion Inhibitors

Water is almost always present in traces and can ingress into the fuel during processing, transfer, and storage by condensation. Moreover alcohols, when present in the fuel, can also lead to corrosion problems either by themselves or because of their water-solubility. Corrosion must be inhibited not only to avoid damage of the tank or the fuel system but also because fine rust particulates can block fuel filters and injectors.

The effect of corrosion inhibitors is based on their ability to cover a metal surface with a monomolecular layer of inhibitor molecules and thus protect it from contact with corrosive components [40]. The structure of the corrosion inhibitors corresponds to this requirement: a polar end of the molecule (e.g., a carboxylate, ester, or amine group) is attached to the metal surface; while a nonpolar, high molecular mass alkyl chain at the other end of the molecule extends into the hydrocarbon phase (Fig. 3.6). An oily covering layer is thus formed that effectively repels water.

High alcohol contents in fuel (e.g., gasohol in Brazil) are a problem because alcohols, in addition to their own corrosiveness, can also dissolve the protective film. Apart from these cases, which require exceptionally high concentrations (e.g., 100 ppm) of corrosion inhibitors, 10–20 ppm, is generally sufficient for effective corrosion protection.

The efficiency of corrosion inhibitors can be tested in the corrosion test according to DIN 51585, ASTM D 665, in which gasoline – water mixtures are allowed to react with steel test objects.

Examples of suppliers are BASF (Kerokorr additives), Octel (DCI series), Lubrizol (LZ series), Bayer AG (RC additives), Infineum, Nalco/Exxon, Baker/Petrolite (Tolad) Betz-Dearborn, Apollo Technologies (PRI), UOP (Unicor), Ethyl (Hitec), and Champion Technologies (RPS).

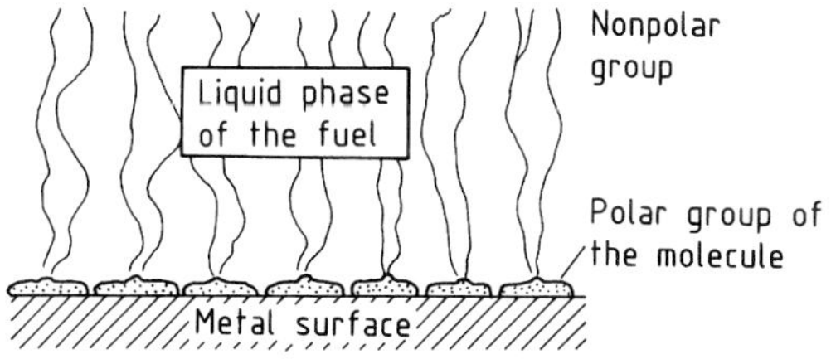

Fig. 3.6 Mode of action of corrosion inhibitors.

3.5.1.2 Detergents

In Otto engines, high molecular mass organic deposits can be formed in various sites in the fuelling and inlet and mixing systems. These deposits can lead to considerable problems in engine operation. Deposits on the *carburetor* and *injectors* prevent the formation of an optimal fuel–air mixture and cause increased fuel consumption and poor exhaust gas emissions [41–43]. In extreme cases the engine can even stall. Severe deposit formation on the *inlet manifolds* and *valves* causes, in addition to increased fuel consumption and deterioration in exhaust emissions, a decrease in power [44–47]. Sometimes even mechanical engine damage can occur.

The formation of deposits is caused by several factors [48–51]: Fuel components with high olefin or heteroatom content (which also have a marked tendency toward gum formation) are relatively prone to deposit formation. This is also true for fuels with very high final boiling points. Engine oil, which enters the inlet system via the valve stem guides, also contributes to deposit formation. Finally, combustion gases play a role in deposit buildup. These gases get into the inlet system either with the recirculated blowby gases from the crankcase or directly from the combustion chamber because of valve overlap. The multiplicity of factors contributing to deposit buildup shows that the problem cannot be solved by individual improvements in single areas but only by the addition of detergents with extensive effectiveness.

Detergents for Carburetor and Injector Cleanliness. Detergent additives were first used in the 1950s to maintain carburetor cleanliness ("keep clean effect"). Later, detergent additives were developed which removed deposits already present in the carburetor ("cleanup effect"). First-generation detergents were specially designed to work in the relatively cold region of the carburetor. These detergents include alkylamines, alkyl phosphates, alkyl-substituted succinimides, imidazolines, and oleylamides, which were added to the fuel in concentrations of 50–100 ppm. These compounds are also suitable, in principle, for keeping injectors clean, although concentrations of 150–300 ppm are then required [52].

To assess the efficiency of a carburetor detergent additive, the CEC developed a test cycle with a Renault 5 engine [53].

Detergents for Inlet Valve Cleanliness. From the mid-1970s, the search for decreasing fuel consumption and increased power led to leaner fuel–air mixtures and temperature increases in the inlet system. This created several problems that could not be overcome by the first-generation detergent additives, because these were effective only in the low-temperature range. In individual cases the carburetor detergents caused even higher deposit accumulation on the inlet valves due to decomposition.

The development of second-generation detergent additives made it also possible to keep the hot inlet valve clean [47], [54–56]. Long-chain, high molecular mass alkyl compounds (molecular mass around 500–1500 g/mol) with a polar end group, are very effective. The mechanism is analogous to that of corrosion inhibitors described in Section Corrosion Inhibitors. Effective, modern detergent additives are based on polyisobuteneamine, polyisobuteneamide/succinimide, or polyetheramine technology.

To meet most stringent environmental legislation these additives have to be produced halogen-free. These detergents, except for the polyetheramines, are normally used in combination with very thermally stable carrier oils that keep the surface of the valve lubricated and ensure continuous flowing off of deposit particles. To keep inlet valves clean relatively high concentrations of detergent (300–400 ppm) and carrier oil (300–400 ppm) are used. Depending on the polar end group modern detergents provide both keep-clean as well as cleanup performance.

A number of tests have been developed to assess the detergency performance of these gasoline additives: the Mercedes Benz M102E test was developed in Europe by CEC [57], [58].

Examples of suppliers are BASF (Keropur multipurpose additives), Infineum (Vectron), Ethyl (Hitec), Octel (OGP), Oronite (OGA series), and Lubrizol (ADX).

3.5.1.3 **Antioxidants (→ Antioxidants)**

As early as 1930 the aging of gasoline was known to be inhibited considerably by antioxidants (oxidation inhibitors). Nevertheless, antioxidants achieved importance only after World War II, when olefin-rich fuel components from cracking processes were increasingly used (→ Oil Refining). Olefins, in particular conjugated dienes, undergo liquid-phase free-radical chain reactions relatively easily at low temperature, which finally lead to the formation of highly polymeric residues, the so-called gum [59]. Other potentially gum-forming components are aromatics and the nitrogen and sulfur compounds present at trace levels. The efficiency of antioxidants is based on their capability to interfere with the mechanism of gum formation. In this mechanism, the initiating step is formation of a radical ($R^1{-}H \rightarrow R^1\cdot$); in the second step, hydroperoxides are formed by reaction with dissolved oxygen from the air and the chain reaction is started (Eqs. 1 and 2):

$$R^1\cdot + O_2 \rightarrow R^1{-}O{-}O. \qquad (1)$$

$$R{-}O{-}O\cdot + R^2H \rightarrow R{-}O{-}OH + R^2\cdot \qquad (2)$$

Gum-forming reactions continue until the chain reaction is terminated by the coupling of two radicals (Eqs. 3 and 4). The chain termination reaction can also be taken over by an antioxidant (AH), which leads to the formation of an inactive radical A· that is not able to continue the chain reaction (Eq. 5):

$$R^1\cdot + R^2\cdot \rightarrow R^1{-}R^2 \qquad (3)$$

$$R^1{-}O{-}O\cdot + R^2\cdot \rightarrow R^1{-}O{-}O{-}R^2 \qquad (4)$$

$$R^1{-}O{-}O\cdot + AH \rightarrow R^1{-}O{-}O{-}H + A. \qquad (5)$$

The aging of gasoline must be inhibited not only because of gum formation, which (1) blocks carburetors and injectors, filters in engines or dispenser pumps; and (2) can lead to dirt accumulation in the inlet system. An additional function of the antioxidant used to be the stabilization of tetraethyllead.

The gum content of gasoline fuels is measured according to EN ISO 6246, ASTM D 381, giving the washed gum evaporation residue. The residue has to be washed with *n*-hexane to remove detergent additives and carrier oils used in high performance fuels.

The induction period (EN ISO 7536) is a laboratory test to measure the storage stability of gasolines. Most antioxidants belong to one of the following two classes of substance.

***para*-Phenylenediamines** have the general formula R^1–NH– C_6H_4–NH–R^2, in which R^1 and R^2 can be *sec*-butyl-, isopropyl-, 1,4-dimethylpentyl-, or 1-methylheptyl-. *para*-Phenylenediamines are very effective at concentrations of 5–20 ppm, particularly for gasolines with high olefin content. Their basicity, however, makes them susceptible to reaction with acidic impurities or fuel components. On the other hand, phenylenediamines offer the advantage of catalyzing the oxidation of mercaptans (which have foul odors even when present in trace amounts) to odorless disulfides.

Hindered Alkylphenols. Alkylphenols are optimally efficient if their hydroxyl group is sterically hindered by alkyl groups in positions 2 and 6, e.g., as in 2,6-di-*tert*-butylphenol.

Alkylphenols are particularly suitable for olefin contents <10 %. Because of steric hindrance, the acidity of the hydroxyl group is lowered to such a degree that the danger of side effects is low. To achieve the same efficiency as phenylenediamines, higher concentrations (5–100 ppm) are necessary. Because of their positive synergism, mixtures of phenylenediamines and alkylphenols are also used sometimes [60].

Examples of suppliers are BASF (Kerobit additives), Octel (AO), and Ciba (Irganox).

3.5.1.4 Metal Deactivators

The aging (oxidation) of gasoline can be accelerated considerably by traces of metals that get into the fuel during transfer. Mercaptans, phenols, and alcohols increase the solubility of metals in fuel. Copper has the highest catalytic activity, accelerating the oxidation of fuel considerably at concentrations as low as 0.01 mg/L.

The function of metal deactivators is based on the fact that they form stable chelate complexes with dissolved metal atoms so that the metal loses its undesirable catalytic activity [40]. The addition of *N,N'*-disalicylidene-1,2-propanediamine is particularly effective,

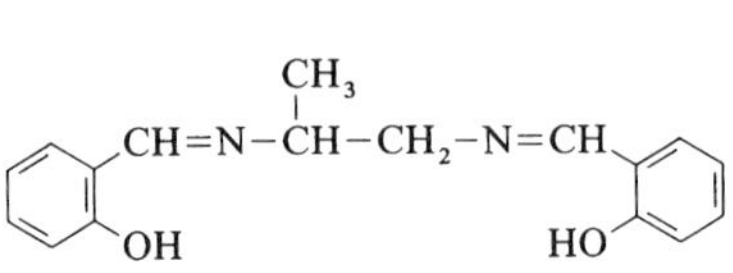

and copper is complexed as follows:

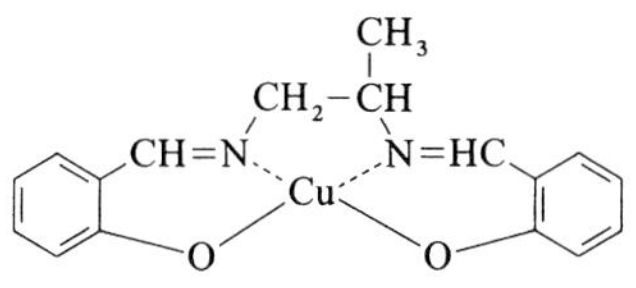

Examples of suppliers are BASF (Keromet additives) and Octel (DMD).

3.5.1.5 **Anti-Icing Additives**

The problem of carburetor icing has virtually disappeared because of the introduction of multipoint fuel injection systems which are practically immune to icing.

During the evaporation of fuel in the carburetor or in single point injection systems, heat is absorbed from the surroundings so that even at ambient temperatures >0 °C, temperatures below 0 °C can occur and lead to freezing of the air moisture there. Besides its temperature, the relative humidity of the air is an important factor for the formation of icing (Fig. 3.7) [54].

Carburetor icing was particularly noticeable during idling (engine stalling because of ice formation) or constant speed driving making itself noticeable by engine power loss, high fuel consumption, unfavorable exhaust gas emissions, and finally, engine stalling. To avoid carburetor icing, two different types of additives were used: (1) surfactants and (2) freezing point depressants (→ Antifreezes).

Surface-Active Substances. Surfactants produce protective films over the metal surface so that ice is prevented from sticking to it. Examples of surface-active compounds are amines, diamines, amides, or glycol esters of fatty acids, which are effective at 10–60 ppm. Other surface-active substances added to fuels, such as detergents, act as anti-icings in addition.

Freezing Point Depressants. Effective freezing point depressants include alcohols, glycols, dimethylformamide, and other water-soluble polar substances. Depending on the type used they were added in concentrations of 0.02–2 vol %.

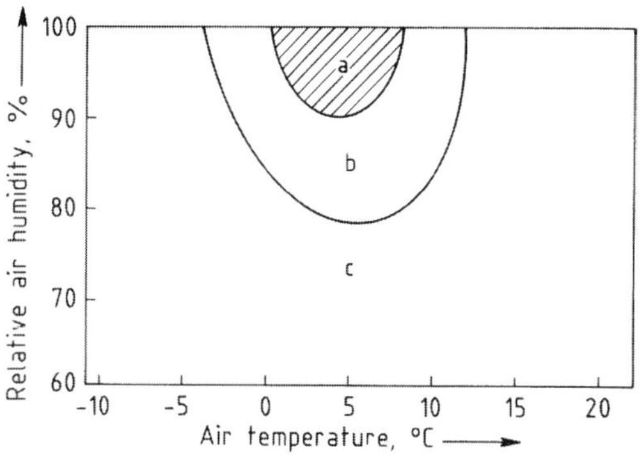

Fig. 3.7 Effect of relative air humidity and air temperature on carburetor icing [54] a) Severe icing; b) Icing; c) No icing.

3.5.1.6 Additives for Combating Combustion Chamber Deposits

Mainly when leaded gasolines were used considerable quantities of lead compounds, but also carbon, and engine oil components could deposit in the combustion chamber and on the spark plugs [61–64]. This could lead to knocking combustion and to spark plug fouling problems.

Organophosphorus and organoboron compounds were therefore added to fuel at an early stage. These additives do not remove the deposits but are at least able to convert them to materials with high glow temperature and low electrical conductivity (*deposit modifiers*). The commercial products contained, for example, tricresyl phosphate, triisopropyl phosphite, borazane, or boric acid esters. These additives are no longer on the market because (1) the phase out of lead compounds reduced the problem and (2) phosphorus compounds are incompatible with catalytic converters because of their tendency toward catalyst poisoning.

Research was devoted to the development of *ash-free* additives, which counteract the buildup of combustion chamber deposits and control any increase in the octane number requirement of the engine [*octane requirement increase control* (ORIC) additives]. These additives have shown some positive effects in test engines. However their effect under actual road conditions seems to be limited. Reduced engine oil consumption, precise fuel metering to obtain stoichiometric air-fuel ratios, the shape of modern combustion chambers and of course the use of unleaded gasoline have reduced the problem of combustion chamber deposit formation. Specific additives to reduce these deposits are therefore not in general use.

3.5.1.7 Valve Seat Recession Protection Additives

Some older cars with "soft valve seats" (e.g., cast iron) relied on leaded gasoline, since lead compounds inhibited corrosion and wear of the valve seats. Unleaded gasoline for older cars can be provided with organic potassium or sodium compounds as lead replacement. These can be added to the finished fuel by the end consumer using a bottle solution. However, valve seat recession occurs predominantly under high-duty (motorway) operation. As vintage cars with soft valve seats are normally not operated under these conditions it is generally not necessary to use these additives.

3.5.1.8 Antiknock Agents (→ Octane Enhancers)

The most important antiknock additives were alkyllead compounds, i.e., tetraethyllead and tetramethyllead. As pointed out the use of these additives was discontinued because of their poisoning effects on the exhaust-gas aftertreatment systems and their toxicity. Other metal-based antiknock additives have either similar unwanted side effects or limited activity. Organic antiknock additives are less effective than ash forming products. Best known are nitrogen-containing aniline derivatives. None of these antiknock additives is used commercially on a larger scale. Instead oxygen-containing blending components, mainly ethers (→ Octane Enhancers; → Methyl *Tert*-Butyl Ether) are used to increase the octane quality of unleaded gasolines.

3.5.1.9 Dehazers and Antistatic Additives

Dehazers and antistatic additives are rarely used in gasoline. They are similar to the corresponding types of additives in diesel fuel (see Sections Dehazers and Antistatic Additives).

3.5.2
Additives for Diesel Fuel

Apart from a few exceptions treatment of diesel fuel with additives was introduced much later than that of gasoline. Several reasons existed for the introduction of additives: (1) a significant progress in diesel engine technology occurred from about the mid-1970s onward which led to a greater market share of diesel engines in passenger cars; (2) fleet operators and car owners required improved reliability and comfort; (3) more stringent emission standards required improved diesel quality.

3.5.2.1 **Ignition Improvers (Cetane Improvers)**

The cetane number (see Section Quality Aspects of Diesel Fuels) is a measure of the ignition quality of a diesel fuel. Because a range of aspects essential to the operation of diesel engines are also connected with the cetane number (cold-start properties, particulate, e.g., white and black smoke emission, noise emission, fuel consumption, engine durability), the increase in cetane number by additives is a valuable means of improving diesel fuel quality [65].

The effect that can be achieved by the use of ignition improvers depends on the base diesel fuel. Unfortunately, base fuels with high natural cetane numbers react better to additive treatment than those with lower cetane numbers, for which an increase in the cetane number is more essential (Fig. 3.8). In view of constantly increasing efforts to reduce exhaust gas emissions, the fact that even small quantities

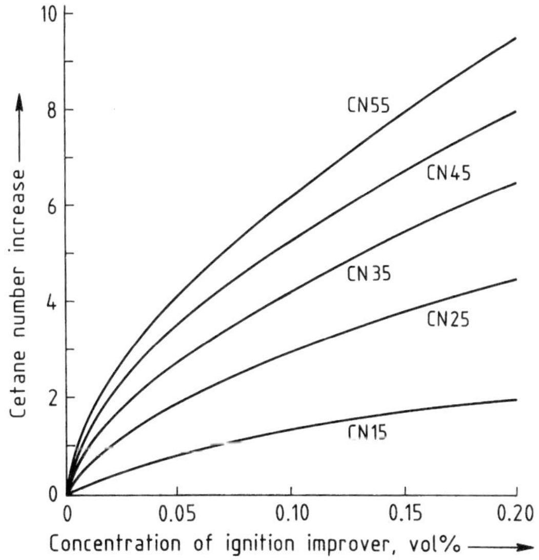

Fig. 3.8 Response of different base fuel types to addition of ignition improvers.

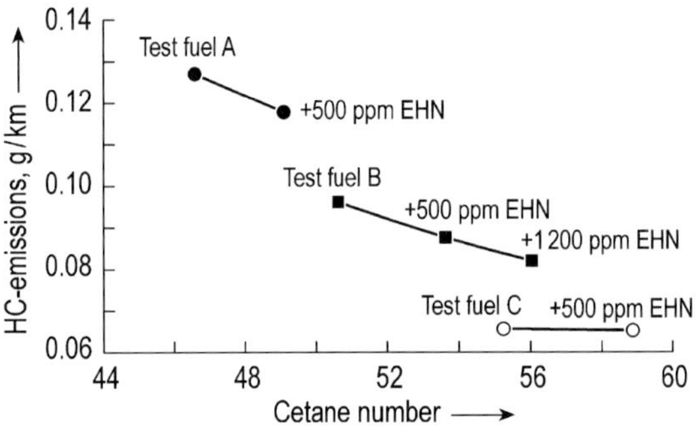

Fig. 3.9 Improvement of exhaust gas emissions by ignition improvers; average emissions of three passenger cars with pre- and swirl chamber diesel engines [67].

of ignition improvers can improve the exhaust gas quality of the diesel engine is important (Fig. 3.9) [66, 67].

The effect of ignition improvers is based on their rapid decomposition, with the formation of free radicals that accelerate the chain reaction of diesel fuel combustion. Suitable compounds are alkyl nitrates, ether nitrates, alkyldiglycol nitrates, and organic peroxides. From a commercial point of view only alkyl nitrates such as 2-ethylhexyl nitrate (EHN) have been successful. With these compounds, the ready cleavage of the $RO–NO_2$ bond leads to radical formation [69].

Examples of suppliers are Octel (CI additives) and Ethyl (Hitec).

3.5.2.2 **Detergent Additives**

Corresponding to the different construction of the diesel engine compared with the Otto engine, the problem of deposit formation occurs in a different form. In diesel engines the injection system is particularly affected, where gum- or carbonaceous deposits are formed on the injectors. This inhibits the ability of the valve needle to move and the formation of a finely dispersed spray of fuel. The mixing system is thus disturbed, and increased emissions (particulates), higher fuel consumption, and lower engine power, result.

A range of substances are suitable as detergents in diesel engines: amines, imidazolines, amides, succinimides, polyalkyl succinimides or amines, and polyetheramines [70, 71]. The following mechanism of action has been proposed: the surface activity of the polar groups ensures the formation of a protective film, whereas the dispersing effect of the polymeric chain prevents agglomeration and deposit formation by smaller particles. In this way a “keep clean effect” can be achieved; however, for already dirty nozzles, a solubilization of formed deposits material is also necessary. The addition of detergents offers other advantages for diesel fuel, such as better storage stability and a certain amount of protection against corrosion [72].

Examples of suppliers are Infineum (Vectron), Octel (OMA), Oronite (ODA series), and Lubrizol.

3.5.2.3 **Cold Flow Additives**

Diesel fuel has the unfortunate property that crystallization of high molecular mass *n*-paraffins (waxes) can occur at low ambient temperature. These wax crystals can clog the fuel filter and lead to engine stoppage after a short period of operation. The tendency toward wax separation depends strongly on the composition of the diesel fuel, which is determined mainly by the crude petroleum used and the refinery structure [73]. In principle, the low-temperature behavior of a diesel fuel could be improved by removing the particularly critical *n*-paraffins. However, this would be unacceptable not only because of the high costs, but also because *n*-paraffins are very valuable diesel fuel components due to their good ignition performance and low smoking tendency [74].

Before the development of highly effective additives for improving low-temperature properties, owners of diesel vehicles were frequently forced to add up to 30 % regular gasoline to diesel fuel, risking engine damage and increased exhaust emissions.

To assess the low-temperature behavior of diesel fuel the following two laboratory test procedures are used:

1. **Determination of the Cloud Point** (EN 23015, ASTM D 2500). The cloud point is the temperature at which diesel fuel begins to become turbid at a defined cooling rate because of the separation of small wax crystals. The cloud point is especially important for the evaluation of refinery components in order to determine type and amount of cold flow additive that has to be added to the components. The cloud point does not define the operability of the final, additivated diesel fuel.
2. **Determination of the Cold Filter Plugging Point (CFPP)** (EN 116) [75, 76]. The CFPP is the limiting temperature at which, under defined conditions, a diesel fuel sample can no longer pass through a metal sieve within a specified length of time. Generally speaking the CFPP gives a reliable indication of the operability limit expected for a diesel fuel in modern vehicles. Since the development of wax antisettling additives, which drastically lower the CFPP, the operability of critical older vehicles (e.g., without heating of the fuel system) is not as low as suggested by the CFPP result.

Flow Improvers. Flow improver additives improve the cold flow properties of diesel fuels in a cost- effective way. Refineries are given the flexibility to use high cetane quality paraffinic product streams for diesel fuel, kerosene, otherwise necessary for achieving good low-temperature behavior of diesel fuel, for the production of jet fuel.

Flow improvers have been used increasingly since 1973. Their mechanism of action is based on two different effects [77]. First, the additive forms crystallization nuclei on reaching the cloud point, to which the wax crystals become adsorbed. Second, further crystal growth is slowed down by adsorption of the additive on the growing crystal surface. Thus, both the crystal size and the tendency to agglomerate

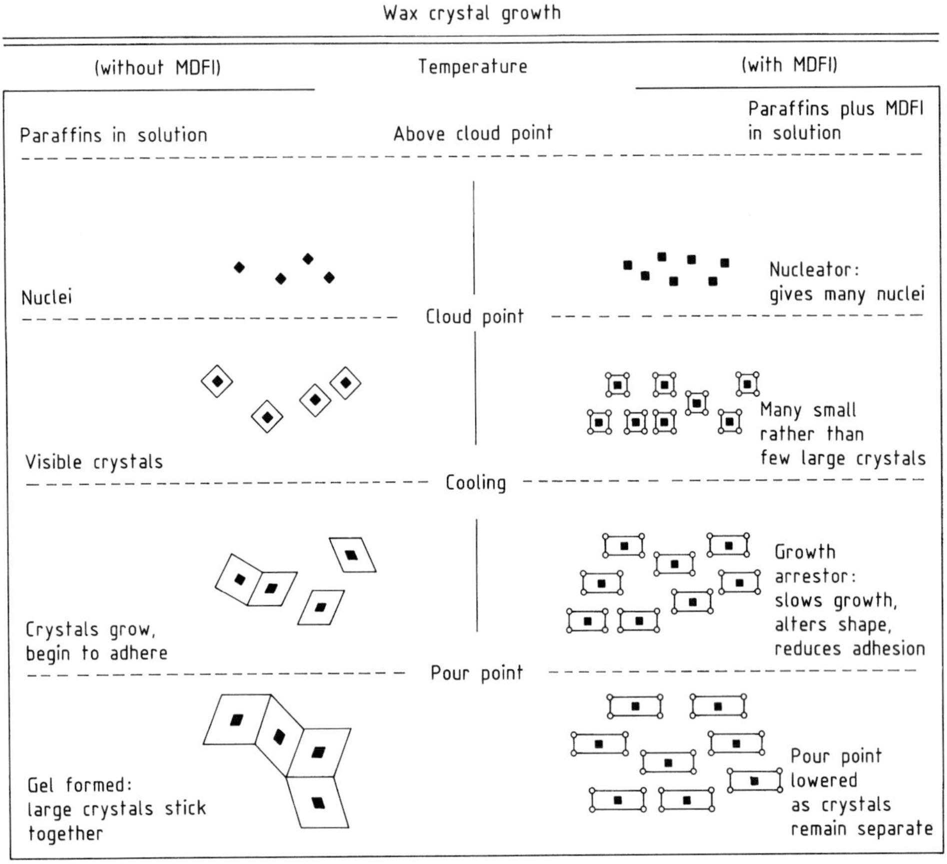

Fig. 3.10 Mode of action of flow improvers [79] MDFI = Middle distillate flow improvers.

are reduced (Fig. 3.10). The small crystals thus formed have a considerably lower tendency to block fuel filters because they (1) pass through the filters to some extent, (2) produce a still penetrable layer on the filter, and (3) dissolve again rapidly when the engine is warmed.

Flow improvers are effective only when they have been added to the fuel at temperatures well above the cloud point of the base fuel, i.e., at about +40 °C, and preferably in the refinery. Addition after crystal formation is not effective [78]. Ash-free polymers are used as flow improvers, in the vast majority of cases ethylene–vinyl acetate (EVA) copolymers. The dosage with which a particular CFPP value can be achieved depends on the base fuel and is normally 50–500 ppm (Fig. 3.11).

Examples of suppliers are BASF (Keroflux additives), Octel (OFI), Infineum, and Clariant (Dodiflow).

Wax Antisettling Additives (WASA). Wax antisettling additives keep small wax crystals in suspension and thus inhibit the formation of wax layers, which can

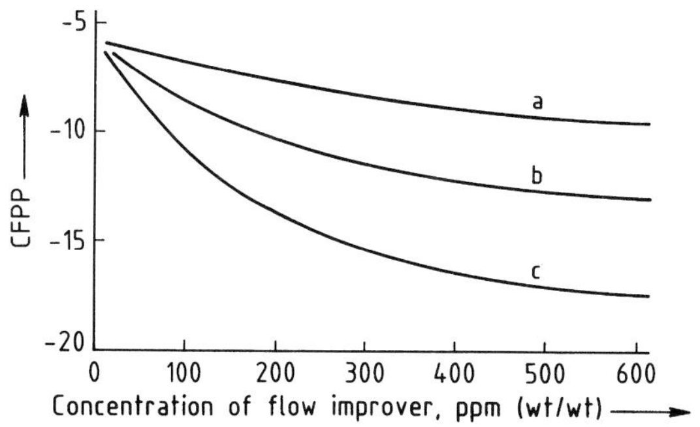

Fig. 3.11 Effect of a flow improver on various base diesel fuels [54] a) Poorest response behavior; b) Normal response behavior; c) Best response behavior

rapidly lead to filter clogging [25]. This is achieved by their ability to slow the growth of wax crystals, which is even greater than that of flow improvers. According to Stokes' law, small particles have a considerably lower settling rate than large ones. By reducing the size of wax crystals by one-fifth, the sedimentation time can be extended from an hour to a day. Wax antisettling additives have been used since 1986—generally in combination with flow improvers. Typical concentrations are 100–500 ppm. They are usually based on modified ethylene–vinyl acetate copolymers.

Examples of suppliers are BASF (Keroflux ES additives) and Infineum [79].

Cloud Point Depressants. The use of conventional flow improvers to lower the CFPP value generally has hardly any influence on the cloud point. However, polymer additives (based on ethylene–vinyl acetate, unsaturated esters, imides, or olefins) have been developed that permit additional lowering of the cloud point. Their mode of action is based on shifting the thermodynamic equilibrium by irreversibly decreasing the crystallization temperature. This particularly increases the solubility of higher molecular mass *n*-paraffins. With a dosage of 500 ppm the cloud point can be lowered by ca. 3 °C. The market importance of cloud point depressants is considerably smaller than that of flow improvers and wax antisettling additives [73].

3.5.2.4 **Lubricity Additives**

The intrinsic, natural lubricity of diesel fuel is reduced during the desulfurization process, because surface active components are destroyed during the hydrogenation. In the case of low and ultra-low sulfur diesel fuels the lubricity of the finished product can be too low, resulting in wear problems in the injection pumps. To ensure safe operation low and ultra-low sulfur diesel fuels have to be additivated with efficient lubricity additives. Examples of surface active substances are fatty acid derivatives which are added in the range of 50–200 mg/L.

The lubricity performance of fuels is evaluated in the high frequency reciprocating wear rig (HFRR) laboratory test (ISO 12156–1). A maximum wear (HFRR value) is defined and specified in EN 590.

Examples of suppliers are Octel (OLI), Ethyl (Hitec), Infineum, and Lubrizol (LZ).

3.5.2.5 Antifoam Additives

When a vehicle tank is filled with diesel fuel, severe foam formation can occur because of intensive mixing with air. This has unpleasant effects: either the time necessary to fill the tank completely is extended or the amount of fuel taken into the tank is considerably lower.

Silicon-containing antifoams, mostly polysiloxanes, have been developed and added to diesel fuel at a concentration of 10–50 ppm to avoid excessive foam formation. An effective antifoam additive not only depresses foam formation but also accelerates foam decomposition.

Examples of suppliers are Goldschmidt (Tegopren series), OSI Witco (MR additives), and Dow Corning.

3.5.2.6 Additives for Increasing Storage Stability — Antioxidants

During the storage of diesel fuel, aging and corrosion effects can occur in a way similar to gasoline (see Sections Antioxidants and Metal Deactivators). The aging of diesel fuel also occurs via oxidation and radical polymerization reactions, which first cause darkening and later sediment formation [80]. The problem of unstable diesel fuel has increased particularly since ca. 1970 because higher proportions of diesel fuel components from cracking processes are in use [81]. These components have a higher content of unsaturated hydrocarbons than straight-run middle distillates. Middle distillates from cracking processes are generally subjected to hydrogenation ("hydrotreating"); however, additional treatment with antioxidants is often necessary.

In diesel fuel aging, sulfur- and nitrogen-containing substances can also be responsible for sediment formation to a considerable extent, although they sometimes have an oxidation-inhibiting effect. In addition, a high acidity (caused by carboxylic acids or phenols) accelerates aging catalytically. Sediment formation brought about by aging can clog the fuel filter; gumlike deposits can affect the flow of the fuel in the injection pump and the injectors.

The additives used to inhibit aging are similar to those used for gasoline (see Sections Antioxidants and Metal Deactivators), namely sterically hindered phenols and phenylenediamines. Trialkylamines are also very efficient. The additive dosage is usually ca. 50 ppm; however, the efficiency depends on the particular diesel base fuel.

Since dissolved metal ions (especially copper) can also catalyze the aging of diesel fuel, ca. 5 ppm of a *metal deactivator* can be added.

Corrosion inhibitors (see Section Corrosion Inhibitors), which generally consist of high molecular mass amines when used in diesel fuel, are also effective against aging because of their basicity.

The *detergents additives* added to diesel fuel principally to keep the injection nozzles clean are often effective in preventing diesel fuel aging. Additives for stabilizing diesel fuel are frequently used in the form of a multifunctional mixture at a concentration of ca. 100 ppm.

3.5.2.7 **Dehazers**

Through contact with water vapor in refinery processes and possible entry of water during transportation and storage, diesel fuel often becomes cloudy due to the presence of finely dispersed water droplets. Since such a cloudy, water-containing fuel would lead to considerable operational problems, the water droplets must have separated before the fuel is distributed. Especially in diesel fuels rich in aromatics, water separation may take a long time so that large tank capacities are necessary. The settling of water can be accelerated considerably by the addition of dehazers. These additives mainly lower the surface tension between water droplets, so that agglomeration to form larger, more rapidly sinking drops is achieved. Quaternary ammonium salts in a concentration of 5–50 ppm are effective dehazers.

Examples of suppliers are Rechem (ER and EXP additives), Petrolite (Tolad additives), and Nalco (Nalco).

3.5.2.8 **Biocides (→ Biocides)**

As explained in Section Dehazers, the contact of diesel fuel with water cannot be ruled out totally. In some cases, water layers are formed at the bottom of the storage tank due to insufficient house keeping. At the interface between water and the diesel fuel, conditions are provided that allow the growth of microorganisms such as bacteria, yeast, and fungi. Infestation with microorganisms leads, on the one hand, to sediment formation which increases the danger of filter blocking. On the other hand, sulfate-reducing bacteria create a corrosive atmosphere that produces rust particles and can finally lead to tank destruction [82]. Through the addition of biocides, microbial infestation can be prevented or already affected tanks can be disinfected.

Microorganisms can develop resistance to biocides under certain circumstances, so biocides should be used only sparingly for specific purposes. Generally careful cleaning of the affected storage tank and consequent removal of water from the tank solve the problem.

A whole range of substances can act as biocides: formaldehyde derivatives, isothiazolones, triazines, *N,N*-methylenebis(5-methyloxazolidine), and boric acid derivatives. Biocides are usually applied in a range of 50–400 ppm.

Biocides are supplied, e.g., by Schülke & Mayr, Norderstedt, Germany.

3.5.2.9 **Antistatic Additives**

When hydrocarbons are pumped rapidly, a buildup of static electricity can occur which, on discharge, may lead to fuel vapor ignition. In comparison to kerosene, this danger is relatively small for diesel fuel because of its lower volatility and its content of polar components that lead to increased conductivity. However, the use of small quantities of antistatic additives (up to 5 ppm) which, for example, contain calcium compounds is sometimes advisable where, very high pumping rates (e.g., at refineries, transfer, or truck refueling) can lead to the danger of static charge buildup. Electrical conductivity can be determined by means of a testing method ASTM D 2624, developed for jet fuel.

Antistatic additives are supplied, e.g., by Octel (Stadis).

3.5.2.10 **Reodorants**

The odor of diesel fuel is often regarded to be unpleasant, with the result that people can reject diesel engines as "dirty". If the skin or clothes are contaminated with diesel fuel, the unpleasant odor can remain for a long time because of its low volatility.

The unpleasant smell is caused particularly by certain sulfur-containing and unsaturated components. Desulfurization with hydrogen removes most of the unpleasant smelling ingredients. Modern low-sulfur fuels have generally a neutral smell so that specific reodorants are no longer required.

Reodorants are based mostly on nature-identical substances that are toxicologically harmless, such as vanillin or terpenes. They usually work by blocking the olfactory nerves and can be added in concentrations of 30–100 ppm.

Examples of suppliers are Haarmann & Reimer (Kompensols) and Bush Boake Allen (LG additives).

3.6
Fuel Standardization and Testing

The engine performance strongly depends on the physical and chemical properties of a fuel (see Chap. Fuel Composition and Engine Efficiency). To guarantee adequate fuel quality for a broad range of engine requirements, minimum specifications must be met, which are defined, for example, in *Europe* in EN 228 (for gasoline) and EN 590 (for diesel) (Tables 3.9 and 3.10).

In these specifications the great climatic differences within Europe (Sweden–Portugal, for example) are taken into account be defining appropriate limits for, e.g., volatility of gasoline and cold-flow performance of diesel fuels for each country.

The ASTM testing methods used in the *United States* are similar to the corresponding European (EN, ISO) methods. For all testing procedures a variation in measured values is unavoidable, and determination of a "true value" is impossible. The terms "true value", repeatability, and reproducibility are precisely defined in EN ISO 4259.

The EN standards for gasoline and diesel fuel are mandatory. The quality of the fuels and their conformity with standards is, for example, tested on a random basis by the appropriate authorities in the Federal Republic of Germany and is followed up by court in the case of an offense. This is all the more important because not conforming to standards can cause damage to engines and provide economic advantages for disreputable suppliers.

Table 3.9 European standard for gasoline; extract from EN 22*

	Regular**	**Super**	**Super Plus****	**Test method**
ρ(15 °C), kg/m³	720–775	720–775	720–775	EN ISO 3675,
Knock rating				EN ISO 12185
RON	min. 91.0	min. 95.0	min. 98.0	EN ISO 5164
MON	min. 82.5	min. 85.0	min. 88.0	EN ISO 5163
Lead content, mg/L	max. 5	max. 5	max. 5	EN 237
Distillation range, total proportion distilled, vol% up to 70 °C:***				
Winter and intermediate	22–50	22–50	22–50	EN ISO 3405
Summer	20–48	20–48	20–48	
up to 100 °C:				
Winter and intermediate	46–71	46–71	46–71	
Summer	46–71	46–71	46–71	
Up to 150 °C	min. 75	min. 75	min. 75	
Final boiling point, °C	210	210	210	
Vapor pressure, kPa*				EN 13016–1
Winter	60–90	60–90	60–90	
Intermediate and summer	45–60	45–60	45–60	
Benzene content, vol %	1.0	1.0	1.0	EN 238, EN 12177 EN 14517
Aromatics content, vol %	max. 35.0	max. 35.0	max. 35.0	ASTM D 1319, EN 14517
Olefin content, vol %	max. 21.0	max. 21.0	max. 21.0	ASTM D 1319, EN 14517
Sulfur content, mg/kg	max. 50 or max. 10	max. 50 or max. 10	max. 50 or max. 10	EN ISO 20846, EN ISO 20847, EN ISO 20884,
Oxygenates, vol %				EN 1601, EN 13132
Methanol	max. 3.0	max. 3.0	max. 3.0	
Ethanol	max. 5.0	max. 5.0	max. 5.0	
2-Propanol (isopropyl alcohol, I((Ida)) PA)	max. 10.0	max. 10.0	max. 10.0	
2-Methyl-2-propanol (*tert*-butyl alcohol, TBA)	max. 7.0	max. 7.0	max. 7.0	
2-Methyl-1-propanol (isobutyl alcohol, I((Ida)) BA)	max. 10.0	max. 10.0	max. 10.0	
Ethers containing 5 or more carbon atoms	max. 15.0	max. 15.0	max. 15.0	
Other mono alcohols and ethers with a finaboiling point below 210 °C	max. 10.0	max. 10.0	max. 10.0	
Oxygen content, mass %	max. 2.7	max. 2.7	max. 2.7	EN 1601, EN 13132
Induction period, min	min. 360	min. 360	min. 360	EN ISO 7536
Copper corrosion visual rating	max. 1	max. 1	max. 1	EN ISO 2160

* To reduce CO_2 emissions a minimum concentration of biofuel components is legally required in Germany, increasing from 1.2% in 2007 stepwise by 0.8% per year to 3.6% in 2010.
((bitte "Sten chenfupnoteu" Sortieser, erst "*", dana "**", dana "***"))
** Regular grade and Super plus are standardized locally, figures shown are valid for Germany.
*** Mandatory limits vary depending on climatic conditions, figures shown are valid for central European climate.

Table 3.10 European standard for diesel fuel; extract from EN 59****

		Test method
ρ(15 °C), kg/m³	820–845	EN ISO 3675, EN ISO 12185
Cetane number	min. 51	EN ISO 5165
Cetane index	min. 46	EN ISO 4264
Distillation range, total proportion recovered vol %		Pr EN ISO 3405
up to 250 °C	below 65	
up to 350 °C	min. 85	
95% recovered	max. 360 °C	
Flash point °C	above 55	EN ISO 2719
Viscosity (40 °C), mm^2/s	2.00–4.50	EN ISO 3104
Filterability (CFPP), °C**		EN 116
Summer	max. 0	
Winter	max. −20	
Intermediate	max. −10	
Sulfur content, mg/kg***	max. 50 and max. 10	EN ISO 20846, EN ISO 20847, EN ISO 20884
Conradsen carbon residue, mass %	max. 0.30	EN ISO 10370
Ash content, mass %	max. 0.01	EN ISO 6245
Copper corrosion visual rating	max. 1	ISO 2160
Oxidation stability, g/m^3	max. 25	EN 12205
Sediment, mg/kg	max. 24	EN 12662
Water content, mg/kg	max. 200	EN ISO 12937
Lubricity HFRR, μm*	max. 460	EN ISO 12156-1
Polycyclic aromatics, mass %	max. 11	EN 12916

* Protection against wear of injection pumps defined in the High Frequency Reciprocating wear Rig (HFRR).

** mandatory limits vary depending on climatic conditions, figures shown are valid for central European climate.

***The specification of 10 mg/kg supports the introduction of "sulfur free" Diesels fuels.

**** To reduce CO_2 emissions a minimum concentration of 4.4% biofuel components is legally required in Germany since 2007.

3.7 Storage and Transportation

To transport fuels from the refinery to a vehicle tank a sophisticated infrastructure is necessary.

Storage. In the refinery or in gasoline depots, fuels that are ready for use are first placed in cylindrical, generally steel tanks for intermediate storage. To store volatile gasoline fixed roof tanks with *internal floating roofs* are generally used. The roof, which is sealed against the wall of the tank, can rise and fall with the liquid level. To avoid hydrocarbon emissions the tanks breathe into a sealed common system (→ Hydrocarbons in Air). For the storage of diesel fuel, *fixed roof tanks* are frequently used.

Small quantities of fuels can be stored in steel drums, steel cans or fuel-resistant plastic containers.

Transportation. Fuels can be transported *from* the refineries via pipelines, tank ships, rail tankers, or road tank trucks. In some countries mineral oil companies exchange base fuels from their refineries. The shortening of the transport routes allows cost saving and reduces safety risks.

In the refineries or depots, vapor recovery plants with a retention efficiency of ca. 99 % are sometimes employed for the filling of gasoline tankers [83]. When underground gasoline station tanks are being filled, displaced gasoline vapors are often returned to the tank trucks (vapor recovery stage 1). In Europe and the United States the emission of gasoline vapors is also controlled when car tanks are being filled by means of special dispensing nozzles (vapor recovery stage 2) [84].

Gasoline belongs to hazard class I, whereas diesel fuels are assigned to hazard class III because their flash point is >55 °C. Hazard classifications for the transport of gasoline and diesel fuel are as follows:

Gasoline	
GGVS, ADR/RID	Class 3, number 3 b
IMDG Code	Class 3.1
UN no.	1203
ICAO/IATA-DGR	Class 3
Diesel fuel	
GGVS, ADR/RID	Class 3.3, number 31 c
IMDG Code	Class 3.3
UN no.	1202
ICAO/IATA-DGR	Class 3

3.8 Alternative Fuels

The search for alternative fuels is not new. During earlier efforts to introduce alternative fuels (e.g., during World War II or after the first oil crisis in 1973), concern about the extent of crude petroleum reserves or too great a dependence on oil-producing countries was predominant. Overcoming these problems was attempted by shifting fuel production from crude petroleum to other fossil energy sources. A true picture of the primary energy consumption can only be obtained by drawing up energy chains that take all conversion steps into account.

3.8.1
Liquefied Petroleum Gas

Stephen M. Thompson, Gary Robertson, Robert Myers

3.8.1.1 Introduction

The term liquefied petroleum gas (LPG or LP Gas) describes hydrocarbon mixtures in which the main components are propane, butane, isobutane, propene and butenes. Most commonly this term is applied to mixtures of propane and butane. These components and mixtures thereof are gaseous at normal temperature and pressure but can be liquefied by cooling, compression or a combination of both processes.

Liquefied petroleum gas is produced from two distinct sources. The first is by *extraction* from crude oil and natural gas streams at or close to the point of production from the reservoir and contains propane and butane. The amounts of LPG in the well-stream fluid vary greatly, depending on the nature of the reservoir. Production may be: (1) of natural gas from a gas reservoir, (2) of gas and light liquid hydrocarbons from a gas condensate reservoir or (3) of crude oil and gas from a combined oil and gas field. The extent of recovery of LPG and heavier hydrocarbons from gas depends on the composition of the gas produced and the quality specifications of the gas to be transported to the consumer. Before storage or transportation of crude oil by tanker, its vapor pressure must be lowered so that it can safely be contained in a ship's tank. This reduction, known as stabilization, is achieved by the removal of LPG and lighter species to yield stabilized crude oil and additional natural gas (see Fig. 3.12). The LPG components in produced oil and gas are predominantly the saturated hydrocarbons propane, butane and isobutane.

Secondly, LPG is obtained by *processing* of crude oil in refineries and in some instances as a byproduct from chemical plants. The remaining LPG in the stabilized crude oil at the refinery is separated in the crude oil fractionation column. The constituents of this LPG are propane, butane and isobutane. Additional LPG is produced in refinery conversion processes, such as catalytic reforming, thermal

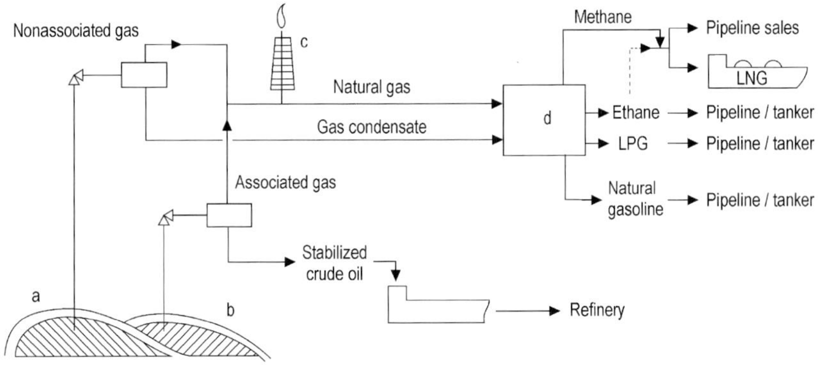

Fig. 3.12 Oil – gas recovery scheme: (a) gas and gas condensate; (b) oil and gas; (c) vent-flare; and (d) gas processing plant

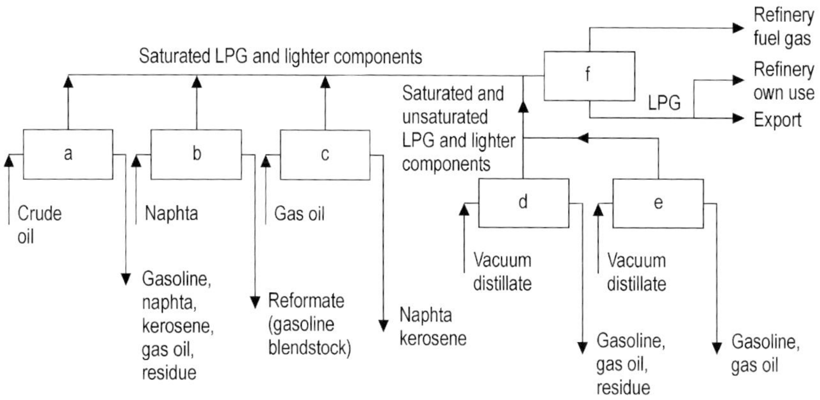

Fig. 3.13 Sources of LPG in the refinery: (a) crude oil fraction; (b) catalytic reforming; (c) hydro-cracking; (d) thermal cracking; (e) catalytic cracking; and (f) LPG recovery.

cracking, catalytic cracking and hydrocracking (see Fig. 3.13). The composition of this LPG depends on the processing configuration at the refinery but typically includes both saturated (propane, butane and isobutane) and unsaturated (propene and butenes) components.

3.8.1.2 Properties

The physical properties of the main constituents of LPG are listed in Table 3.11. In addition to these components, other species may be present in trace amounts. Typically, sulfur compounds, water and occasionally residual oils and tars may occur. Depending on the use of the LPG, these contaminants must be reduced to an acceptable level, commensurate with the applicable LPG specifications in the country of use.

Table 3.11 Physical properties of the main LPG components [87]

Component	Boiling point (101.3 kPa), °C	Vapor pressure (37.8 °C), kPa	Liquid density (at saturation pressure; 15.6 °C), kg/m	Gross calorific value (25 °C), kJ/kg
Propane	−42.1	1310	506.0	50 014
Propene	−47.7	1561	520.4	48 954*
n-Butane	−0.5	356	583.0	49 155
Isobutane	11.8	498	561.5	49 051
1-Butene	−6.3	435	599.6	48 092
cis-2-Butene	3.7	314	625.4	47 941
trans-2-Butene	0.9	343	608.2	47 878
Isobutene	−6.9	435	600.5	47 786

3.8.1.3 Production and Processing

Recovery from Natural Gas Natural gas consists largely of methane (see Section 3.8.2) and smaller amounts of ethane, propane, butanes and heavier hydrocarbons, together with varying amounts of water, carbon dioxide, nitrogen, sulfur compounds and other nonhydrocarbons. Depending on the specifications of the gas supplied to the consumer, a portion of the ethane and heavier component content is recovered in gas processing plants, which yield additional products such as ethane, LPG and higher boiling hydrocarbons (natural gasoline). When no incentive to recover ethane exists, only the LPG and natural gasoline are recovered from natural gas. Increasingly, however, ethane is being considered as a valuable product in the petrochemical industry, and plants are designed to recover not only LPG but also ethane. Potentially all of the ethane, LPG and natural gasoline may be obtained as products from natural gas; however, some ethane must normally remain in the gas to meet sales specifications.

Recovery and Manufacture in the Refinery After crude oil has been stabilized for transportation to the refinery, small but significant amounts of LPG and lighter components (methane, ethane) remain in the oil. In the refinery, crude oil is first separated into a number of fractions by distillation (fractionation). Gas, which is the lightest fraction, is produced as the top product from the atmospheric fractionation column; it consists of LPG, ethane and methane. Other fractions that contain the main refinery products are, in order of ascending atmospheric boiling points, naphtha, kerosene, gas oil and residue.

In practice, the yield of each fraction does not match market demand. Consequently, the less valuable heavy fractions must be converted into lighter ones, and the lighter fractions must be chemically modified to improve their properties. Examples of these conversion processes are catalytic reforming, catalytic cracking, thermal cracking, hydrocracking and alkylation.

The products of *thermal* and *catalytic* cracking include LPG and lighter components (ethane, ethylene, methane, hydrogen), both saturated and unsaturated. Conversely, during *catalytic reforming* and *hydrocracking* only saturated LPG and lighter components are formed. Generally, the products of these conversion processes are separated in a fractionation column to yield gas and the main products. Thus, gas streams consisting of LPG and lighter components are produced from direct fractionation of crude oil and from fractionation of the products of various refinery conversion processes.

These gas streams are used as fuel gas for refinery furnaces. However, LPG normally has a significantly higher economic value than fuel gas and recovery of LPG is economically attractive. On the other hand, the most effective use of ethane in the refinery is as fuel gas because normally no incentive exists to recover ethane other than for the petrochemical industry.

Depending on the circumstances, various techniques may be used for the recovery of LPG, which can be applied in gas plants or refineries. These methods are: (1) recontacting – compression, (2) refrigeration, (3) absorption, (4) adsorption or (5) a combination thereof.

(1) Recontacting – Compression The overhead vapor product from the fractionator is compressed to 1.2 – 1.5 MPa and combined with the liquid top product.

The combined stream is cooled to ca. 40 °C by air or water and directed to the separator. Because of the higher pressure, more LPG is obtained in the liquid phase of the separator than in the liquid phase in the overhead accumulator vessel of the fractionator. Components heavier than butane are also recovered from the fractionator vapor top product by recontacting. The liquid phase from the separator passes through a de-ethanizer from which the vapor product is recycled to the separator to improve LPG recovery. The vapor phase from the separator, which still contains some LPG, is used as fuel gas. The liquid product from the de-ethanizer is LPG, which may be further separated into propane and butane products in a de-propanizer.

Recontacting – compression achieves ca. 75 % recovery of the LPG in the fractionator overheads. In gas processing plants, this technique is hardly used today because better recovery of LPG is almost always required.

(2) Refrigeration The recovery of LPGs from a gas stream can be accomplished by a variety of techniques. The most common technique applied in new LPG recovery plants is to refrigerate the gas stream to condense the LPG fraction. The recovered liquids are then fractionated to separate the LPG components.

Some of the most commonly employed processes are summarized below.

(a) Low-temperature Separation (LTS) Refrigeration of an LPG containing gas stream can be accomplished by indirect heat transfer with an external refrigerant stream. In a conventional gas conditioning plant, refrigeration is normally provided by a closed-loop propane refrigeration circuit (see Fig. 3.14). Treated and dried feed gas to the plant is first contacted with cold LPG lean gas exiting the plant (a). The pre-cooled gas is then further cooled by propane chilling to condense the LPG fraction in the feed gas (b). The condensed liquids are then recovered in the HP separator

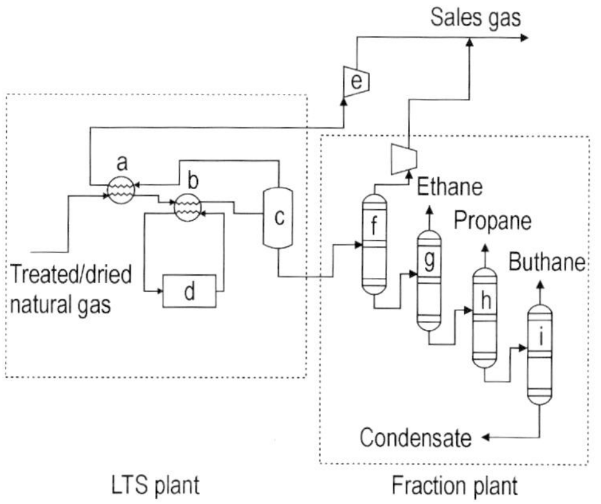

Fig. 3.14 Schematic of an LPG plant: (a) feed gas – gas heat exchanger; (b) feed gas – refrigerant heat exchanger; (c) HP separator; (d) refrigeration unit; (e) sales gas compressor; (f) de-methanizer; (g) de-ethanizer; (h) de-propanizer; and (i) de-butanizer.

(c) and fed to a downstream fractionation plant. Cold from the lean gas exiting the HP separator is then used for cooling the feed gas.

The liquids from the HP separator are fed to a conventional fractionation plant comprising a de-methanizer, de-ethanizer, de-propanizer and de-butanizer for separation of the LPG components. If no extra ethane outlets are available, then the de-methanizer and de-ethanizer columns can be combined into a single column. The chief advantages of this type of process are its simplicity and low pressure drop.

Alternatively, chilling of the gas can be provided by a cascade refrigeration circuit. The refrigerant in this case is either a propane – ethane, a propane – ethylene or a propane – ethane – methane – nitrogen mixture (commonly known as *mixture refrigerant*). A cascade ethane – propane circuit produces a lower temperature than a single propane circuit. Accordingly, the method selected depends on the ethane and LPG recoveries desired. Recoveries of LPG >99 %, corresponding to ca. 85 % ethane recovery, are achievable in practice by using mechanical refrigeration. Mechanical refrigeration is more usual in gas processing plants than in refineries because higher LPG recovery is normally required in a gas processing plant to meet the hydrocarbon dew point specification of the sales gas.

In a cascade refrigeration plant, impurities such as water, carbon dioxide and sulfur compounds are removed from the inlet gas, whereupon it is chilled against product gas and external refrigerant. Removal of water and carbon dioxide avoids possible freezing problems in the cold sections of the process. The liquid produced by chilling is fed to the de-methanizer, where methane and some ethane are stripped off. The vapor separated from the separator is used to chill the incoming feed before being commingled with the overhead vapor from the de-methanizer. The liquid product from the de-methanizer consists of ethane, LPG and natural gasoline, from which LPG is separated by fractionation. The vapor phase from the de-methanizer (mainly methane with some ethane) is heated against the incoming feed before being compressed prior to sale.

Liquid petroleum gas can also be recovered in a liquefied natural gas (LNG) plant, in which natural gas is fully liquefied by mechanical refrigeration. Generally, the external refrigerants employed are propane and a mixed refrigerant consisting of nitrogen – methane – ethane and propane. In the liquefaction process the partially condensed natural gas stream is passed to a distillation column in which ethane, LPG and the heavier natural gasoline are separated from natural gas. The natural gas (predominantly methane) leaves the distillation column overhead as a vapor product and is subsequently fully liquefied. Some or all of the ethane and propane products may be used as refrigerants for the LNG plant, with any excess exported.

(b) Expander Plants Expander plants generate refrigeration through expansion of the feed gas. By dropping the pressure via a control valve or an expander, a sufficiently low temperature can be generated to condense the LPG fraction in the feed gas. Treated and dried gas supplied to the plant is first cooled via indirect heat exchange with the cold lean gas exiting the plant. Any liquids formed during this first stage of cooling are then recovered in the HP separator and fed to the bottom section of the de-methanizer column. The overhead vapor from the HP separator is then expanded either via a rotating expander or via a control

valve (commonly termed a JOULE – THOMSON valve). The cold gas/liquid from the expansion is then fed to the top of the de-methanizer. The de-methanizer strips the lighter components (methane, nitrogen) from the liquids. The overhead vapor is subsequently fed to the feed gas – gas heat exchanger for recovery of the cold before being compressed for export as sales gas. Coupling the first stage of the compression to the expander minimizes compression power.

A recent development of the conventional expander process has been the gas subcooled process. This process is an enhanced expander process in which a rectifying section is added to the de-methanizer column. The purpose of adding this section is to reduce the loss of ethane and LPG in the overhead vapor by refluxing with cold liquefied feed gas. The reflux is obtained by taking a small portion of the HP separator gas that is then liquefied by chilling against the overhead vapor of the de-methanizer. The high-pressure cold liquid is then de-pressurized and fed to the top tray of the de-methanizer column. The gas subcooled process is capable of recovering >99 % of the LPGs in the feed gas and is also ideally suited for recovery of >95 % of the ethane in the feed.

(c) Combined Processes. When high recoveries of LPG and/or ethane are desired, it is possible to assist the conventional expander process with external refrigeration. External propane or cascade refrigeration is applied to cool the feed gas entering the HP separator of the expander plant. The cold gas is then expanded and fed to a conventional de-methanizer column.

With this type of process, more than 95 % ethane recovery and essentially 100 % recovery of LPG can be achieved.

(3) Lean Oil Absorption The lean oil absorption process employs a hydrocarbon oil to recover light components from either a natural gas or a refinery fuel gas stream. The process is used in refineries worldwide, as well as in natural gas processing plants, especially in the United States, where the vast majority of plants used the oil absorption process until the first expander plant started up in 1963. However, since 1963 expander plants have dominated natural gas processing.

Hydrocarbon oil may be brought in or components of natural gasoline produced in the gas processing plant may be used. In the refinery, an available fluid such as naphtha is often employed. Despite the different terminology, these streams from the natural gas processing plant and oil refinery are very similar in terms of molecular mass. Typical absorption oils have a molecular mass of 100–180.

The process applied in natural gas processing plants is, in principle, the same as that used in the refinery, although LPG recoveries of around 98 % are often needed to ensure that gas supplied to the customer satisfies the hydrocarbon dew point specification. This high LPG recovery is achieved by cooling the lean oil to −40 °C, usually by propane refrigerant. So that the absorber column can operate at this temperature, the natural gas feed must be dried to avoid problems caused by freezing.

(4) Adsorption In adsorption, gas molecules are reversibly bonded to the surface of a solid material, the adsorbent. Examples of adsorbents include silica gel, activated carbon and alumina. Adsorption plants are generally used to obtain natural

gasolines from natural gas to satisfy sales gas specifications for heavy hydrocarbon content. The LPG recovery is significantly lower than that achieved by refrigeration or lean oil absorption processes.

Whichever method is used to obtain LPG from natural gas or refinery fuel gas, LPG may be produced as a mixed product or as separate propane – propene, known as C_3 product, and butanes – butenes, known as C_4 product. The relative amounts of saturated and unsaturated species depend on the types of processes in the refinery that are responsible for LPG production (see Section 3.8.1.1). Further fractionation is sometimes employed to recover isobutane from the remainder of the C_4 product and propene from the C_3 product. Isobutane is used in the refinery in the alkylation process, in which it reacts catalytically with either propene or butene to produce high-octane gasoline blendstocks. Propene can also be a valuable chemical feedstock. Thus, depending on the configuration of each refinery, the LPG pool may contain various blends of the saturated and unsaturated components that comprise LPG.

3.8.1.4 **Purification**

Whether produced in the refinery or a natural gas processing plant, LPG contains contaminants. Most common are water and sulfur compounds, including hydrogen sulfide, carbonyl sulfide, mercaptans and elemental sulfur. Removal of these contaminants is necessary to satisfy locally applicable quality specifications, for example, DIN 51 622 in the Federal Republic of Germany, BS 4250 in the United Kingdom and the Gas Processors Association (GPA) specification in the United States. As an example, Table 3.12 summarizes the requirements for different LPG products according to GPA specifications. Various processes are available to remove contaminants to the prescribed level, some of which are described below.

The first stage of drying LPG is achieved normally by separating *undissolved* water in a coalescent. If the LPG is commercial-grade *butane*, which is stored at ambient temperature under its vapor pressure, this is sufficient to meet the water specification. However, if the butane is to be held in refrigerated storage at ambient pressure, further drying is necessary to remove *dissolved* water. This must be performed before the butane is chilled, otherwise problems caused by water freezing will ensue. Water levels in refrigerated LPG are typically less than 10 ppm (mol) to avoid freezing.

Irrespective of the method of storage, commercial-grade *propane* requires the removal of dissolved water to satisfy the specification.

Adsorptive Purification Dehydration of LPG can be effected by using a solid adsorbent such as a molecular sieve, alumina, silica gel or calcium chloride. With the exception of calcium chloride, these are regenerative processes operating on a cyclical basis. Dried LPG is taken from the outlet of one of the beds and vaporized by heating. It is then passed through the bed which is to be regenerated, where it heats the bed, thereby desorbing the water. The outlet stream is cooled to condense water and LPG, and the LPG is separated from the water before being returned to the inlet of a bed which is on adsorption service. The regenerated bed is then cooled, filled with LPG and returned to service, adsorbing water from the LPG.

Table 3.12 The LPG specifications of the U.S. Gas Processors Association [89], [90]

Product characteristics	Commercial propane	Commercial butane	Commercial propane–butane mix	Propane HD-5[a]
Composition	mainly propane and propene	mainly butane and butene	mainly mixes of propane – propene and butane – butene	not less than 90 % propane; not more than 5 % propene
Vapor pressure (max.) at 100 °F, psig[b]	208	70	208	208
Temperature of volatile residue: at 95 % evaporation (max.), °F[b]	−37	36	36	−37
Butane and heavier, vol %	2.5	–	–	2.5
Pentane and heavier, vol %	–	2.0	2.0	–
Residual matter, residue on evaporation of 100 mL (max.), mL	0.05	–	–	0.05
Oil stain observation	pass[c]	–	–	pass[c]
Corrosion, copper strip, (max.)	no. 1	no. 1	no. 1	no. 1
Volatile sulfur (max.), grains per 100 ft^3 [e]	15	15	15	10
Moisture content	pass[d]	–	–	pass[d]
Free water content	–	none	none	–

[a] HD-5 is the specification of propane for use as automotive fuel.
[b] Conversion factors: 208 psig = 1434 kPa (g); 70 psig = 483 kPa (g); −37 °F = −38 °C; 36 °F = 2 °C.
[c] An acceptable product shall not yield a persistent oil ring when 0.03 mL of solvent residue mixture is added to a filter paper in 0.1-increments and examined in daylight after 2 min as described in test method ASTM D 2158.
[d] Passes the GPA Propane Dryness Test (cobalt bromide) or ASTM D 2713-70.

Molecular sieve units can also handle the combined removal of water, hydrogen sulfide, carbonyl sulfide and small amounts of mercaptans. Adsorption and regeneration occur in the same manner as for drying only, although water recovered during the regenerative cooling step, as well as the regeneration gas itself, is "sour" (i.e., it contains sulfur compounds). Molecular sieves tend to be favored when the amount of contaminant to be removed is small.

Absorptive Purification The LPG can be brought in contact with an *amine solution* in an absorber column primarily to remove relatively large amounts of hydrogen sulfide. However, amines are not particularly effective for the removal of mercaptans. The sour amine solution is regenerated continuously in a regenerator column at low pressure and high temperature before being recirculated to the absorber column. *Caustic soda treatment* is used primarily to remove mercaptans and smaller amounts (<100 ppm) of hydrogen sulfide. In an absorber column, LPG is contacted countercurrently with sodium hydroxide solution. The sour caustic solution is then passed to a regenerator column where air stripping oxidizes the mercaptans to disulfides, which are removed from the regenerated sodium hydroxide solution. Regenerated caustic solution is recycled to the absorber. The removal of hydrogen sulfide by sodium hydroxide renders the latter nonregenerable and it must be discarded; hence, bulk hydrogen sulfide removal is normally performed upstream of the caustic unit. Variants of the regenerative caustic process are marketed with additives to enhance the process, for example, the Merox system from UOP.

Depending on the type and amount of contaminants present in the feed and product specifications, one or more of these processes may be employed for LPG treatment.

3.8.1.5 Storage and Transportation

The storage of LPG can be categorized as follows: pressurized storage at ambient temperature; refrigerated storage at ambient pressure; and semi-refrigerated, partially pressurized product storage. Further, LPG may be stored either above or below ground.

Aboveground Storage The design pressure of aboveground *pressure storage* tanks is normally 1724 kPa for propane vessels and 860 kPa for butane vessels. These values correspond to the vapor pressures of both compounds at the maximum expected ambient temperature. For smaller inventories (up to ca. 100 tonne), the vessels are usually horizontally or vertically mounted cylinders with hemispherical heads. Larger inventories tend to be stored in spherical vessels.

The largest pressurized spheres hold ca. 1500 t; thus, for storage of larger amounts than this, either several spheres are needed or *refrigerated storage* must be used. This is designed to store the LPG product at its atmospheric boiling temperature (i.e., ca. −42 °C for propane and −4 °C for butane). The tank design pressure is typically ca. 110 kPa. Tanks are insulated; vapors liberated by heat leaking in are compressed, liquefied and returned to the storage tank. Usually, the boil-off vapor is liquefied by the refrigeration system provided to chill the LPG product rundown from the producing plant. Sometimes, the aboveground storage facility is mounded with earth.

Semi-refrigerated storage can also be used: propane is stored at −10 °C, at which its vapor pressure is ca. 320 kPa.

Underground Storage The LPG may also be stored underground in salt or quarried caverns that can withstand its vapor pressure at subterranean temperature. These have been used in the United States, Canada, Europe and Australia. In frozen-earth storage, the ground surrounding the excavation is frozen and an insulated roof is constructed over the cavity. Propane is more suitable for storing by this method because its boiling point at atmospheric pressure is considerably lower than the freezing temperature of the soil, thus causing moisture in the soil to freeze.

Refrigerated propane product at ca. −42 °C is loaded into the cavity where it is held at atmospheric pressure. Propane vapors liberated by heat leaking in are compressed, cooled and returned to the cavity storage as liquid.

Transportation Depending on the location of LPG-producing plants in relation to the markets, LPG may be transported by pipeline, sea, road or rail. Traditionally, LPG has been carried in pressurized ships and in trucks and rail cars, except for bulk refrigerated deliveries. In the last 20 years or so a large international market for LPG has developed requiring an extensive fleet of refrigerated ships for seaborne trade. Pipelines have been used for inland transportation of large volumes, particularly in the United States. For transport to the consumer, LPG is delivered in cylinders and bulk vehicles of various sizes. The maximum capacity of a *pressurized* LPG ship is limited to about 2000 m^3 due to high manufacturing costs. *Semi-refrigerated* ships usually hold 2500–5000 m^3, with *fully* refrigerated ships having capacities up to 100 000 m^3. The larger the LPG cargo, the more economical it is to transport fully refrigerated. The LPG can be unloaded into refrigerated storage tanks or, after heating, into pressurized tanks which supply conventional pressurized distribution facilities. Alternatively, cargo may be transferred into smaller coastal tankers for delivery of refrigerated products to smaller terminals that cannot receive large oceangoing ships.

3.8.1.6 Uses

LP gas has a long history of successful use in the transportation sector dating from the 1920s. Today more than 8×10^6 vehicles worldwide use LP gas. The term usually used to distinguish LP gas from other conventional motor fuels is "autogas", however, other terms such as "LPG", "GPL" or "motor fuel" are also used depending on the country. Regardless of the name, the fuel is comprised mainly of a mixture of propane and butane with lesser amounts of propene and ethane. In 2001, 7.7 % of worldwide consumption of LP gas was in the transport sector.

The countries with the greatest number of LP gas vehicles in 2001 were:

• South Korea	1 428 000
• Italy	1 250 000
• Turkey	975 000
• Poland	700 000
• Australia	600 000
• Russian Federation	500 000
• Mexico	400 000
• The Netherlands	360 000
• US	277 000
• Bulgaria	165 000

In the past 30 years the number of autogas vehicles has increased steadily in response to global concerns of vehicle emissions and their impact on air quality. LP gas is considered both as an alternative fuel and as a clean fuel by many governmental agencies, thus it has experienced favorable tax treatment as an incentive to promote its use.

LP gas offers some distinct advantages over gasoline and diesel, including no evaporative emissions, no contamination of soil or aquifers as it is non-toxic, very low particulate matter and greatly reduced hydrocarbons (HC) and carbon monoxide (CO) contents. Also, HC emissions are less reactive than gasoline by at least one-half. Being a gas, LP gas mixes well with air before entering the engine, resulting in low carbon monoxide emissions assuming the air/fuel ratio is kept lean.

In recent years policy makers and the scientific community have adopted life-cycle analysis (LCA) as the appropriate method of assessing the transport sector's contribution to air quality degradation and the production of greenhouse gases (GHG). This method encompasses the entire range of vehicle and fuel production, distribution and use from "well head" to "wheels". LCA shows LP gas contribution to GHG is approximately 20 % less than gasoline.

Refueling Infrastructure LP gas refueling infrastructure requires specialized equipment as the fuel is stored and transferred to the vehicle as a liquid under modest pressure. Compared with some other alternative fuels, however, the cost for refueling equipment is relatively modest depending on the degree of sophistication desired. Thus, it is possible for even small fleets to have their own refueling system on-site, consisting of a tank, pump, meter, hose and hose nozzle. More complex systems matching the appearance and features of common gasoline and diesel dispensers with features such as card lock readers and fleet data recorders are also widely used. Pumping capacity of LP gas refueling systems are equal to or better than gasoline and diesel. The operational and maintenance costs of LP gas dispensers are minimal. In 2001 some 32 000 LP gas refueling facilities were operational worldwide.

Vehicle Conversions to LP Gas Equipping a vehicle to operate on LP gas is done either by retrofitting after the vehicle has been purchased from the manufacturer (conversions) or by the manufacturer prior to sale. In Europe, Japan, Australia and to some degree in North America, vehicle manufacturers offer LP gas vehicles through their regular distribution network (factory equipped). Japan, for example, offers more than 90 models factory equipped; Europe offers some 26 different models. Most of these are passenger cars other than North America where the market evolved mainly with converted fleet vehicles. Light and medium duty vans, pickups, taxis and delivery trucks typify this fleet market. In other countries most of the LP gas vehicles are conversions.

The typical retrofit system, whether a conversion or factory equipped, consists of the LP gas fuel tank, a vaporizer/converter and a carburetor (or air mixer). The vaporizer reduces tank pressure to a negative pressure and converts the liquid fuel into a gas. The carburetor or air mixer proportions the fuel with air in response to engine demand. The LP gas fuel system interfaces and uses the standard vehicle computer so that all emission control systems are fully employed. These systems may be either bi-fuel (capable of using either gasoline or LP gas) or dedicated (only LP gas). The traditional carbureted systems are giving way to more sophisticated injection systems comparable to that of gasoline and diesel in response to more stringent emission standards.

The availability of heavy-duty engines suitable for large trucks and urban buses are currently limited. However, development work is underway in several countries to bring larger engines to the market.

Outlook The ability of the LP gas industry to sustain the growth rate in the transport sector of recent years is highly dependent on government policy. As discussed, LP gas requires a separate on-board fuel system that ultimately adds to the vehicle cost. Further, there is approximately 10 % less energy in a liter of LP gas than gasoline. Consumers look for an acceptable economic payback to offset these additional costs. Most of the economic incentive is currently provided by lower fuel taxes compared with gasoline and diesel. Giving up revenue is always difficult for governments, but these revenue concessions must be weighed against the social and environmental costs of progressively deteriorating air quality in most urban areas.

3.8.1.7 Safety Aspects

Throughout the process of LPG manufacture, storage, transportation and consumption, high safety standards are necessary because LPG is a hazardous substance. It is flammable and, in the event of leakage, can release large amounts of flammable gas which may migrate considerable distances from the source of the leak before being ignited. Being heavier than air, escaping LPG readily accumulates at low points, including open and closed drains and sewers.

Generally, precautions must be taken to minimize the likelihood of uncontrolled leakage and eliminate ignition sources. When controlled leakage of LPG from storage or plant has to be allowed for, it must be piped to a safe location for disposal in a controlled manner, normally by flaring. To this end, equipment used in the storage and processing of LPG is fabricated according to procedures laid down in the appropriate codes. Ignition sources are also kept out of plant or storage areas where the risk of LPG release exists. Accumulation of static electricity is prevented by ensuring that all equipment is earthed.

Occupational Health LPG is an asphyxiant and can cause suffocation by displacing oxygen in the air supply. It is, however, not a toxic substance. In the event of an unexpected discharge it does not contaminate aquifers or soil. The precautions required center on making sure that people are not exposed to a suffocating atmosphere. For example, in a tank manufacturing plant personnel must not enter a vessel unless that vessel is suitable for entry, i.e., it has been physically disconnected from other plant equipment and freed of flammable substances by purging with inert gas (normally nitrogen) followed by replacement of the inert gas with a breathable oxygen atmosphere.

3.8.1.8 Economic Aspects [91]

The major producers and exporters of LPG are the Middle East, Canada, North Africa and the Far East. Reduced flaring of natural gas, particularly in the Middle East, is expected to contribute significantly to the amounts of LPG available for export. Prospects for international trade are also improved by the reduction in freight costs associated with the use of large-capacity refrigerated LPG ships.

The total world consumption of LPG in 2001 was approximately 200×10^6 t, the main consumers being the United States, Japan and Western Europe.

United States In the United States, consumption in 2001 totaled about 50×10^6 t. Historically, ca. 55 % of the LPG production in the United States has been supplied by gas processing plants, with oil refineries providing the remaining 45 %. The volumes produced internally have not quite satisfied local demand; therefore, some imports (ca. 10 % of the total demand) have been necessary. Most of the imports have come from Canada and the Middle East. Future production of LPG from natural gas is predicted to decrease as greater proportions of the gas are taken from wells containing less LPG. At the same time refinery production is expected to increase. This is due to cleaner grades of gasoline and diesel in response to environmental regulations. More LP gas production is the result. Any substantial growth in the United States beyond current demand will require imports.

Japan In 2001 LPG consumption in Japan was 19×10^6 t. Japan has little indigenous natural gas, which strongly influences the LPG market. The lack of natural gas has resulted in a public gas distribution system which is less extensive than in other industrialized countries. Consequently, on the demand side, the market for LPG is large with two-thirds of the total demand typically being from the residential and industrial sectors. Another important feature of the Japanese market is the demand for clean burning fuels, for example, in the automotive and power generation sectors. On the supply side, the lack of natural gas means that the only sources of LPG produced locally are oil refineries. The LPG produced by refineries is insufficient to meet the demand, to such an extent that Japan was the largest single importer of LPG in 2001, importing 14.3×10^6 t.

Western Europe In Western Europe, LPG consumption was 30×10^6 t in 2001. Traditionally, the LPG supply in Europe has come from refinery production, to a market restricted to some extent by the availability of pipeline natural gas. In 2001 most imports to Western Europe came from the North Sea and Algeria.

3.8.2
Natural Gas

Klaus Reders, Margret Schmidt

Natural gas is, like crude oil, a fossil fuel. It has been known to mankind for several 1000 years. Natural wells appear to have existed in the Middle East and in China. The first practical use of natural gas was reported in China, with bamboo pipelines transporting the gas over short distances. While the production of crude oil on an industrial scale started in the 19th century, the search for natural gas and further larger-scale production only began about 100 years later after the middle of the 20th century, when it was realized that the crude oil reserves would not last forever.

3.8.2.1 Occurrence

Natural gas occurs either associated with crude oil (in solution or as gas caps) or non-associated. There are also large reserves of gas hydrates, a specific form of clathrates, where the gas molecules are enclosed in ice crystals. Gas formations of this nature are known to exist under the sea (American and Siberian shores) as well as under permafrost soil (Siberia). However, these deposits currently have no commercial value.

For a long time the associated gas was flared when crude oil was produced, because no market for the gas existed in the areas of oil production and transport to industrial areas was not worthwhile. Meanwhile, there are various techniques available to use the associated gas, such as pumping it back to increase the pressure in the cavities. Today non-associated gas is mainly used for industrial purposes (energy and heat production), the gas is mostly transported by large pipeline networks from the major sources worldwide (Russia, USA, Latin America, Algeria, partly the Near-East) to the industrial customer countries. Non-associated gas that occurs in areas with no pipeline access is named stranded gas. Stranded as well as associated gas is finding growing use as liquefied natural gas (LNG), which is produced today in a few plants next to the gas wells. LNG is mainly transported by marine tankers to countries that do not have their own gas resources and fed into their gas pipelines. Another strong developing business is the conversion of natural gas to GtL (gas-to-liquid) where synthetic diesel and waxes are produced by a Fischer–Tropsch reaction and subsequent refining processes (see Section 3.8.3.1.).

While about 65 % of the known crude oil reserves are located in the Middle East, natural gas is more evenly distributed around the world. The largest known reserves (about 35 %) are located in the areas of the former Soviet Union, followed by the Middle East with about the same magnitude. The remaining 30 % are in Asia, Africa, America and Western Europe (Fig. 3.15).

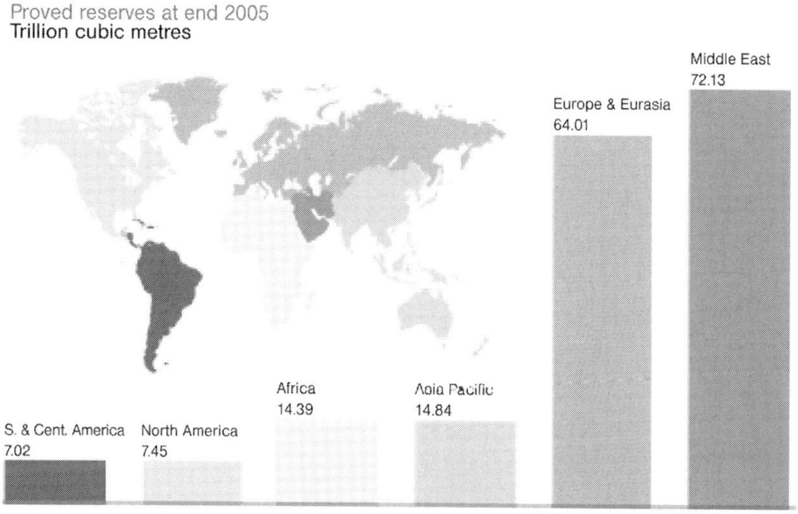

Fig. 3.15 Map of proven natural gas reserves at the end of 2005, www.bp.com.

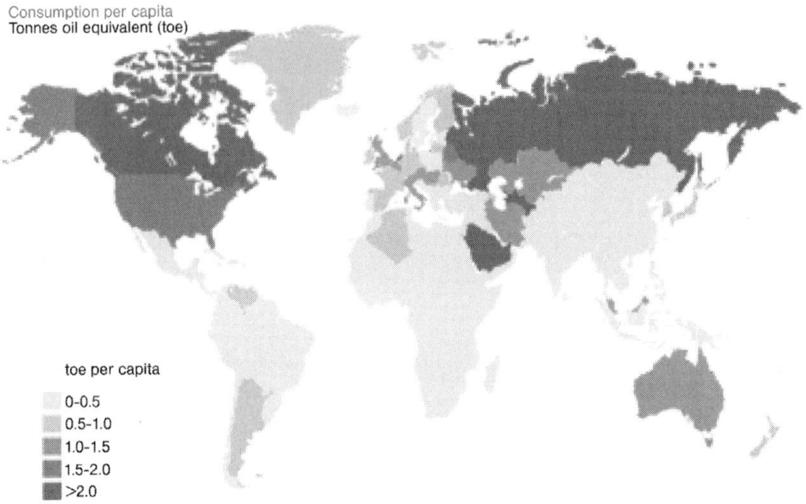

Fig. 3.16 Natural gas consumption worldwide (status 2005), www.bp.com.

Several countries, such as Germany and Japan, rely almost completely on gas imports, others have their own reserves, which are used for the national demand and/or export. It can be seen from this that only a global view of reserves and demand makes sense. Based on the annual gas production for the year 2000 of 2400×10^9 m^3, the natural gas reserves should last for about 100 years, however, international demand is increasing greatly (see Fig. 3.16.)

3.8.2.2 **Composition**

The composition of natural gas depends on its source, untreated gas contains:

- mainly methane–CH_4
- heavier hydrocarbons, such as ethane, propane, butane and their isomers, generally content decreasing with increasing density
- inert gases such as nitrogen and carbon dioxide
- sulfur compounds such as hydrogen sulfide (H_2S), carbonyl sulfide (COS) and mercaptans
- traces of noble gases
- water

Examples of various gas compositions are presented in Table 3.13.

Natural gas is processed before transport and use only as far as is necessary. Therefore, the main properties such as gas composition, and hence energy content, are fairly constant for the final users, who are supplied by pipeline from constant gas sources. However, in some countries, e.g., Germany the gas quality is different locally because natural gas is supplied from various sources and pipeline streams.

Table 3.13 Natural gas compositions from different sources

Source	Methane, mol %	Ethane, mol %	Propane, mol %	Nitrogen, mol %	Carbon dioxide, mol %	Wobbe index, MJ/N m³	Methane number (MN)
Russia 1	96.2	1.20	0.30	1.80	0.30	52.5	96
Netherlands (Ommen)	82.3	2.85	0.44	12.90	1.28	44.4	94
Germany							
–North Sea	85.6	8.79	2.52	0.59	1.67	54.3	74
–Netherlands	83.3	3.47	0.66	11.00	1.27	45.9	89
–Russia 2	98.3	0.52	0.21	0.81	0	53.0	95
Algeria	89.0	8.15	1.03	0.48	0	55.0	74
Singapore (Indonesia)	90.9	5.11	1.47	0.30	1.34	53.8	n.a.
California sources							
–San Joaquin	86.0	8.90	2.70	0.50	1.90	n.a.	n.a.
–South Central	88.5	5.20	3.80	0.50	2.00	n.a.	n.a.
Pakistan							
–Sindh field	82.5	7.05	3.11	3.05	2.08	n.a.	n.a.
–Punjab field	94.4	0.96	0.28	≈4	0.02–0.06	n.a.	n.a.
Argentina, (Dock sud)	95.4	2.28	0.32	1.23	0.65	49.9	n.a.
Italy, (Po Valley)	99.6	0.06	0.02	0.26	0.02	53.5	99

While domestic and industrial burners can easily be adjusted to the energy content of the local gas quality, this is a challenge for mobile use in vehicles and the consequence is different driving distances for a compressed natural gas (CNG) tank filling. Natural gases are classified depending on the heating properties as gas group H and group L, for details see Section 3.8.2.5.

3.8.2.3 Processing

Depending on its properties, wellhead natural gas, i.e., as it comes out of the ground, often has to be processed, to avoid health, pipeline and other equipment problems. Sulfur compounds, water and higher boiling hydrocarbons are removed as required. Various processing methods are available.

A. Dehydration to avoid condensation of water and corrosion
   - compression with subsequent cooling
   - cooling below required dew point
   - adsorption with dry dessicants (silica or alumina based materials)
   - absorption with liquids [various glycols, diethylene glycol (DEG) or triethylene glycol (TEG)]

B. Hydrocarbon extraction to avoid condensation of higher boiling hydrocarbons

In general, higher boiling hydrocarbons in the natural gas increase the heating value, on the other hand, the unwanted tendencies towards condensation and smoking also increase.

C. Cooling and/or compression
D. Removal of toxic and acid gases

The removal of the highly toxic hydrogen sulfide is most important. Carbonyl sulfide, mercaptans, and also carbon dioxide, are reduced as required to avoid corrosion. A large number of physical and chemical absorption solvents are available. A typical example is the Shell Sulfinol process, which uses a combination of physical and chemical solvents (tetrahydrothiophene dioxide mixed with secondary or tertiary alkanolamines and water) removing all sulfur containing compounds and also carbon dioxide. It is generally combined with a sulfur recovery plant based on the Claus process.

3.8.2.4 Transport/Distribution/Local Blending

A large proportion of natural gas is produced far away from industrial or populated areas. Typical examples are the gas production fields in Siberia and in Indonesia and the surrounding territories. Two main methods of transport are in use for long-distance transport. The natural gas from Siberia is usually transported via large-scale high-pressure pipelines about 1 m in diameter with compressor stations at regular intervals to compensate for friction losses (typically every 150 km). Of course the energy required for transport increases with distance (length of pipeline and number of compressor stations). The energy loss for large-scale pipelines is quoted to be about 1.25 % per 1000 km. The other method of transport–liquefaction to LNG and transport at low temperature in specialized marine tankers–is being used with increasing frequency. LNG transport is today mainly utilized in the Far East for the supply of the Japanese and US market from the gas fields in, e.g., Indonesia, Brunei and Australia as well as from Africa to the US. In addition to the approximately 10 % energy loss for liquefaction, about 0.15 % of the energy is lost per day of marine transport because of evaporation (keeping the gas at constant low temperature at ambient pressure is required). The evaporated fraction of natural gas is generally used in the vessel's propulsion system. Another method of long-distance transport is the chemical conversion of natural gas into a liquid near to the production site, e.g., methanol. However, because of the higher energy losses this is only viable when the specific product produced is required at its destination.

After arrival at the local gas supplier and decompression or regasification, the gas may be treated before it is fed into the local low-pressure gas supply system.

Local gas suppliers may add other, mainly gaseous components, for various reasons, e.g., to meet specifications by feeding in LNG into natural gas pipelines, or to increase the availability of gas during periods of high demand by LPG injection (peak shaving) or by use of other locally available gases, such as processed biogas from natural sources. With all these blendings gas suppliers finally have to ensure that the quality specifications for natural gas supplied from the local gas grid are met.

For safety reasons the natural gas received from the local gas grids is always odorized. Odorization is done worldwide mainly through injection of low amounts of mercaptans either at the point where it is fed from the well into the large pipeline systems or later in the gas grids. However, because mercaptans increase the sulfur content of natural gas, some sulfur-free odorants are already on the market.

3.8.2.5 Properties and Specification

As described earlier, the composition and hence the properties of natural gas vary substantially according to the source of the gas. Various classifications are used:

In Europe	LL-gas	very low energy content
	L-gas	low energy content
	H-gas	high energy content
In USA	I high inert	low energy content
	II high methane	medium energy content
	III high Btu	high energy content

Each country where natural gas for industrial and household uses is delivered via pipelines has certain quality specifications. In Germany the DVGW Arbeitsblatt G 260 describes the quality requirements for natural gases supplied from the gas grids to industrial and household users (Table 3.14).

Specification of Natural Gas The main quality parameters are the energy content ranges (expressed with the Wobbe index = ratio of heating value to relative density of the gas). With this parameter the gas qualities are differentiated into L- and H-gases. Furthermore, dew points for higher hydrocarbons, water, dust and sulfur content, etc., are limited. Most important for industrial and household users is the energy content on a volume basis, e.g., in megajoules per cubic meter. With increasing concentration of higher-density hydrocarbons (ethane, propane, butane) there is a noticeable increase in energy content on a volume basis resulting from the increase in total gas density. However, higher contents of inert gases such as nitrogen and carbon dioxide also result in higher gas density but reduced energy content. In any given area a change in the energy content brought about by changes in the gas composition is undesirable and may be adjusted by the local gas supplier either by the addition of nitrogen or air to reduce the energy content, or by supplementation with, e.g., propane, to increase the energy content.

3.8.2.6 Natural Gas as Automotive Fuel

While gasoline and diesel fuels are complicated mixtures of many different types of hydrocarbons, containing paraffins, aromatics, naphthenes, olefins, oxygenates, etc., the main component of CNGs is methane. These differences in composition and other physical properties between CNG and gasoline and diesel influence the combustion behavior of CNG and the resulting exhaust emissions and performance. Table 3.15 gives a brief comparison of the physical properties of CNG and other fuels.

Worldwide there are only a few specifications dedicated to CNG as an automotive fuel, a fixed specification in California, in addition to a US SAE recommendation

Table 3.14 German national standard for natural gas, (technical regulation G260 of the German Society of Gas and Water Branch)

Combustion properties	dimension	L-gas	H-gas
Wobbe Index			
total range	kWh/m³	10.5 to 13	12.8 to 15.7
	MJ/m³	37.8 to 46.8	46.1 to 56.5
Nominal	kWh/m³	12.4	15.0
	MJ/m³	44.6	54.0
Range in local distribution network	kWh/m³	+ 0.6 − 1.4	+ 0.7 − 1.4
Higher calorific value	kWh/m³	8.4 to 13.1	8.4 to 13.1
	MJ/m³	30.2 to 47.2	30.2 to 47.2
Relative density (air = 1.0)		0.55 to 0.75	0.55 to 0.75
Distribution network pressure	mbar		
Total range		18 to 24	18 to 24
nominal		20	20
Attending substances (impurities)		**max. concentrations**	**max. concentrations**
Hydrocarbons: dewpoint	°C	Ground temperature at given pipeline pressure	Ground temperature at given pipeline pressure
Water: dewpoint	°C	Ground temperature at given pipeline pressure	Ground temperature at given pipeline pressure
Fog, dust, liquids		technically free	technically free
Oxygen			
In dry distribution systems	vol%	3	3
In wet distribution systems	vol%	0.5	0.5
Total sulfur content			
Yearly average (without odoration substances)	mg/m³	30	30
short term	mg/m³	150	150
Mercaptans	mg/m³	6	6
short term	mg/m³	16	16
Hydrogen sulfide	mg/m³	5	5
short term in exceptional situations	mg/m³	10	10

and a German regulation (in preparation, see Table 3.16). For type approval of CNG engines and vehicles according EU exhaust emissions standards two CNG types must be tested. These two CNG fuels are specified in the regulation and correspond approximately to gas types L and H.

For the use of natural gas as CNG automotive fuel some properties other than those for industrial/household uses are important. These properties must be specified to ensure problem-free vehicle operation, such as high antiknock performance (methane number), sufficient heating value, low dew point (this relates to the higher hydrocarbon, water and oil dust content) and sulfur content.

The most important factor for CNG as an automotive fuel in a spark-ignition engine is its antiknock performance. It describes the ability of a fuel to avoid

Table 3.15 Properties of CNG in comparison with other fuels

Properties	CNG	LPG	Gasoline	Diesel
Chemical	CH_4	C_3H_8 (main comp.)	C_4–C_{12}	C_8–C_{32}
Avg. boiling range at 101.3 kPa, °C/°F	−162/*−259*	−42 to 31/*−44 to 0*	30–220/*80–420*	160–380/*320–720*
–Octane number range (R+M)/2 [a]	120 (pure CH_4)	104	87–93	n.a.
–Methane number	100 (pure CH_4)	–	–	–
–Cetane number	–	–	5–20	40–55
Lower heating value, MJ/kg/Btu/lb	avg. 47–59/avg. *21300*	46.1/*19800*	18–19/*7740–8170*	18–19/*7740–8170*
Autoignition temperature, °C/°F	760/1004	454–510/*850–950*	232–482/*450–900*	204–315/*400–600*
Flashpoint, °C/°F	−184/*−300*	−73 to −101/*−100 to −150*	−42/*−45*	74/*165*
Flammability limits (vol % in air)				
lower/higher	5.3/14	2.4/9.6	1.4/7.6	0.7/5.0
Stoichiometric air/fuel ratio, weight	17.2	15.7	≈ 14.7	≈ 14.7

[a] R = research octane number, M = motor octane number.

Table 3.16 E DIN 51624, Draft March 2006, CNG for the use as automotive fuel, requirements and test methods

Property	Test method	Dimension	Min.	max.	Comment
Lower calorific value, H-gas	DIN 51857 DIN ISO 6976	MJ/kg	46		
Lower calorific value, L-gas	DIN 51857 DIN ISO 6976	MJ/kg	39		
Density (absolute)	DIN 51857, DIN 1871, DIN EN ISO 6976	kg/m³	0.72	0.91	
Methane no. (calculated)	AVL-Method		(70/75)		min. requirement not yet defined
Methane content	ISO 6974-6	vol%	80		
C_2 Hydrocarbons	ISO 6974-6	vol%		(10/16)	max concentr. not yet defined
> C2 Hydrocarbons	ISO 6974-6	vol%		(8.5 ?)	max concentr. not yet defined
Propane	ISO 6974-6	vol%		(5)	max concentr. not yet defined
Butane	ISO 6974-6	vol%		(2)	max concentr not yet defined
Pentane	ISO 6974-6	vol%		(1)	max concentr. not yet defined
Hexane and higher hydrocarbons	ISO 6974-6	vol%		(0.5)	max concentr. not yet defined
Oxygen	ISO 6974-6	vol%		3	
Nitrogen	ISO 6974-6	vol%		15	
Carbon dioxide	ISO 6974-6	vol%		2	
CO_2 plus N_2	calculation	vol%		15	
Hydrogen sulfide (H_2S)	DIN 51855, DIN EN ISO 6326-5	mg/m^3		5	
Mercaptans (CH_3SH and others	DIN 51855, DIN EN ISO 6326-6	mg/m^3		6	
Sulfur, total	DIN EN ISO 6326-5	mg/kg		15	
Sulfur, total from 1.1.2009	DIN EN ISO 6326-5	mg/kg		10	
water content	EN ISO 18453, DIN EN ISO 10101-3	mg/kg		40	
Compressor lubricants and suspended particles					test method and limits to be defined

uncontrolled combustion (knock, pre-ignition) and is measured as octane numbers (see Section 3.1.1.). Typical automotive gasolines have octane numbers in the range of 92–100 RON. Owing to its high physical–chemical stability, methane has a much higher octane number of about 135 RON and 122 MON. More recently a different scale for the antiknock performance of CNG has been introduced, the methane number (MN), where methane has been given the methane number 100 and hydrogen the methane number 0. The methane number can be measured in an adapted test engine or can be calculated from the MON octane number or the gas composition, because CNG has far less but more well defined constituents compared with gasolines. Increasing concentrations of ethane, propane and butane reduce, whereas nitrogen increases, the antiknock performance of CNG. Marketed CNG has methane numbers in the range of 69–99.

As methane and most CNG fuels have noticeably higher antiknock performance than gasolines, they would allow higher compression ratios of engines compared with gasoline.

As the majority of CNG vehicles are bi-fuel aftermarket converted gasoline concepts, allowing the use of either CNG or gasoline, and designed for the lower antiknock performance of gasolines, the potential of CNG can not be utilized in this type of vehicles (an increase in engine compression ratio would require practically impossible changes of pistons and/or cylinder heads). However, the share of OEM (vehicle manufacturer) CNG vehicles in the marketplace is rising in countries with stringent emission regulations. Euro 4 and 5 emissions limits with CNG passenger cars and heavy-duty engines can only be met by combustion strategies dedicated to CNG as a fuel. This is done by taking advantage of the high compression capability of CNG, in addition to stoichiometric and lean-burn concepts, which are different for light- and heavy-duty applications. Furthermore, special lambda sensors and three-way catalysts are necessary to cope with methane. Such advanced CNG vehicles are usually equipped with a small reserve tank containing gasoline in case no CNG refueling station can be reached. Because of the CNG oriented combustion process, these vehicles only run poorly when in the safety mode on gasoline.

Also important for the combustion in engines is the fuel dependent air:fuel ratio. While there are few differences between gasolines, the ratio varies considerably between different CNG fuels. When vehicles move between areas of different CNG sources and are accordingly fueled with different gas qualities, the engine fuel metering system must cope with a fairly wide range of gas injection rates to achieve the correct air:fuel ratio. Tests have shown that the modern closed-loop combustion control systems of emission controlled vehicles can cope with this situation.

The relative high thermal stability of methane and hence of CNG, as indicated by the relatively high autoignition temperature and low flame speed, requires some special measures to be taken for engine management, e.g., the ignition system (timing and spark energy) may need adaptation, and exhaust catalysts with increased reactivity have to be fitted if the same level of unburnt hydrocarbon emissions, as compared with gasoline, is required.

Ignitability limits are dealt with in the safety section.

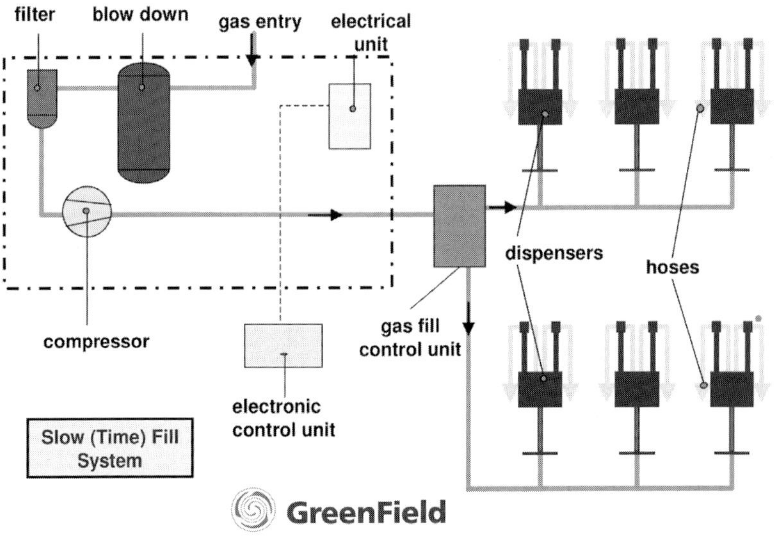

Fig. 3.17 Slow-fill CNG refueling scheme.

Vehicle Refueling Systems CNG is stored in the vehicle tanks at a high pressure of about 200 bar. For this reason the refueling systems have to increase the gas pressure from the low pressure local gas supply system in urban areas, typically between 18 and 24 mbar to pressures higher than the required 200 bar. Depending on the number of vehicles to be refueled in a given period and the time available for the refill there are basically two different systems in use; a low-cost slow-fill system and a higher cost fast-fill system (see Figs. 3.17 and 3.18).

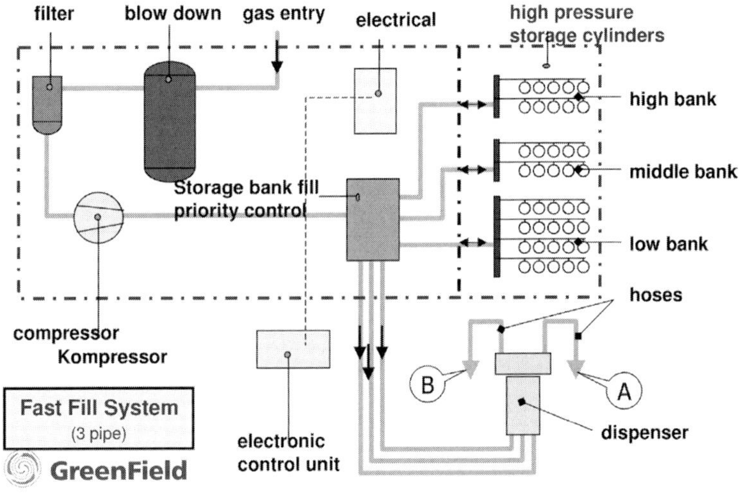

Fig. 3.18 Fast-fill CNG refueling scheme.

Slow-Fill Refueling In slow-fill and in fast-fill systems the NG flows through a blow-down vessel for pressure equalization, which is normally combined with a dryer. Slow-fill refueling of vehicles operates directly from the compressor through a special slow-fill nozzle connector to the vehicle (about 8–10 h refueling time). Slow-fill refueling works due to CNG flowing as a result of overpressure from the nozzle to the vehicle tank with only a slight overpressure (relative to the vehicle tank pressure) and requires only a lower power (lower cost) compressor. Slow-fill stations are mainly used for CNG buses, which can be refueled in depots overnight. 'Sim' versions of the slow-fill units are the home fill stations, which work only with a small compressor taking the gas from the grid and feeding the compressed gas directly into a passenger car (used in teh USA).

Fast-Fill Refueling Fast-fill CNG refueling allows refueling times comparable to those of liquid fuels (4–6 min), whenever a vehicle arrives. Fast-fill systems are therefore common at all public CNG refueling stations and also at bus depots (there often in combination with slow-fill). Figure 3.18 shows a typical scheme of a fast-fill system.

Fast-fill systems use CNG storage cylinder cascades so that they always hold a sufficient amount of CNG in stock for refueling in a very short time. CNG must therefore be stored at a significantly higher pressure in these cylinders, approx. 250 up to 350 bar, depending on the filling level of typically three cylinder banks. As the CNG is fuelled by the pressure difference between the storage banks and the vehicle, fast filling can only be achieved if there is sufficient pressure in the banks with a switching control between the banks during refueling. With the three-bank system about 25–60 % (average 30 %) of the entire stored gas volume can be used for refueling, depending on the specific system. Compared with a slow-fill system, clearly a larger and more powerful compressor is needed to fill up a typical storage volume of at least 2000 L (the sum of all gas cylinders) at all demanded times. For standard fast-fill design, each of the banks is connected with a separate line to the dispenser where the choice of the actual bank for the CNG stream is done by an internal control system, depending on the CNG pressure difference between the vehicle and dispenser and the gas flow speed.

Vehicle and Engine Concepts Compared with conventional diesel and gasoline vehicles, vehicles running on NG are characterized by different engine settings and the presence of larger volume fuel storage tanks. Owing to its high antiknocking capacity, CNG cannot be used on its own in engines with compression ignition. It is preferable to use it in the Otto cycle, i.e., spark-ignition engines–the majority of today's light-duty and heavy-duty CNG engines operate with spark ignition. There are also dual-fuel heavy-duty diesel engines that use NG as the primary fuel source and a small amount of diesel fuel injected separately as a pilot charge, which then enables NG to ignite under compression. The dual-fuel engine still works on a diesel cycle (compression ignition), therefore retrofitting a heavy-duty diesel engine to dual fuel is possible. However, the difficult management of switching between diesel and natural gas combustion and potential problems with particulate emissions due to the use of diesel fuel are reasons for the fairly limited number of duel-fuel vehicles on the market.

Dedicated CNG vehicles operate exclusively on NG without the option of switching to another fuel. The target of dedicated CNG vehicles is the full utilization of the NG advantages, such as high octane number and knock resistance, i.e., higher compression ratio and therefore higher efficiency and energy output, which can be achieved when the engine is optimized to CNG combustion only. However, the majority of today's NG vehicles worldwide are bi-fueled vehicles converted from gasoline passenger cars (there could be potential confusion between dual-fuel engines that operate on a combined diesel/CNG strategy and bi-fuel engines that are *either* in gasoline or CNG operation). OEM-built bi-fueled CNG cars have a relatively low market share compared with the aftermarket converted bi-fuel natural gas vehicles. Conversion from gasoline or diesel (heavy-duty) vehicles to dedicated CNG vehicles plays a minor role in aftermarket conversions. The reasons for the major spread of bi-fueled NG vehicles (even most of the OEM-offered CNG passenger cars are bi-fuel) are flexibility in fuel selection in terms of either assured mobility, where CNG outlets are sparse, or based on price fluctuations.

Special optimization of a bi-fuel vehicle to both gasoline and NG combustion requires high technical effort, realistically found only in OEM-built equipment. As a result of optimization on gasoline, bi-fuel vehicles cannot utilize the full potential of NG for lower exhaust emissions.

Vehicle Fuel Supply System The entire fuel system up to the engine inlet pressure regulator operates under the same pressure as the gas storage cylinder, i.e., 200 bar.

The fuel system of a CNG vehicle consists of the following main components:

- CNG fuel filling valve, which works as the coupling system with the CNG refueling nozzle, the CNG valve includes a device for emergency and maintenance open/shut off.
- Gas storage cylinders combined with a pressure regulator, temperature sensor and an internal emergency shut-off valve.
- Manual shut-off valve, pressure relief device (PRD) of emergency venting of the NG.
- Multiple stage pressure regulation which reduces the pressure from 200 bar down to about 8 bar (low pressure value might vary for different CNG engine technologies).
- Air and fuel mixing and metering unit.
- Gas injection system.

Figure 3.19 shows the fuel supply and entire engine system, typical for a stoichiometric light-duty vehicle.

Combustion and Emissions The main factor that affects mixture preparation and combustion is the fact that natural gas requires no vaporization prior to ignition, which has a number of implications. Firstly, when injected into the port it displaces part of the incoming charge, thus reducing the engine's volumetric efficiency and power output by about 10 % compared with gasoline. However, NG does not need to be vaporized and, thus, requires no cold start enrichment.

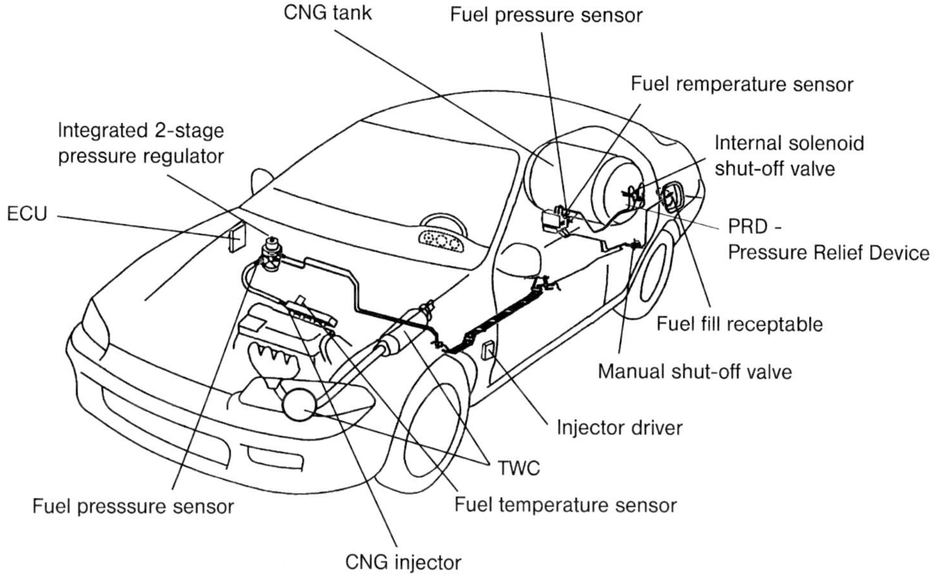

Fig. 3.19 Fuel supply, engine combustion system and exhaust gas treatment of NG passenger car.

Secondly, NG is less prone to cyclic variability, because its combustion depends only on mixing and not on the rate of fuel vaporization. As a result of these improvements, i.e., more stable combustion, the emissions of unburned hydrocarbon upstream of the catalysts are generally lower. However, the majority of these emitted hydrocarbons in CNG engines are methane, which can only be oxidized by specialized (e.g., palladium-based) catalysts. This is due to the fact that the methane molecule is fairly stable and thus more resistant to catalytic oxidation, in addition, the air:fuel ratio margins for catalyst functioning are smaller for natural gas than for gasoline. This necessitates better control of the air:fuel ratio.

Another key characteristic for CNG engine operation is the high octane number of natural gas (average 120). It allows higher compression ratios to be used compared with a standard gasoline spark-ignition engine, leading to higher thermal efficiencies (compression ratios: approx. 10:1 for a gasoline compared with approx. 13:1 for a dedicated CNG engine). All currently available light-duty CNG OEM-built vehicles run with stoichiometric-burn engines, they are equipped with three-way catalysts and lambda sensors to meet the light-duty emissions regulations. In addition, stoichiometric feedback control of the air:fuel ratio enables optimization of engine performance across a range of CNG gas qualities using conventional gasoline engine technology.

However, compared with diesel engines CNG spark ignition engines have a noticeably lower efficiency, partly because of the higher compression ratio of diesel engines. Accordingly, bus fleets have shown considerably higher fuel consumption when operated with CNG compared with diesel engines, typically in the range of about 20 % and sometimes even higher.

Table 3.17 Combustion characteristics of CNG, gasoline, and diesel

Property	CNG	Gasoline	Diesel
Flammability limits, vol% in air	5–15	1.4–7.4	0.6–5.5
Autoignition temperature, °C	450	300	230
Minimum ignition energy in air, 10^{-6} Btu	0.27	0.23	0.23
Peak flame temperature, °C	1883	1977	2054

With regard to exhaust emissions, CNG has the advantage of noticeably lower emissions of air toxics and, compared with diesel engines, also of noticeably less particulates. Because of the higher hydrogen content of CNG there are also lower emissions of carbon dioxide. With optimized exhaust gas treatment the standard regulated exhaust gas constituents CO, HC and NO_x are in the same range as those of gasoline.

Well to wheel efficiency, i.e., the overall efficiency including production, transport, refining and vehicle, of CNG is better than that of gasoline but only the same or even worse than that of diesel fuel.

Safety Aspects CNG is, in some respects, safer than gasoline and other liquid fuels. Flame luminosity is similar to other hydrocarbon fuels and noticeably better than that of alcohol fuels. Natural gas is less toxic than all other fuels apart from hydrogen. There is no possibility of ground water pollution and there is slightly lower air pollution when CNG is used in engines, see above. The flammability limits are narrower and the autoignition temperature is higher than that of gasoline and diesel fuel (see Table 3.17).

However, the risk of explosion remains when mixed with air should a leakage occur, especially in closed rooms. Outside in the open air this risk is much smaller, because methane has a lower density than air and will rapidly disperse and disappear. Only in the case of a major leakage could the cooling by rapid expansion possibly temporarily cause a cold and ignitable gas layer heavier than air on the ground. The storage under high pressure at service stations and in vehicle tanks of course requires attention. High standards for the equipment and regular control ensure reliable storage.

3.8.3 Synthetic Fuels

H. P. Calis, Wolfgang Lüke, Ingo Drescher

3.8.3.1 Introduction

This chapter describes synthetic transportation fuels (synfuels), in particular gasoil, produced from carbon-containing feedstocks (natural gas, coal, biomass) via the following three-step process:

1. conversion of the feedstock into synthesis gas (syngas, a mixture of CO and H_2);

2. conversion of the syngas, via low-temperature FISCHER–TROPSCH (FT) synthesis, into primarily a distribution of unbranched paraffinic (and olefinic) hydrocarbons, ranging from as light as CH_4 (methane) to as heavy as $C_{200}H_{402}$ (heavy paraffinic wax);
3. hydroprocessing of the primary FT products to produce the final product slate, with naphtha, kerosene and gasoil as the main products in terms of volume.

This manufacturing process produces synthetic gasoil product that is virtually free of sulfur, nitrogen and aromatics, clear as water and practically odorless and colorless. It is a superior transportation fuel for diesel engines, with exceptionally high cetane numbers (over 75) as well as improved engine performance, particularly in the area of regulated emissions. By tuning the product specifications, fuels required for the next generation of internal combustion engines, known as Combined Combustion System (CCS) or Homogeneously Charged Compression Ignition (HCCI) engines, can be produced. This future fuel – engine combination allows the high fuel economy of diesel engines to be combined with the favorable emission characteristics of Otto (gasoline) engines.

The synfuels manufacturing process was pioneered in Germany in the first half of the previous century, mainly based on coal as the feedstock. From the 1950s onwards, Sasol in South Africa further developed the coal-based process. It is still in commercial use today in South Africa, although targeting (via *high*-temperature FT synthesis) gasoline rather than gasoil as the primary product. With coal being the most abundant fossil feedstock, the coal-based FT process may find a much wider application in the near future, not just for gasoline but also gasoil and new generations of fuels.

In the 1980s the route from natural gas as feedstock was also commercially developed, to provide an alternative monetization route for natural gas resources remote from the market. The first commercial GTL plant applying low-temperature FT synthesis is a relatively small plant (14 700 barrels per day, bpd) operated by Shell in Malaysia, and in 2007 a larger plant (34,000 bpd) was commissioned in Qatar by Sasol. New plants are under construction by Shell in Qatar (140,000 bpd) and Sasol/Chevron in Nigeria (34,000 bpd) and other projects are under development.

Currently the route from *renewable carbon-containing feedstocks*, such as biomass or even organic waste, is in an early stage of development. However, concerns over the security of supply of fossil energy feedstocks, as well as concerns over climate change, may result in policy changes that favor the commercialization of this route. The German company Choren (partly owned by Shell) has demonstrated synfuel manufacture from biomass, and is in the process of commercializing this process.

The different routes from carbon-containing feedstock to synfuels derive their name from the type of feedstock: “Gas to Liquids” (GTL), “Coal to Liquids” (CTL) and “Biomass to Liquids” (BTL). A generic term for all these processes, where “X” can be a range of feedstocks, is “XTL”. The different feedstocks of the various XTL processes can produce identical product slates, as all XTL processes use syngas as the universal intermediate product.

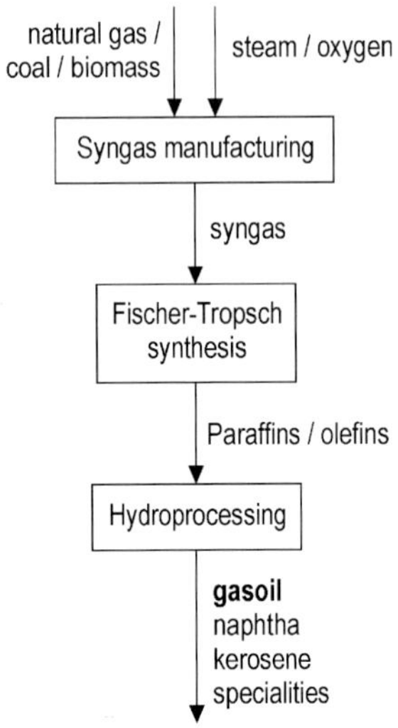

Fig. 3.20 Principle of the XTL process.

The remainder of this section will first address XTL technology and describe the process concept, mentioning the main industrial players. Next, the nature and the performance of XTL derived gasoil as a transportation fuel are discussed; fuel properties, combustion performance when used as blending component or as a neat fuel, and options for dedicated engine development. Finally, the environmental impact of XTL is compared with crude-based refineries on the basis of life-cycle assessment.

3.8.3.2 **XTL Technology**

The core elements of the XTL process (based on low-temperature FISCHER – TROPSCH synthesis and aiming for gasoil as the desired product) are shown in Fig. 3.20.

Not shown in this diagram are the process units for the production or management of utilities such as oxygen, steam, water, power and fuel gas. The utilities section is a very important part of an XTL plant. In GTL, net water is produced, whereas in CTL net water is consumed.

Although the XTL process is highly complex, the current description is kept concise. No references are provided other than a global reference to STEYNBERG and DRY [92], who have compiled an excellent set of detailed reviews with many references to professional literature.

Syngas Manufacturing In the first step, a carbon-containing feedstock (natural gas, coal, biomass) is converted into syngas, a mixture with CO and H_2 as the desired main components (see Section 4.4.1.). A wide range of syngas manufacturing technologies is commercially available. For solid feedstocks such as coal and biomass, these are generally referred to as *gasification* technologies. Examples relevant for XTL are moving-bed, fluidized-bed and entrained-flow systems. For natural gas feedstock (GTL), the syngas manufacturing technologies of current industrial relevance comprise partial oxidation, steam reforming and hybrids thereof, such as Autothermal Reforming (ATR, see Section 4.4.1.).

The syngas manufacture for XTL typically takes place at elevated pressures (10 bar or higher; 1 bar = 10^5 Pa), not just to reduce the syngas reactor volume, but also because the subsequent FT synthesis reaction benefits from elevated pressures. An important differentiator between the syngas technologies is the source of oxygen (to produce CO), which can be air, oxygen or steam. A main quality indicator of the produced syngas is the H_2/CO ratio, which may range from 0.6 for gasification of coal to 5 for steam reforming of natural gas. In the FT synthesis step the ratio in which H_2 and CO are consumed may range from below 1 to just over 2, depending on the type of catalyst. Hence, the hydrogen balancing of an XTL plant is of key importance in the design and economics of the plant.

Fischer–Tropsch Synthesis The second step of the XTL process is the FISCHER – TROPSCH synthesis:

$$CO + 2H_2 \rightarrow -CH_2- + H_2O \tag{3.8.3.1}$$

where $-CH_2-$ represents a typical hydrocarbon unit element of the FT product, as is present in, for example, cetane ($C_{16}H_{34}$, the model component for a diesel fuel).

The performance of the FT synthesis and its primary product slate are governed by the combination of catalyst, reactor and process conditions, and the integration of the FT unit with the rest of the XTL plant.

The traditional catalyst for CTL is based on iron, as this catalyst is inexpensive and moderately robust against poisoning by contaminants present in coal-derived syngas, such as sulfur species. Furthermore, it enables the use of syngas with a low H_2/CO ratio (typical for coal gasification), because it favors the water – gas shift reaction ($CO + H_2O \rightarrow H_2 + CO_2$) that produces the required additional hydrogen *in situ* (see Section 4.4.1.). The drawbacks of the iron catalyst are, however, a short lifetime (typically up to several months), low activity and relatively poor selectivity to heavier hydrocarbons. Hence, this catalyst allows only a moderate yield of synthetic gasoil.

In the world-scale GTL projects under development in Qatar, the various projects (Sasol, Shell) will use cobalt-based catalysts, as does Shell's current commercial GTL plant in Malaysia. This catalyst is more expensive and more sensitive to poisoning, but offers superior activity and selectivity towards heavy wax products, which (via mild hydrocracking) allows the highest yield of synthetic gasoil (and high-value specialty products).

The most relevant reactor types for an XTL plant aimed at producing synthetic gasoil are slurry bubble and multitubular fixed-bed reactors. As a slurry reactor allows on-line replacement of catalyst, it is particularly suitable for catalysts with a short lifespan, such as the iron catalyst. For the cobalt-based catalyst, with lifetimes up to several years, multitubular fixed-bed reactors are feasible. These allow higher reaction rates and higher selectivity towards heavier products, favoring the gasoil yield of the plant. The production of relatively more gasoil is also more attractive from a market perspective as the middle distillates market is supposed to grow more rapidly in major markets.

Typical FT reaction conditions for XTL aimed at gasoil as the primary product are a temperature window from 180 to 270 °C ("low-temperature FT") and a pressure window from 20 to 80 bar. Lower temperatures favor heavier products (beneficial for the gasoil yield), whereas higher pressures allow higher reaction intensity (less catalyst required) and higher selectivity.

The primary product of low-temperature FT synthesis with a cobalt-based catalyst is a distribution of unbranched paraffinic (and olefinic) hydrocarbons, ranging from CH_4 (methane) to $C_{200}H_{402}$ (heavy wax). In terms of mass distribution, only a small fraction (around 10 %) of the FT product consists of gaseous hydrocarbons (butane or lighter), the rest is pentane or heavier.

Hydroprocessing The third step of the XTL process is the conversion of the mix of primary FT product into the desired product slate, including gasoil. The heavier fraction of the FT primary product can be converted into middle distillates via a catalytic, highly selective, mild hydrocracking step (see also Section 1.3.5.). In this step the heavier, long-chained, unbranched wax molecules are cracked selectively to the desired shorter and branched molecules; while olefins and oxygenates are hydrogenated. Following the hydrocracking, distillation provides the correct boiling range for the desired gasoil fraction. For maximum gasoil production, the bottom end comprising the heavier fraction is recycled to the hydrocracker.

The C_{10}^{+} fraction in the primary product of the low-temperature FT synthesis consists primarily of unbranched paraffins. The longer "tails" (longer paraffin chains) in the gasoil cut have a natural tendency to crystallize at temperatures just below ambient, potentially resulting in clouding of the product and plugging of the fuel filter under cold conditions. These properties, measured via the cloud point and the cold filter plugging point (see Section 3.5.2.3.), are subject to specifications for automotive gasoil. The branching of the paraffins that is introduced via the hydrocracking step is an important means of decreasing the cloud point and cold filter plugging point to the desired low temperatures (to enable use under winter conditions). The extent of branching depends on the type of hydrocracking catalyst and on the process conditions (expressed in terms of "severity"), including the degree of recycling of the distillation bottom product to the hydrocracker. Hence, the cold flow properties of GTL gasoil can be tuned to a large extent via the GTL process, and further adapted via additives.

3.8.3.3 **Performance of Synthetic Fuels**

Introduction The diesel combustion process is undergoing constant development in an effort to further reduce emissions and increase efficiency as well as driving fun. One further overlying challenge is the question of energy sources for future fuels and drive systems.

Synthetic fuels have the potential to meet all of the challenges in question. Thanks to the special manufacturing processes, synthetic fuel is in any case free from sulfur and aromatics, which leads to considerable reductions in emissions. Furthermore, synthetic fuels can be optimized for new engine requirements that allow emissions to be reduced further and also increase efficiency. The extremely high and easily reproducible fuel quality does not depend on the primary energy carrier (natural gas, coal or biomass) and thus allows a significant step to be made in the direction of reliable supplies. In addition, CO_2 emissions are almost completely avoided when biomass is used to produce synfuel (BTL) (see Section 3.8.3.4.).

Advantages of Synthetic Fuels for Emission Control Sulfur compounds and aromatics contribute to the formation of soot and nitrogen oxides when fuels are combusted in engines. The automobile industry has therefore been calling for the reduction of these fuel components for decades. Today low-sulfur fuels (<10 mg/kg) have established themselves in several regions. Synthetic fuels go one decisive step further, however. They are not only completely free of sulfur, but also free of aromatic compounds.

When Shell GTL fuel is used in the Golf TDI, the NO_x emissions in the New European Driving Cycle (NEDC) could be reduced by 6 %, the soot emissions by 26 %, the hydrocarbon emissions by 63 % and the carbon monoxide emissions even by 91 %. No changes to the engine hardware and software were necessary. Further emissions reductions can be achieved by fine tuning the engine control program to the slightly lower density of the fuel. In today's engines, NO_x emission reductions in the NEDC of over 30 % are possible with full compatibility – in a continuous test program in 2003, 25 Golf TDIs demonstrated that Shell GTL could be used in practice by customers.

As the synthetic fuel is free of aromatics, its density is slightly lower (785 kg/m^3 instead of 830 kg/m^3) than commercially available diesel, which contains around 20 % aromatic compounds. The density of synfuel is therefore outside the European specification for diesel fuel (EN 590).

Accordingly, synfuel has to be blended with diesel to meet this specification. By the time the production capacity for GTL has been set up, fuel specifications across the globe will very probably have been adjusted to this new generation of highly clean fuels.

In cities with air-quality problems, for example, Bangkok and Athens, the addition of Shell GTL to diesel fuels has helped reduce summer smog problems since 2002 and 2003. GTL alone can significantly improve the air quality in cities without expensive and sophisticated technical modifications to vehicles and infrastructure.

DaimlerChrysler investigated the emissions reduction potential of synfuel additions to diesel. The soot and nitrogen oxides emissions were measured on a bench dynamometer at a characteristic partial load point. The rate of exhaust gas recycling was varied and the start of the pilot injection phase and main injection phase were kept constant during the test. From a 20 % proportion of synfuel, considerable improvements could be achieved. Once the synfuel fraction in the diesel reached 50 %, the reduction in emissions almost reached the level of pure GTL (see Fig. 3.21).

Combined Combustion System A significant reduction in diesel engine emissions requires intensive, continued development of the diesel combustion process and

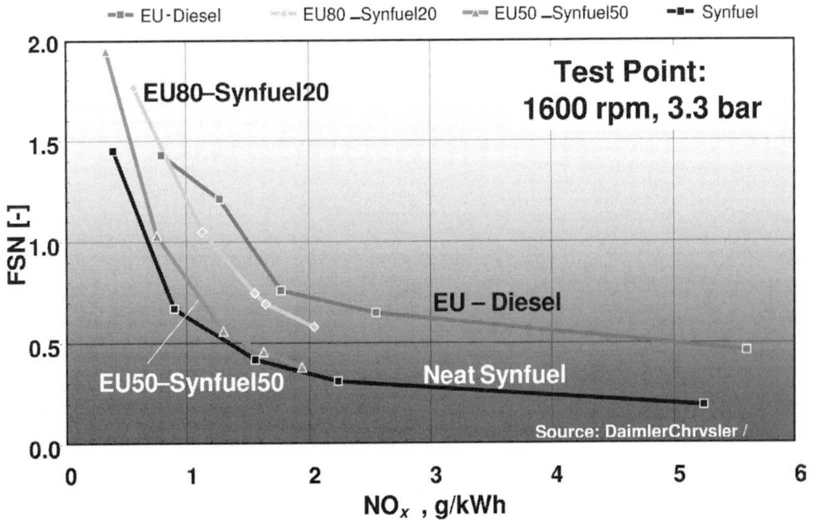

Fig. 3.21 Influence of GTL additives on soot – NO_x trade-off (FSN = filter smoke number). Source: Daimler Chrysler –SASOL

the use of homogenization effects. Zones with oxygen deficiency, which cause soot formation, and local peak temperatures, which are responsible for NO_x emissions, are then avoided or reduced. Thus both the NO_x and soot emissions can be reduced with internal engine measures.

Further development of both the engines and the fuel is necessary. The diesel combustion process has great potential for reducing emissions if the engines and fuel are tailored to each other. Synfuel will make this possible to a significant extent for the first time.

The new generation of Combined Combustion System (CCS) engines, which combine the low emissions of Otto (gasoline) engines with the low consumption of diesel engines, is based on the use of homogenization effects that can be achieved with engine and fuel measures. Carefully directed changing of fuel properties is a key issue to master homogenization in diesel engines. One fuel that has been optimized for the CCS combustion process is CCS fuel.

Figure 3.22 clearly shows the emission reduction potential of the CCS combustion process at an average constant load point. The dark circular spot represents the emissions of a standard TDI engine. The CCS engine is designed via the hardware and the engine control management system to obtain good homogenization of the fuel – air mixture. This allows both emissions to be reduced considerably when diesel fuel (curve with closed symbols) is used. A further reduction of both emissions to almost zero can be achieved by optimizing the fuel, i.e, the CCS fuel. The basic special features of the CCS fuel are its lower boiling range as well as it being free from sulfur and aromatics. This allows very good homogenization with extremely low emissions.

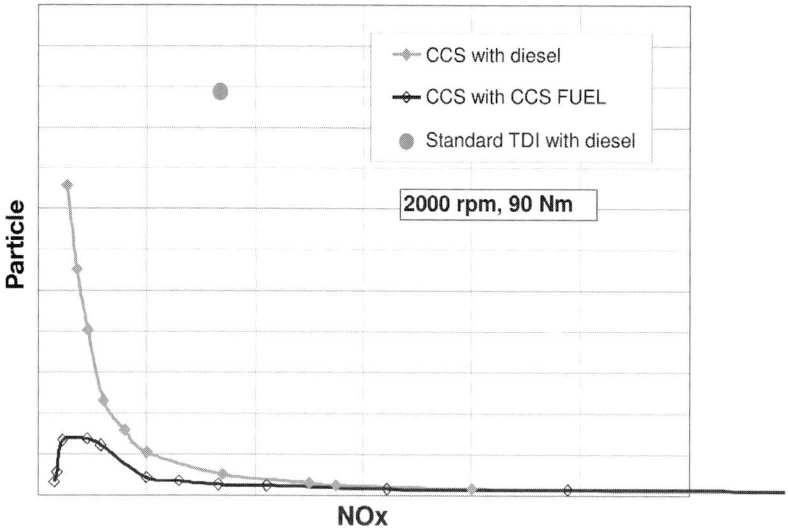

Fig. 3.22 Emission reduction potential of CCS combustion process and CCS fuel. Engine operating point: 90 Nm at 2000 rpm.

From the viewpoint of the automobile industry, synthetic fuels provide solutions for the three big challenges of future mobility:

- further emissions and CO_2 reductions
- reliable supplies, i.e., diversification of primary energy
- use of existing infrastructure.

3.8.3.4 **Well-to-Wheel Comparisons and Life-Cycle Analysis**

Well-to-Wheels (WTW) and Life-Cycle Analysis (LCA) are the most frequently used methodologies for comparing the Greenhouse Gas (GHG) emissions associated with transportation fuels. Both aim to quantify the environmental impact of producing and consuming one or more (alternative) oil products, in terms of grams of CO_2 equivalents produced (including during the fuel production phase) per kilometer driven by a standard vehicle.

WTW focuses on production of automotive fuels only, such as gasoline and diesel. However, the production process of a certain fuel often involves coproduction of several other products as well. This implies that fractions of the total amount of GHG produced need to be allocated to the individual products.

LCA considers the full systems including all products of the two technologies, comparing for example the GTL process with a refinery process. However, the GTL and refinery processes are not directly comparable as a result of their different product slates, which serve different functions. In the case of the refinery, all products are not equal in terms of market desire and economic value. In contrast, the GTL system only provides products valued by the market.

To enable an objective like-for-like comparison, the GTL system has to be expanded to meet the same range of market functions as products from the refinery system.

This implies that products and processes should be added, together with their associated energy and environmental burdens. The expansion of the GTL system is performed on the basis of existing complementary technologies. The choice of product for fulfilling the missing functions is based on market analysis and industry trends.

This LCA approach is different from the WTW where only the diesel and gasoline streams from a refinery are taken into account. The WTW comparison ignores the fact that a refinery has a range of coproducts associated with the diesel stream, some of which are not so desirable from an environmental and economical point of view, specifically the heavier carbon-rich residues, whereas GTL produces only desirable products.

Both WTW and LCA require numerous assumptions in addition to the factual data. The result is a wide scatter of published results. The most extensive study published is a joint effort of EUCAR (European car industry), CONCAWE (European oil industry) and JRC (the Joint Research Centre of the EU Commission) [93]. CONCAWE/EUCAR/JRC are finalizing an updated report that was published in 2006.

An LCA to compare GTL, CTL and BTL with crude-based refineries was commissioned by the U.S. Department of Energy [94]. The main results of this and similar industry studies is that GTL produces similar amounts of GHG as crude-based fuels; CTL produces about twice the amount. BTL produces less than 20 % of the amount, illustrating the significant potential of BTL for reducing GHG emissions.

GTL has additional environmental benefits over crude-based fuels. GTL is very selective in producing only desired products such as gasoil, whereas crude-based refineries inevitably produce much less desired "bottom-of-the-barrel products" such as heavy fuel oil and residue. In addition, GTL has considerably lower local emissions in the product application phase (compared with crude-based diesel fuel), owing to a number of key properties, including its very high cetane number and to being virtually free from sulfur and aromatics. This has important benefits for air quality in urban environments, in particular congested "megacities".

The CO_2 emissions during the manufacturing process of GTL are considerably higher in comparison with crude-based refineries. However, most of the CO_2 produced in the GTL process can effectively be obtained in pure form, suitable for CO_2 sequestration. Application of CO_2 sequestration provides GTL with the potential to outperform crude-based fuels in terms of GHG emissions on an LCA basis.

The GHG advantage of GTL is anticipated to increase in the near future. The crude-oil feedstocks of refineries tend to become heavier, implying higher GHG emissions. In contrast, continuous technical improvements in the still young GTL industry are likely to result in further improvement of the energy efficiency of GTL, implying even lower GHG emissions.

If CO_2 sequestration is applied to CTL, similar GHG emissions (on an LCA basis) can be obtained as for crude-based refineries.

Applying CO_2 sequestration to a BTL plant results in a process that has an overall *negative* GHG production. The rationale of this feature is that biomass captures CO_2 from the atmosphere during its growth. During the BTL process, the solar energy stored in the biomass is further concentrated into liquid fuel, with CO_2 as a pure byproduct. Sequestering this CO_2 stream is an effective means of actually *reducing* the CO_2 levels in the atmosphere.

3.8.4
Bioethanol
Lutz Guderjahn

3.8.4.1 General

Utilizing ethanol as a fuel is not a recent invention, but has been practiced for almost a century and a half. Early on, the pioneers of the automotive industry were inspired by this "fuel of the future" as HENRY FORD called it. Use of ethanol played a crucial role in two major developments that helped to determine the future of the automotive industry. As early as the 1860s, NIKOLAUS AUGUST OTTO used ethanol as the fuel to drive the prototypes of his combustion engine and almost 50 years later, when HENRY FORD designed the car that revolutionized production in the automotive industry – his legendary Model T – it was his understanding that ethanol would be used to fuel this "car of the people". As for its commercial use, ethanol has been sold in Germany since 1925 as an octane enhancer.

The rediscovery of ethanol as a fuel in recent decades stems from the need to explore alternative sources of energy to forestall the finiteness of fossil fuel sources such as oil, the production and refining of which are coupled with an ever-growing energy demand, and to attain a degree of independence from these shrinking resources. Moreover, the production of ethanol not only provides the opportunity of sustainable development and guarantees the availability of fuel in the future, it also makes an important contribution to limiting CO_2 emissions.

The "oil crises" in the 1970s inspired the governments of various countries to promote the use of fuels from renewable sources. Since then, bioethanol has been applied particularly in Brazil and the USA for transportation.

The EU supports the use of such fuels, for example, by enacting the so-called "biofuels directive" in which clear targets for the share of biofuels in total fuel consumption are defined (5.75 % as per energy equivalent by 2010) [95].

3.8.4.2 Production

Raw Materials Bioethanol can be produced from any carbohydrate-containing material by means of fermentation techniques. The sources in the various regions of the world vary according to their availability. The main carbohydrates and their feedstock are:

- starch
  - cereals (corn, wheat, barley, sorghum/milo, etc.)
  - potatoes
  - cassava
- sucrose
  - sugarcane
  - sugarbeet
- lactose
  - whey
- cellulose
  - wood
  - straw

In Brazil, the preferred feedstock is sugarcane, production in the USA focuses on corn, whereas in Europe the utilization of wheat and sugarbeet dominates. In Asian regions, such as Thailand, the preferred feedstock for bioethanol production is the manioc root (cassava, tapioca).

Cellulose-based production from wood or agricultural byproducts such as wheat or corn straw has recently attracted considerable attention because of the general possibility of thereby increasing the biomass yield per hectare. Furthermore, the cellulose-based process is considered to deliver the highest savings in CO_2 emissions of all industrial-scale bioethanol processes. During the past decade numerous initiatives have been started, for example, steps have been taken to improve the industrial applicability of lignocellulosic raw material-based processes. However, so far only pilot-scale plants have been built [96], [97]. The transfer to an industrial-scale procedure requires further process optimization, but might be competitive in terms of production costs compared with sucrose- or starch-related processes within the next five to ten years, and is therefore a very important pathway to be followed.

Production from Cereals Industrial-scale production plants have a capacity in the range of 150–300 000 m^3 bioethanol per year. As the compositions of the various feedstocks applicable for bioethanol production are significantly different (see Table 3.18), the set-up for an industrial scale bioethanol process has to be adapted.

The production process of bioethanol from cereals such as wheat is, in most cases, based on the so-called dry-milling technology, which consists of the following steps:

- cereal reception and storage
- milling of the dry grain
- liquefaction/saccharification
- fermentation
- distillation
- ethanol: drying and storage
- stillage: drying, pelletizing, and storage

The production process starts with the milling of the grain. In most cases this is a dry-milling process, in which the grain is preferably milled by hammer mills. The resulting grind is mashed in water. In some cases wet-milling

Table 3.18 Composition (%) of raw materials for bioethanol production

Source	Starch	Fat	Ash	Protein	Fiber
Rice (broken)	90.0	0.4	0.4	8.4	0.2
Corn	71.6	4.8	1.4	9.8	2.6
Potato	76.0	0.6	4.8	8.0	6.4
Wheat	67.0	2.5	2.0	13.0	12.0
Rye	58.0	3.0	2.0	10.0	15.0
Cassava (chips)	78.6	1.0	2.1	2.0	4.0
Milo/sorghum	68.5	3.5	1.9	14.0	2.0

processes are also applied. This process is applied if the starch has to be extracted from the cereals.

As starch cannot be fermented directly by yeasts, it must first be hydrolyzed to fermentable glucose. This is done in two steps usually in a cascade of stirring vessels by means of industrial enzymes. Acid catalyzed hydrolysis could generally also be applied, but this is less common nowadays.

The first step is called liquefaction or the dextrinization phase. The mash is treated with "liquefying" enzymes (usually bacterial a-amylases) at approximately 85–90 °C. This leads to the formation of dextrins.

In the second step ("saccharification"), the dextrins are further hydrolyzed by means of fungal glucoamylases, in which the dextrins are hydrolyzed to glucose. The temperature is adjusted for the saccharification step and is usually in the region of approximately 60 °C [98].

The achievement of a high yield of bioethanol necessitates a complete saccharification of the starch. Using different cereals might require additional enzymes to accomplish the complete saccharification, rye for instance requires the additional application of enzymes such as xylanases and cellulases that are able to hydrolyze hemicellulosic material.

After saccharification, the mash is cooled and transferred to the fermentation unit, in which the hydrolyzed starch is fermented by means of yeasts (*Saccharomyces cerevisiae*). Industrial yeast fermentation to bioethanol is usually conducted at temperatures between 30 and 35 °C. The fermentation can be conducted either batchwise or continuously in a cascade of stirring vessels. The reaction is exothermic and can be summarized by the following equation:

$$C_6H_{12}O_6 + 2\ P + 2\ ADP \rightarrow 2\ C_2H_5OH + 2\ CO_2 + 2\ ATP\ \Delta H = -156\ kJ$$

The conversion of 1 kg of glucose leads to the formation of 511 g bioethanol, 489 g CO_2 and 867 kJ of heat, which has to be removed by cooling devices [99]. At the end of the fermentation, the ethanol content is in the range of 11.5–12.5 vol %. The fermentation is followed by the distillation step in which the ethanol is removed from the mash by evaporation.

The setup of the distillation unit can vary significantly with the type of process, raw material supplied, and the desired product quality. Usually, the unit consists of at least two steps:

- removal of the ethanol from the mash (distillation)
- purification of the raw alcohol (rectification)

The distillation and rectification is conducted in columns with a variable number of trays, the construction and design of which varies significantly. The setup of the distillation unit is directly linked to the quality requirements for the ethanol produced. The distillation yields a raw alcohol with an alcohol content of approximately 85–87 %. The rectification is necessary to increase the alcohol content and to remove so-called fusel oils, which are higher alcohols (C_3–C_5). In modern distillation units, these two basic steps are integrated to optimize the energy requirement for the process.

Table 3.119 Typical composition of DDGS from wheat (time period: August – December 2005)

Dry matter (DM)	**95.00 %**
Ingredients per kg dry matter (DM)	
Crude ash	58 g
Crude protein	385 g
Crude fiber	71 g
Crude fat	65 g
Crude starch	30 g
Crude sugar	30 g
N-free extracts	421 g
Neutral-detergent fiber	311 g
Acid-detergent fiber	143 g

Source: Südzucker

For the production of fuel-grade ethanol, the ethanol has to be "dried". Anhydrous ethanol cannot be produced by simple distillation because ethanol forms an azeotropic mixture with water. The maximum ethanol content achievable by distillation is approximately 97.2 vol %, which is usually not sufficient for the application of ethanol as a fuel component. The residual water can be removed either by azeotropic distillation by the addition of, e.g., cyclohexane or by the application of molecular sieves.

Today, state-of-the-art plants operate with molecular sieves which provide considerable advantages in terms of investment and operating costs.

The stillage is usually dried and mainly sold as protein-rich feed product (DDGS, Distillers Dried Grain with Solubles, see Table 3.19). It contains all the non-starch components of the applied cereals as well as yeast resulting from biomass propagation in the fermentation step. The byproduct plays an important role in the economics of a bioethanol plant.

Quality of Fuel Ethanol To date, there is no globally adopted specification for fuel ethanol, thus the quality varies considerably [100], [101], depending partly on the final application of the ethanol. In Europe, a typical specification for direct blending as well as ethyl *tert*-butyl ether (ETBE) production is as follows (Table 3.20).

Table 3.20 Typical specification for fuel ethanol

Property	**Unit**	**Min.**	**Max.**
Ethanol	wt %	99.7	
Higher alcohols	wt %		0.2
Methanol	mg/kg		300
Water	mg/kg		2000
Acidity, calculated as acetic acid	mg/kg		30
Nonvolatile matter	mg/100 mL		10
Appearance		clear and bright	

Source: Südzucker

Table 3.21 Characteristics of liquid fuels and hydrocarbons

Fuel	Density (20°C), kg/L	Boiling temperature, °C	Specific vaporization heat, kJ/kg	Specific calorific / fuel value, MJ/kg	Ignition point, °C	Air requirement theor., kg/kg
Automotive gasoline						
Regular	0.715–0.765	25–215	380–500	42.7	ca. 300	14.8
Super	0.730–0.780	25–215		43.5	ca. 400	14.7
Aviation gasoline	0.720					
Kerosene	0.770–0.830	40–180		43.5	ca. 500	
Diesel	0.815–0.855	170–260		43	ca. 250	14.5
		180–360	250	42.5	ca. 250	14.5
Ethanol	0.790	78	904	26.8	420	9
Methanol	0.790	65	1110	19.7	450	6.4

Source: Südzucker

Properties of fuel ethanol in comparison with hydrocarbon-based fuel are listed in Table 3.21.

3.8.4.3 Application of Bioethanol as Fuel

Bioethanol can be used in the transportation sector in the following forms:

- direct blending with gasoline
- direct blending with diesel
- stand-alone fuel ("E85", "E95")
- precursor for the production of oxygenates
- hydrogen carrier for fuel cells

Direct Blending with Gasoline The primary applications of ethanol in the fuel sector are blends with gasoline. In Brazil the proportion of bioethanol in gasoline (often referred to as "gasohol") is between 20 % and 25 vol % (the proportion of bioethanol varying according to availability and cost of the bioethanol), and in countries such as the USA, China, Thailand, or Australia the usual blend level is 10 vol %.

In the EU, the composition and parameters of motor fuels are regulated by the fuel quality directive (directives 98/70/EC and 2003/17/EC), which currently tolerate a bioethanol blend level of 5 vol % in gasoline [95].

Ethanol is fully miscible with hydrocarbons. However, the blending of ethanol with gasoline requires the preparation of the fuel logistics, for the following two reasons in particular:

- ethanol has a higher water affinity than hydrocarbons and less polar fuel components
- ethanol and hydrocarbons form nonideal mixtures, which leads to a nonlinear behavior of the physicochemical mixtures of the blends

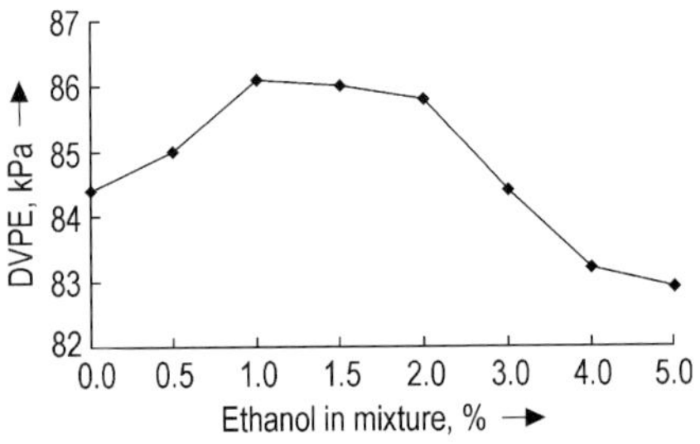

Fig. 3.23 Effect of blending of pure gasoline and a gasoline with 5% ethanol added (source: Südzucker).

The blending of ethanol with gasoline increases the water tolerance of the mixture. This means that ethanol can act as a "solvent" for water, keeping the water molecules in solution. If the water content exceeds a critical limit, the water phase will separate from the hydrocarbon phase. The high water affinity of ethanol will lead to the migration of the ethanol molecules into the water phase. Because ethanol works as an octane booster, the residual hydrocarbon phase loses its knock-resistance properties, which leads to a significant loss in fuel quality. Therefore, the application of directly blended ethanol requires the careful avoidance of water in the fuel distribution system. This behavior has recently been investigated by the German DGMK (Deutsche Wissenschaftliche Gesellschaft für Erdöl, Erdgas und Kohle) [102].

The nonlinear physicochemical behavior of ethanol – gasoline mixtures is particularly significant in blends with a small ethanol share. Although pure ethanol has a low vapor pressure (dry vapor pressure equivalent, DVPE = 15.5 kPa) compared with gasoline, the vapor pressures of mixtures with gasoline increase significantly at low ethanol concentrations (up to 5 %, see Fig. 3.23).

To meet certain fuel specifications, the basic gasoline has to be adapted in such a way that the most volatile hydrocarbons (C_4/C_5 products) are removed from the mixture.

However, at higher blend levels the nonlinear behavior is greatly reduced. Depending on the composition of the gasoline used for blending purposes, the vapor pressure of the blend returns to the starting value at ethanol concentrations between 20 and 50 % (Fig. 3.24).

In many countries the oil companies commonly share fuel depots to optimize the costs for fuel logistics. Therefore, blending of ethanol- and non-ethanol-containing

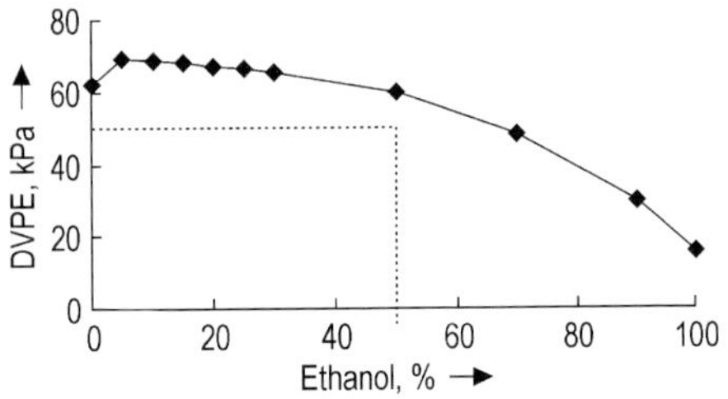

Fig. 3.24 Behavior of vapor pressure of ethanol – gasoline blends (0–100 % ethanol in the mixtures) (source: Südzucker).

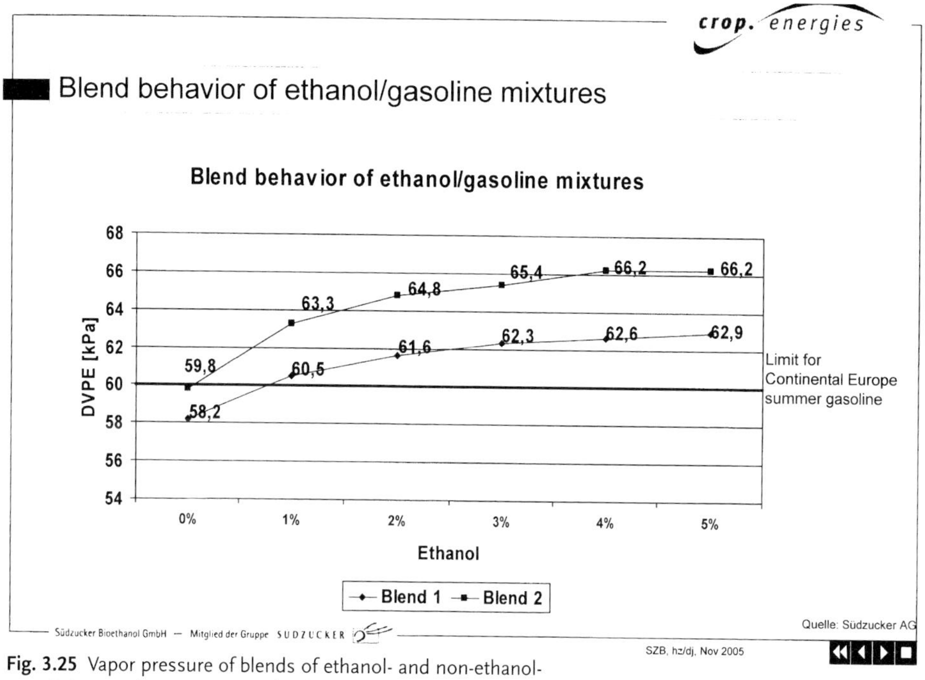

Fig. 3.25 Vapor pressure of blends of ethanol- and non-ethanol-containing gasoline (source: Südzucker).

gasoline might occur, which could cause an increase in vapor pressure. This effect has been recognized as the "commingling effect" (see Fig. 3.25).

This could be avoided by either introducing ethanol as a fuel component on a wide scale or by matching the vapor pressure of ethanol-containing fuel at lower values.

Direct Blending with Diesel Blending of bioethanol with diesel is currently not widely applied, although applicants who conducted trials in the USA claim that an addition of ethanol to diesel would reduce the emission of particulate matter in heavy-duty vehicles by up to 40 %. In the EU, the blending of bioethanol with diesel is not very common, because there are no generally accepted specifications for such a fuel. Furthermore, there are safety matters to be investigated due to the different chemical properties of the fuels to be blended [103].

The addition of ethanol to diesel lowers both the flash point and the cetane number of the mixture. The decrease in cetane number could technically be compensated by the addition of cetane-enhancing additives. The strong reduction of the flash point requires checking and possibly the adoption of safety devices in the vehicles.

Bioethanol as a Stand-alone Fuel

E85 The recent development of so-called "flexible fuel vehicles" (FFVs) have led to the possibility of using blends with a higher percentage of ethanol – namely up to 85 %. These cars are able to run on any mixture of ethanol and gasoline. Sensors

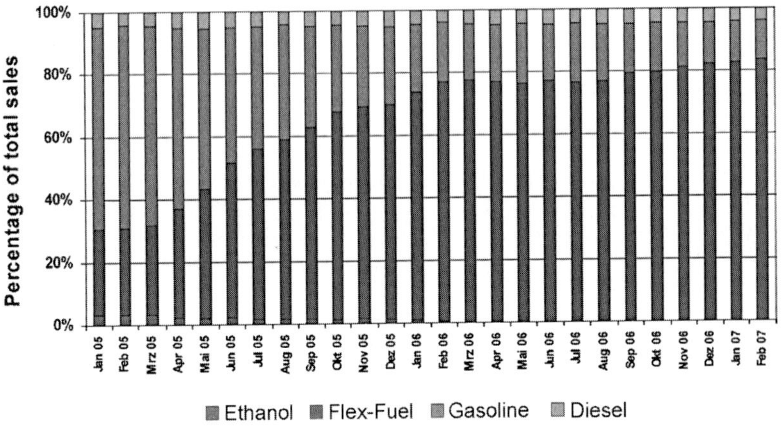

Fig. 3.26 Sales of new cars in Brazil (source: Carta da Anfavea, no. 250).

detect the oxygen content of the fuel and the parameters for the engine management are adjusted accordingly.

E85 – a blend of 85 % ethanol and 15 % regular gasoline – has been successfully introduced into a number of markets across the world. In March 2006, almost 80 % of the vehicles sold in Brazil were FFVs (see Fig. 3.26). A large automotive manufacturer in Brazil has stoped manufacturing vehicles with gasoline-only engines for the Brazilian market. FFVs can run on E85 or, for that matter, on any blend of ethanol and gasoline up to a level of 85 % ethanol. This is one of the great advantages of these vehicles – they are truly flexible in that they do not depend on a certain type of fuel, as many other vehicles powered by alternative sources of energy do.

As ethanol is chemically more reactive towards certain polymers and metals than gasoline, those parts of car engines that have direct contact with the ethanol have to be made from specific ("suitable") materials. This refers to seals, gaskets, tubes, parts of the injection system and the fuel pump. Furthermore, an enlarged carbon canister is necessary to absorb possible increased emissions during a cold start. In addition, for cold starts, heating of the engine block might be necessary.

In Brazil, these cars have gained significant importance in the market. Almost all automotive manufacturers offer FFVs. The share of new cars with FFV technology has reached some 80 %, creating a huge potential for ethanol as a fuel in Brazil (see Fig. 3.26).

In Europe, a workshop agreement was drawn up in 2005 by the European Committee for Standardization (CEN) to address matters of standardization and quality (Table 3.22).

Vehicle manufacturers such as Ford and General Motors (Saab) have also introduced FFVs into the European markets. Sweden is the pioneer of Europe, where currently some 16 000 FFVs have been sold.

E95 Sweden is also home to an example of heavy-duty vehicle application for ethanol. Scania has developed a heavy-duty compression-ignition engine that is

Table 3.22 Requirements and test methods for ethanol E85

Property	Units	Limits Min.	Limits Max.	Test method See 2. Normative references
Research octane number, RON		95.0	–	EN ISO 5164
Motor octane number, MON		85.0	–	EN ISO 5163
Sulfur content	mg/kg	–	20[a]	EN ISO 20846 EN ISO 20884
Oxidation stability	minutes	360	–	EN ISO 7536
Existent gum content (solvent washed)	mg/100mL	–	5	EN ISO 6246
Appearance This shall be determined at ambient temperature or 15°C whichever is higher		Clear and bright, visibly free of suspended or precipitated contaminants		Visual inspection
Higher alcohols (C3–C8)	% (V/V)	–	2.00	EN 1601 EN 13132
Methanol	% (V/V)		1.0	
Ethers (5 or more C atoms)	% (V/V)		5.2	
Phosphorus	mg/L	not detectable		ASTM D 3231
Water content	%V/V		0.3	ASTM E 1064
Inorganic chloride content	mg/L		1	ISO 6227
pHe		6.5	9.0	ASTM D 6423
Copper strip corrosion (3 h at 50 °C)	rating	class 1		EN ISO 2160
Acidity, (as acetic acid CH_3COOH)	% (m/m) (mg/L)	–	0.005 (40)	ASTM D 1613

[a] In line with Directive 2003/17/EC the limits will be aligned with those for petrol. In accordance with this Directive, by no later than 1 January 2005 unleaded petrol with a maximum sulfur content of 10 mg/kg should be available on an appropriately balanced geographical basis within the territory of the Member States. By 1 January 2009 all unleaded petrol marketed in the territory of the Member States should have a maximum sulfur content of 10 mg/kg.

able to run on ethanol. These engines have been utilised in buses that have been in use for more than 15 years by the municipal transportation companies in some cities in Sweden, most of them in Stockholm. The number of ethanol-fuelled buses in the country is around 400.

The main arguments for using ethanol in heavy-duty vehicles are:

- the replacement of diesel provides a stronger reduction of NO_x and particle emissions than the replacement of gasoline by ethanol
- in Europe the demand for diesel fuel is increasing greatly, so the replacement of diesel fuel by ethanol is more in line with the development in the demand for fuels
- to make use of biofuels clearly visible to the consumers

The latest trials conducted for the Swedish Bioalcohol Fuel Foundation (BAFF) resulted in a significant reduction in CO, NO_x and particles emission [105].

As ethanol itself has no lubricant properties, additives have to be applied to ensure lubricity of the fuel [106].

Bioethanol as a Precursor for Ether-Based Octane Enhancers The directives quoted above also allow the blending of up to 15 vol % of ethers with five or more C-atoms. These ethers can be produced by etherification of alcohols with either isobutene or higher olefinic components (see Sections 6.1 and 6.2).

Methanol + isobutene → MTBE (methyl *tert*-butyl ether)
Ethanol + isobutene → ETBE (ethyl *tert*-butyl ether)

Methanol + methylbutene (isoamylene) → TAME (*tert*-amyl methyl ether)
Ethanol + methylbutene (isoamylene) → TAEE (*tert*-amyl ethyl ether)

These ethers, especially MTBE, have been introduced as fuel components since the phase-out of lead-based octane boosters and have gained significant importance in the gasoline pool worldwide (see Chap. 6). The installed European etherification capacity is in the range of 5.7×10^6 t/a [107].

These ethers are less polar than ethanol, therefore form ideal mixtures with gasoline (Fig. 3.27), and have a lower affinity for water. Both properties make them very easily applicable in the fuel sector. In Europe, the production of ETBE is currently the main application for fuel ethanol. For this reason, producers have switched existing MTBE plants to ETBE facilities, with some changes to the technical set-up of the etherification plant.

Ethanol as an H_2-Carrier for Fuel Cell Applications It is important to emphasize that ethanol is not only applicable in internal combustion engines, but also for fuel cells (treated in Chap. 5). There are currently various initiatives aimed at investigating the possibilities for ethanol in high-temperature fuels cells and direct ethanol fuel cells. Ethanol in fuel cells provides the following advantages:

- low toxicity
- easy handling

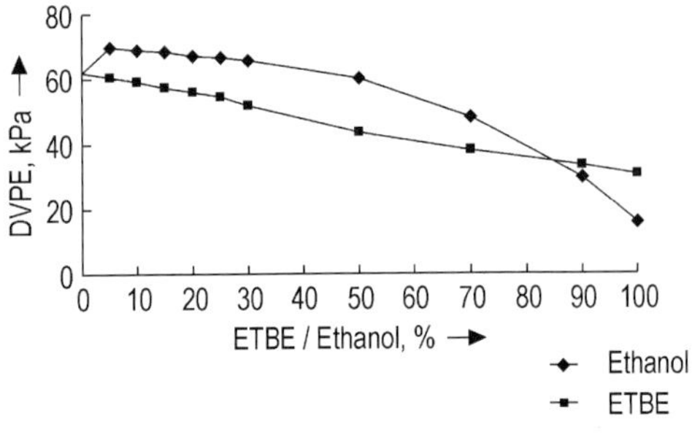

Fig. 3.27 Vapor pressure of ETBE – gasoline and ethanol – gasoline blends (source: Südzucker).

Table 3.23 Energy density of potential H_2-carrier for fuel cells

Product	kWh/kg	kWh/L
H_2	39.4	0.4
Methanol	5.5	4.4
Ethanol	7.4	5.9
Diesel	12.6	9.8
Li-polymer	~0.2	~0.3

- in comparison with methanol higher energy density (see Table 3.23)
- very good availability
- good process security

3.8.4.4 The Bioethanol Fuel Market

The worldwide production of ethanol in 2005, more than 95 % of which was of biogenic origin, was 45.6×10^6 m^3 [108].

The largest producers of ethanol are Brazil and the USA, which together manufacture approximately two thirds of the total worldwide production. The EU contributes only 6.5 % of the worldwide production.

Some 80 % of the ethanol produced worldwide is applied in the fuel sector. Brazil and the USA together produce about 90 % of the fuel ethanol worldwide. The share of the EU is currently only ca. 2 % or 750 000 m^3 per year (see Fig. 3.28).

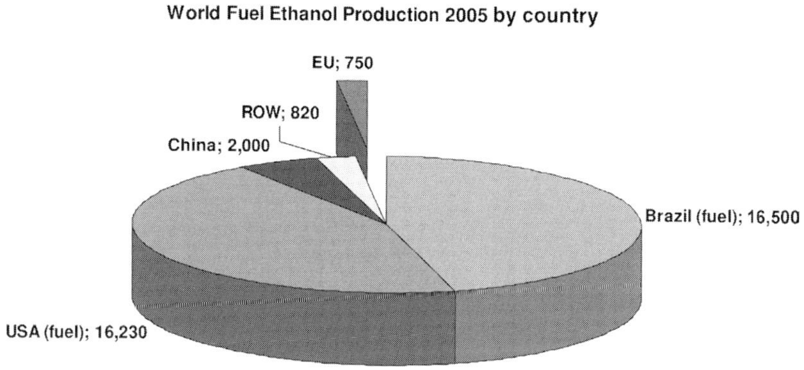

Fig. 3.28 Worldwide fuel ethanol production, 10^6 L; ROW = rest of the world [108].

3.8.4.5 Outlook

Bioethanol is by far the most important biofuel worldwide. This is mainly a result of the proven and established process technology in addition to the wide range of raw materials suitable for its production.

The main forces behind further successful market development are:

- bioethanol assists countries in the development of strategies to become more independent of crude oil
- bioethanol enables countries to achieve their targets for CO_2 emission reduction
- bioethanol helps to create jobs in rural areas

The production and application as a fuel results in a significant reduction in the CO_2 emission, one of the major topics in the discussion on climate change. Bioethanol therefore enables countries to achieve their CO_2 emission reduction targets. The potential for CO_2 emission reduction strongly depends on the raw material, process technology, and the sources for the primary energy supply. There have been numerous publications on this issue [109–111].

Based on the discussions above, F.O. LICHT has projected a steady increase in the worldwide ethanol market and expects a worldwide production of approximately 123×10^6 m^3 by 2020 (Fig. 3.29), of which the EU would contribute some 20×10^6 m^3 [106].

Applying this perspective also to fuel cells and extending the raw material supply with cellulose-based products such as wood or straw, bioethanol has the potential for further strong growth and to become established in the pool of sustainable fuels.

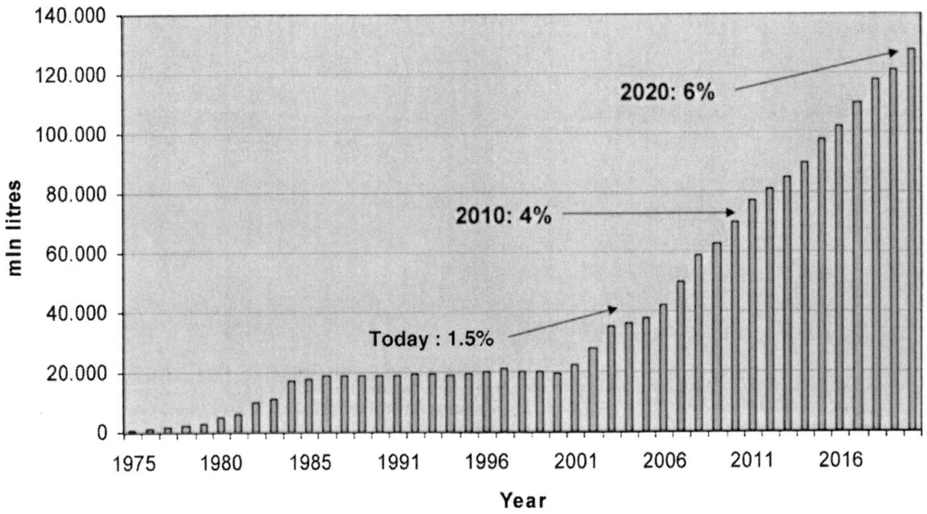

Fig. 3.29 Development of world ethanol production.

3.8.5 Fatty Acid Methyl Esters (FAME)

Klaus Reders, Margret Schmidt

Fatty Acid Methyl Esters (FAME) used as alternative diesel fuels are regarded as one of the "first generation biofuels" [112]. FAME is produced by chemical conversion processes mainly from vegetable oils but it is also possible to use animal based greases. With the currently available technology use of FAME allows the immediate substitution of petroleum-based diesel fuel. Generally only a small proportion of the crop is used, e.g. the oil pressed out of the seeds of the rape plant. In the longer term it is planned to produce second- and third-generation biofuels from "biocrude" using the whole crop by new technologies and in new "biorefineries". The new processes currently under development will increase the availability of biomass for fuel production and will also require less energy for the conversion processes.

In principle vegetable oils could be used directly – without conversion – as alternative fuels in the diesel combustion process. However, some of the properties of vegetable oils cause problems in their handling and application in vehicles and engines. These problems result from their high viscosity, their poor low-temperature fluidity, natural impurities and their limited temperature and storage stability, e.g., their susceptibility to attack by microorganisms.

These disadvantages of vegetable oils can be overcome by esterification with methanol (Fig. 3.30). However this process requires additional (fossil) energy. Esterification, therefore, adds to carbon dioxide emission and is a move away from the desired but practically impossible closed-loop carbon dioxide circle attributed to the cultivation of plants and their use as an energy source (Figs. 3.31 and 3.32). Because of this energy input (fertilizer, harvest, oil production and chemical conversion) the use of neat FAME leads only to about 50% reduction of CO_2 emission compared to petroleum-based diesel fuel. However, advanced biodiesel fuels currently under development will lead to about 90% reduction of CO_2 (Fig. 3.33).

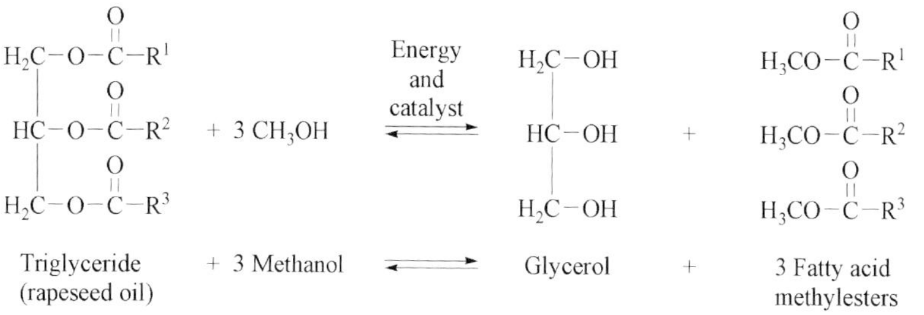

R^1, R^2, R^3:

Various fatty acid rests e.g. —$C_{17}H_{35}$ (Stearic acid)

—$(CH_2)_7$–CH=CH–$(CH_2)_7$–CH_3 (Oleic acid)

—$(CH_2)_7$–CH=CH–CH_2–CH=CH–CH_2–CH=CH–CH_2–CH_3 (Linoleic acid)

Fig. 3.30 Esterification of rapeseed oil with methanol.

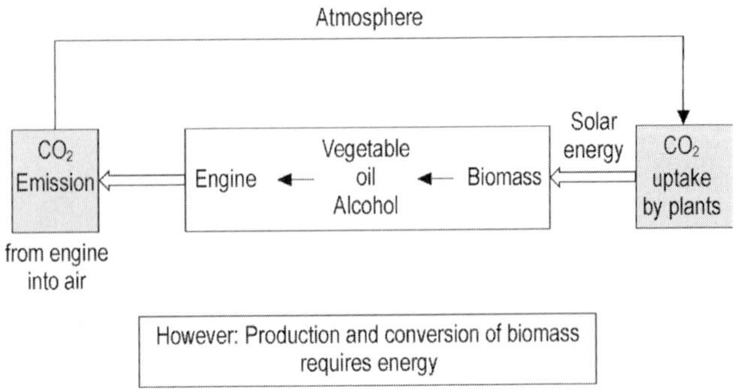

Fig. 3.31 Fuels from biomass, idealized closed CO_2 loop.

The production cost of FAME using currently available technology (usually from rape seed oil) is according to the EU Commission in the range of 500 € /1000 L [113] (spot market price in Germany 2006 was 0.73 €/L). This cost depends on a number of factors, particularly on the price of the raw material (rape seed oil), the size and type of the production plant and the yield and the value of the byproducts (protein, glycerol). In the past with crude oil prices in the range of $ 20 to 30/barrel FAME produced from rape seed oil was not competitive and had to be subsidized. With the currently observed crude oil prices well above 50 $/barrel the production cost of FAME from rape seed oil is now in the same range as that of conventional diesel fuel. When comparing cost of fuels the lower energy content of FAMEs in the range of 6 to 8% on a volume basis has to be taken into account. This applies mainly when neat FAMEs are offered as diesel fuel replacement. However, it has little relevance when FAMEs are used as blending component in conventional diesel fuels at low concentrations.

Esterification brings the critical properties of vegetable oils into a range suitable for diesel engines (specifically the injection systems) and the remaining fuel

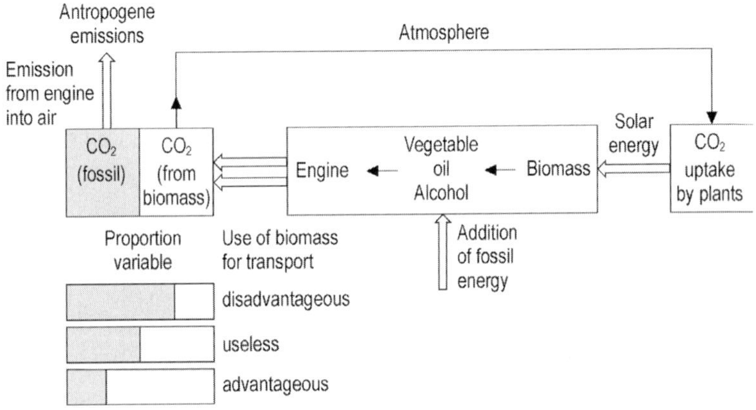

Fig. 3.32 Fuels from biomass; in reality CO_2 emission is larger compared to that consumed by plants, proportion depending on type of plant and biomass conversion process.

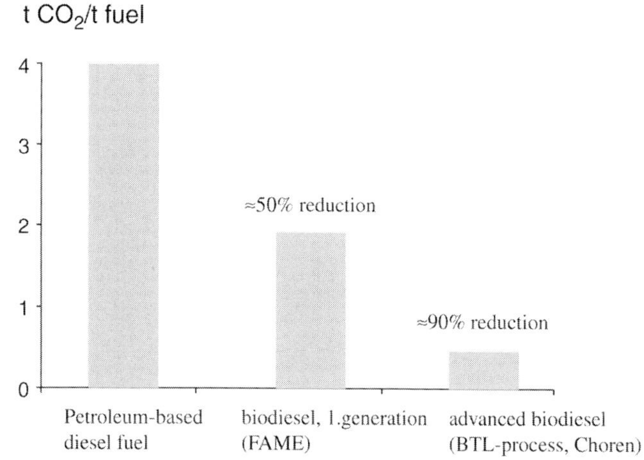

Fig. 3.33 CO_2 reduction with FAME and advanced biodiesel.

systems of the current vehicle population. However, some differences compared with conventional diesel fuels remain (see Table 3.24). The slightly higher densities of FAMEs increase the transported fuel weight to a minor extent. The higher viscosities may lead to larger fuel droplets during the injection process and therefore to somewhat poorer mixture preparation. However, this is compensated by the better ignition quality (higher cetane number) of vegetable oil methyl esters. The lower energy content on a mass basis is to some extent compensated by the higher density. The remaining difference of typically 6–8 % on a volume basis (diesel injection systems meter by volume), leads to an increase in volumetric fuel consumption, a reduction of maximum power output of a diesel engine adjusted for conventional diesel fuel and consequently to a reduction of mainly full throttle particulate and PAH (polyaromatic hydrocarbons) emissions. Higher emissions of aldehydes observed with vegetable oil esters can be controlled by exhaust gas catalysts. In general, fuel systems must be equipped with specific elastomers to allow the use of FAME and care must be taken to keep the contents of glycerol in the fuel at a low level to avoid elastomer incompatibility. The quality requirements of neat FAME diesel fuel are standardized in Europe and other areas (see Table 3.25), taking into account additional specific and relevant properties, for example impurities, such as methanol, glycerol, and phosphorus.

The use of FAME as a blending component for diesel fuel is meanwhile well proven and fuels of this type, mainly rape seed oil esters, are in commercial use with increasing tendency. At the same time, however, neat FAMEs have a very small market share and cover so far only niche markets. The worldwide limited amount of landscape available for the cultivation of plants for fuel production will limit the proportion of biomass fuels. However, new types of "energy plants" and new conversion technologies will allow a diesel biofuel market share of probably 20% or even higher.

Table 3.24 Composition and properties of various FAME's compared to diesel fuel

FAME produced from	No. of double bonds	Composition: Chemical C	Chemical H	Chemical O	Mass C	Mass H	Mass O	Mol weight (g/Mol)	Iodine No. (g/l)	Density at 15 °C (kg/m³)	Lower calorific value (MJ/kg)	Lower calorific value (MJ/l)	Viscosity at 40 °C (mm²/s)	Cetane No.
Coconut oil	0.08	13.4	26.7	2	73.3	12.2	14.5	220.4	10	872	35.3	30.8	2.7	62.7
Palm kernel oil	0.13	13.9	27.7	2	73.7	12.3	14.1	227.6	15	–	35.5	–	–	–
Palm oil	0.59	18.0	34.9	2	76.3	12.4	11.3	283.7	52	872–877	37.0	32.4	4.3–4.5	64.3–70.0
Peanut oil	1.10	19.0	35.0	2	77.0	12.2	10.8	296.3	94	–	37.2	–	–	–
Cotton seed oil	1.20	18.5	34.6	2	76.9	12.1	11.1	289.2	105	–	37.0	–	–	–
Rapeseed oil (rich in erucic acid)	1.25	21.0	39.5	2	77.8	12.3	9.9	324.2	98	–	37.7	–	–	–
Rapeseed oil (low in erucic acid)	1.33	19.0	35.4	2	77.2	12.0	10.8	296.0	114	882	37.2	32.8	4.2	51.0–59.7
Sunflower oil	1.49	18.9	34.8	2	77.2	11.9	10.9	294.1	129	885	37.1	32.8	4.0	61.2
Soybean oil (rich in erucic acid)	1.51	18.8	34.6	2	77.2	11.9	10.9	293.0	131	–	37.1	–	–	56
Linseed oil	2.12	18.9	33.6	2	77.5	11.6	10.9	292.1	183	891	37.0	33.0	3.7	52.5
Typical Diesel fuel	–	1	1.85	0	86.6	13.4	0	120–320	–	830–840	42.7	35.5	2–3.5	51

Quelle: Mercedes-Benz

Table 3.25 FAME-100 specifications worldwide

Standard Date Application*		Europe EN 14214 Oct-03 FAME	Aus ON C1191 Sep-03 FAME	Phi DPNS Draft Feb-02 CME	Brz ANP 255 01-Sep VOME/FAME	USA ASTM D6751-02 Jul-99 S-500	 /01 S-15	Test Method
Density	g/cm³	0.86–0.90	0.86–0.89	–	0.82–0.88	–	–	ISO 3675, ASTM 1298; 4052
Viscosity	mm²/S	3.5–5.0	3.5–5.0	1.9–6.0	2.5–5.5	1.9–6.0	1.9–6.0	ISO 3104. D445
Flash point	°C	>120	>120	>100	>100	>130	>130	pr ISO 3679 CD 3679 D93
CFPP (max)	°C	(1)	No data	0	3	–	–	
Cetane No.		>51	–	>40	>45	>40	>40	ISO 5165, D613
Sulfur	ppm	<10	<50	<500	<0.001%	<500	<15	D5453, pr EN ISO 20846
Sulfated ash	% m/m	<0.02	<0.02	<0.02	<0.02	<0.02	<0.02	ISO 3987, D874
Water	mg/kg	<500	<0.05%	<500		<0.05%	<0.05%	ISO 12937. D 2709
Methanol	% m/m	<0.2	–	–	<0.5	–	–	EN 14110
Monoglyceride	% m/m	<0.8				–	–	EN 14105, D 6584
Diglyceride	% m/m	<0.2				–	–	EN 14105, D6584
Triglyceride	% m/m	<0.2				–	–	EN 14105, D6584
Free glycerol	% m/m	<0.02	–	–	<0.02	<0.02	<0.02	EN 14105, D6584
Total glycerol	% m/m	<0.25	–	–	<0.38	<0.24	<0.24	EN 14105, D6584
Alkaline metals	mg/kg	<5	–			–	–	EN 14108
Phosphorus	mg/kg	<10	<10			–	–	EN 14107, D4951
Ester content	% m/m	>96.5	>96.5			–	–	EN 14103
Carbon residue	% m/m	<0.3	<0.3			–	–	ISO 10370, D4530, 189
Cu corrosion	rating	class 1	Class 3	3	1	Class 3	Class 3	ISO 2160, D130
Ox. St. 110°C	hours	>6	–	–	>6	–	–	EN 14112
Acid value	mg KOH/g	<0.5	<0.8	<0.8	<0.8	<0.8	<0.8	EN 14104, D664
Iodine value	gI/100 g	<120		–		<125	<125	EN 14111
Linolenic acid methyl ester	% m/m	<12				–	–	EN 14103

(1) cold flow properties identical to diesel Fuel, EN590

In Germany a mineral oil tax reduction for diesel fuels containing biofuel components was in place since January 2004 which ended in December 2006. From January 2007 onward a minimum concentration of 5 vol % FAME in conventional diesel fuel is required legally. In several other countries such as France and Italy blending of FAMEs into diesel fuel has also been common practice for several years.

3.9 References

1. O. von Fersen: *Ein Jahrhundert Automobiltechnik*, VDI-Verlag, Düsseldorf 1985.
2. U. Seiffert, P. Walzer: *Automobiltechnik der Zukunft*, VDI-Verlag, Düsseldorf 1989.
3. W. Dabelstein, A. Reglitzky, K. Reders, N. Lucht: *100 Jahre Kraftstoffe für den Strabßnverkehr*, Shell Technischer Dienst, Hamburg 1986.
4. J. S. Meurer, *VDI-Ber.* **714** (1988) 1–30.
5. F. Pischinger, *Verbrennungsmotoren*, RWTH Aachen, 1980.
6. F. Wankel, *Motortech. Z.* **25** (1964) no. 12, 489.
7. F. Huf, *Automobil Rev.* **49** (1961) 23.
8. K. Groth: *Grundzüge des Kolbenmaschinenbaues*, Groth, Hannover 1975.
9. A. Pierburg: *Vergaser für Kfz-Motoren*, 4th ed., VDI-Verlag, Düsseldorf 1970.
10. H. Scholl, *Motortech. Z.* **34** (1973) 4.
11. R. Krebs, B. Stiebels, L. Spiegel, E. Pott: *"FSI Gasoline direct injection Engine for the Volkswagen Lupo"*, Lecture held at the 21. Wiener Motorensymposium, 2000.
12. Bosch: *Kraftfahrtechnisches Handbuch*, 20th ed., VDI-Verlag, Düsseldorf 1987.
13. K. Reders, W. Dabelstein, A. Reglitzky, W. Lange, *Mineralöltechnik* (1989) no. 12, ibid. no. 13, 38.
14. G. Sitkei: *Kraftstoffaufbereitung und Verbrennung im Dieselmotor*, Springer Verlag, Berlin–Heidelberg 1964.
15. M. Gairing, M. Fortnagel, F. Scherer, Lecture held at the Technische Arbeitstagung Hohenheim (1984), Mineralöltechnik (1984) no. 12–13.
16. P. Heinze, *Mineralöltechnik*, (1987) no. 4, 5.
17. H. Waldmann, G. H. Seidel: *Kraft- und Schmierstoffe*, Walter de Gruyter Verlag, Berlin 1979, pp. 986–993.
18. R. L. Allen, A. Reglitzky, *Qualität von Ottokraftstoffen*, Shell Kraftstoff-Symposium, Hamburg Sept. 1985, Shell Technischer Dienst, Hamburg.
19. W. W. Lange, A. Schütze, H. Krumm, K. Reders: *Ottokraftstoff Lexikon*, Shell Technischer Dienst, Deutsche Shell, Hamburg.
20. H. Krumm: *Fortschritte der Fahrzeugtechnik, Stadtauto/Mobilität Ökologie, Ökonomie, Sicherheit*, Vieweg Verlag, Braunschweig 1995.
21. K. Groth: *Brennstoffe für Dieselmotoren heute und morgen*, Expert-Verlag, Ehingen 1989, pp. 195 ff.
22. H. May, M. Regitz, "Studie Dieselkraftstoffe," Forschungsvereinigung Verbrennungskraftmaschinen Heft 430, Frankfurt 1989.
23. *CONCAWE Rep.* 1987, no. 10.
24. K. Weidmann, H. Menrad, K. Reders, R. Hutcheson, SAE-Paper **881 649**, 1988.
25. D. Rickeard, A. Ramsdem, N. Thompson, *SAE-Paper* 890 048, 1989.
26. B. Riediger: *Die Verarbeitung des Erdöles*, Springer Verlag, Berlin–Heidelberg 1971.
27. W. Dabelstein, A. Reglitzky: *Umweltschutz neue Determinante für die Energiepolitik*, Verlag TüV, Rheinland, Köln 1988, pp. 167–221.
28. W. Keim, A. Behr, G. Schmitt: *Grundlagen der Industriellen Chemie*, Verlag Salle + Sauerländer, Frankfurt 1986, p. 133.

29. DEUTSCHE BP: *Das Buch vom Erdöl*, 5th ed., Reuter und Klöckner, Hamburg 1989, pp. 161.
30. F. KIENERT: "Ein Beitrag zur Untersuchung von Reformingkatalysatoren im integralen Rohrreaktor und einem neuen Kreisgasreaktor," Dissertation RWTH Aachen, 1989.
31. Refining Process Handbook, *Hydrocarbon Process.* **63** (1984) 67.
32. G. ZÜRN, K. KOHLHAASE, K. HEDDEN, J. WEITKAMP, *Erdöl Kohle Erdgas Petrochem.* **37** (1984) 62 ff.
33. H. BRUDERRECK, *Erdöl Erdgas* **102** (1986) 463 ff.
34. D. ROHE, *Chem. Ind.* **7** (1986) 570–572.
35. W. DANAHER, L. PALMER, *Fuel* **67** (1988) 1441–1445.
36. M. J. VAN DER BURGT, I. VAN KLINKEN, S. T. SIE, "The Shell Middle Distillate Synthesis Process," Lecture Washington D.C., Nov. 11–13 (1985).
37. J. FALBE, W. HASSERODT: *Katalysatoren, Tenside u. Mineralöladditive,* Thieme Verlag, Stuttgart 1978, pp. 221.
38. INGENIEUR CONSEIL BEICIP, *Fuel Additives in Western Europe,* Nov. 1988.
39. E. BAMBERG, K. REDERS, "Verbesserung von Ottokraftstoffen durch Additive," Lecture held at the Technische Arbeitstagung Hohenheim, Shell Technischer Dienst, 1986.
40. J. LODWICK, *Chem. Rundsch.* **20** (1967) no. 1, 1–3; no. 2, 5–7; no. 3, 8–9.
41. R. TUPA, *SAE-Paper* 872 113, 1987.
42. B. TANIGUCHI, R. PEYLA, G. PARSONS, K. HOEKMANN, D. VOSS, *SAE-Paper* 861 534, 1986.
43. J. BESON, P. YACCARINO, *SAE-Paper* 861 533, 1986.
44. B. BITTING ET AL., *SAE-Paper* 872 117, 1987.
45. F. CARRACIOLO, R. STEBAR, *SAE-Paper* 872 111, 1987.
46. K. STARKE, "Ablagerungen in den Ansaugsystemen von Ottomotoren," Lecture held at the Technische Arbeitstagung, Hohenheim 1977, Shell Technischer Dienst, Hamburg 1977.
47. C. JEWITT, G. BOSTICK, V. KERSEY, *SAE-Paper* 872 114, 1987.
48. J. FABRI, "Chemische Untersuchungen zum Mechanismus und den Einflußparametern der Ablagerungsbildung auf den Einlaßventilen von Ottomotoren," Dissertation RWTH Aachen, 1988.
49. G. LEPPERHOFF, J. SCHOMMERS, O. WEBER, H. LEONHARDT, *SAE-Paper* 872 115, 1987.
50. E. DIMITROFF, A. JOHNSTON, *SAE-Paper* 660 784, 1966.
51. E. DIMITROFF, J. MOFFIT, R. QUILLIAN, *Trans. ASME J. Lubric. Technol.* **91** (1969) no. 7, 406–416.
52. R. TUPA, D. KOEHLER, *SAE-Paper* 861 536, 1986.
53. Coordinating European Council, Tentative Test Method No. CEC F-03-T-81, (1981).
54. K. OWEN (ED.): *Gasoline and Diesel Fuel Additives, Critical Reports on Applied Chemistry,* **vol. 25**, JOHN WILEY AND SONS, New York 1989, pp. 19 ff.
55. H. RATH, K. STARKE, H. VOGEL, "Wirkung optimierter Kraftstoffadditive," Lecture held at the Technische Arbeitstagung, Hohenheim 1988, published.
56. R. LEWIS, H. NEWHALL, *SAE-Paper* 830 938, 1983.
57. J. SCHOMMERS, "Ursachen der Ablagerungsbildung auf den Einlaßventilen von Verbrennungsmotoren," Dissertation RWTH Aachen, 1986.
58. M. GAIRING, *Mineralölrundschau* **34** (1986) no. 11, 209–215.
59. P. POLSS, *Hydrocarbon Process.* **52** (1973) 61–68.
60. J. JOHNSON: *Antioxidants,* Noyes Data Corporation, New York – London 1975.
61. Y. Yonekawa, Y. Nakamura, N. Okamoto, Sekiyu Gakkaishi 24 (1981) no. 2, 85–91.
62. P. MEYERS, O. UYEHARA, R. DEYOUNG, DOE/CS/50 020; NTIS-NO. DE 82 012 576, 1981.
63. J. MOKITA, B. STURGIS, *Proc. World Pet. Congr.* (1955) Sect. 4 F.
64. I. SPILNERS, J. HEDENBURG, *Prepr. Am. Chem. Soc. Div. Pet. Chem.* **26** (1981) no. 2, 632–638.

65. H. Giere, B. Nierhauve, *Mineralöltechnik* **6** (1988) no. 6, p. 5, pp. 14–19.
66. D. Sutton, *SAE-Paper* **860 222,** 1986.
67. R. Olree, D. Lenane, *SAE-Paper 840 108,* 1984.
68. W. W. Lange, H. Krumm, K. Reders: *Dieselkraftstoff Lexikon,* Shell Technischer Dienst, Deutsche Shell, Hamburg.
69. Ting-Man Li, R. Simmons, Presentation at the 21st International Symposium on Combustion, München 1987.
70. X. Montagne, D. Herrier, J. Guibet, *SAE-Paper* 872 118, 1987.
71. D. Sutton, M. Rush, P. Richards, *SAE-Paper* 880 635, 1988.
72. T. Russel, Jugoma Professional Publication, Dubrovnik 1987.
73. T. Coley, F. Rossi, M. Taylor, J. Chandler, *SAE-Paper* 861 524, 1986.
74. C. v. Eberan-Eberhorst, L. Geldern, F. Pischinger, G. Seidel, *Erdöl Kohle Erdgas Petrochem.* **37** (1984) 69 ff.
75. T. Coley, L. Rutishauser, H. Ashton, *J. Inst. Pet.* **52** (1966) no. 510, 173–189.
76. Coordinating European Council Report P-01–74, 1974.
77. J. Zielinski, F. Rossi, A. Stevens, *SAE-Paper* 841 352, 1984.
78. British Technical Council (BTC), Publication BTC/F1/79, 1979.
79. Paramins Corp.: *Fuel Additives Manual-Wax Crystal Modifiers,* Abingdon 1988.
80. R. C. Tupa, C. J. Dorer, *SAE-Paper* 861 179, 1979.
81. H. Bechert, H. Hauschildt, G. Heilmann: "Künftige Dieselkraftstoffqualitäten und ihre Auswirkungen auf das motorische Betriebsverhalten," *Motortech. Z.* **43** (1982) no. 6, 237–241.
82. R. N. Smith (ed.): Papers presented at Microbiology Committee Conference Oct. 1986, Institute of Petroleum, 1986.
83. DGMK-Tagungsbericht 8902, Fachbereichstagung "Umwelt," Köln 1989.
84. Shell Briefing Service: *Liquefied Petroleum Gas,* Shell Int. Pet. Co., London, 1979.
85. Royal Dutch/Shell: *The Petroleum Handbook,* Elsevier, Amsterdam, 1983.
86. J. J. L. Ma, Ching-Shien Wu: *Natural Gas Liquids,* SRI International Report No. 135, Oct 1979.
87. A. F. Williams, W. L. Lom: *Liquefied Petroleum Gases, Guide to Properties, Applications, and Uses,* 2nd edn., Ellis Horwood, Chichester, 1982.
88. R. L. Rowell: (Hi-Cryo Process for Recovering Liquids from Refinery Fuel Gas), *Nat. Pet. Refin. Assoc. Tech. Pap.* (1982.
89. Gas Processors Suppliers Association: *Engineering Data Book,* GPSA, Tulsa, Oklahoma, 1987.
90. Gas Processors Association: *LPG Specifications and Test Methods,* publication 2140–80.
91. World LP Gas Association: *Statistical Review of Global LP Gas,* 2002.
92. A. P. Steynberg, M. E. Dry (eds.): *Fischer-Tropsch Technology, Studies in Surface Science and Catalysis,* **vol. 152,** Elsevier, Amsterdam, 2004.
93. Joint Research Centre (European Union): *Well-to-Wheels Analysis of Future Automotive Fuels and Powertrains in the European Context,* 2004, http://jrc.cec.eu.int/Download/eh.
94. J. J. Marano, J. P. Ciferno: *Life-Cycle Greenhouse-Gas Emissions Inventory For Fischer-Tropsch Fuels,* prepared for U.S. Department of Energy, National Energy Technology Laboratory. Energy and Environmental Solutions, LLC, 2001.
95. European Union: "Directive 2003/30/EC of the European Parliament and the Council of 8 May 2003 on the Promotion of the Use of Biofuels or other Renewable Fuels for Transport".
96. Iogen Corporation: www. iogen.ca.
97. Etek Etanolteknik AB: www.etek.se.
98. P. W. Madson, *Ethanol Distillation: the Fundamentals,* in *The Alcohol Textbook,* ed. K. A. Jacques et al., 4th edn., Nottingham University Press, Thrumpton, 2003.

99. T. SENN, *Ethanolerzeugung und Nutzung*, in *Energie aus Biomasse: Grundlagen, Techniken und Verfahren*, eds. M. KALTSCHMITT, H. HARTMANN, Springer-Verlag, Berlin, Heidelberg, New York, 2001.

100. ASTM International, Designation: D 4806 – 01a Standard Specification for Denatured Fuel Ethanol for Blending with Gasolines for Use as Automotive Spark-Ignition Engine Fuel.

101. CEN TC19 / WG 21 Ethanol Task Force Report to WG 21, April 2004.

102. D. WALTHER, DGMK – Projekt 645: Laboruntersuchungsprogramm zur Ethanolbeimischung zu Ottokraftstoffen, Schwedt, 2005.

103. O2Diesel Inc. www.o2diesel.com.

104. European Committee for Standardization (CEN), Automotive Fuels – Ethanol E85 – Requirements and Test Methods, 2005.

105. K. E. EGEBÄCK, *A Clean Ethanol Fuelled Compression Ignition Bus Engine*, Bioalcohol Fuel Foundation Report, Stockholm 2004.

106. Svensk Etanolkemi AB, www.sekab.se.

107. European Fuel Oxygenates Association, www.efoa.org.

108. C. BERG, *2005 – the year when everything went right*, FO Licht conference, Amsterdam, November 2005

109. M. QUIRIN, et al., *CO_2-neutraleWege zukänftiger Mobilität durch Biokraftstoffe – Eine Bestandsaufnahme*, Heidelberg, May 2004.

110. R. IGELSPACHER, *Ganzheitliche Systemanalyse zur Erzeugung und Anwendung von Bioethanol im Verkehrssektor*, Munich, July 2003.

111. European Commission Joint Research Centre, *Well- to Wheels Analysis of Future Automotive Fuels and Powertrains in the European Context*, Version 2a, December 2005.

112. Bio fuels in the European Union, A vision for 2030 and beyond. Final draft report of the bio fuels advisory councel.

113. Commission of the European Communities, COM (2001) 547, Communication from the Commission to the European Parliament, the Council, the economic and social committee of the regions on alternative fuels for road transportation and on a set of measures to promote the use of biofuels.

4
Hydrogen

Harald Klein, Harald Ranke

4.1
History

Hydrogen was first separated and identified in the second half of the 17th century by BOYLE. He produced it from diluted sulfuric acid and iron. In 1766 CAVENDISH published precise values for the specific weight and density. LAVOISIER demonstrated the splitting of water in 1785 and gave the "inflammable air" the name hydrogen. NICHOLSON and CARLYLE, in 1800, produced hydrogen and oxygen by electrolysis. Then in 1898 DEWAR liquefied hydrogen using the Linde process.

In transportation the buoyancy of hydrogen quickly led to the use of this gas in balloons (CHARLES, 1783), but its use for aeronautics was ended by the Hindenburg disaster at Lakehurst in 1937. The potential of hydrogen as an aviation fuel was re-established 1957 with the lift-off of a hydrogen-powered B-57 twin-engine jet bomber. From 1963 onwards, liquid hydrogen – liquid oxygen propelled rockets were launched. For the Apollo moon flights 12×10^6 L of liquid hydrogen were necessary to fill the tanks of the Saturn carrier rockets.

CECIL had already carried out experiments using hydrogen as an engine fuel in 1820. These experiments were followed by investigations in car engines up until the Second World War and partly afterwards (RICARDO, BURSTAL, ERREN). Various transport systems using hydrogen were tested, e.g., in the United States by BILLINGS or in the Federal Republic of Germany (Daimler Benz, BMW) in the 1960s and 1970s. In 1966 CRAIG MARKS built the world's first fuel cell-driven car, the General Motors Corp. Electrovan. Mounted on a modified Chevrolet Corvair, it contained two pressurized hydrogen tanks, generated 32 kW of continuous power and traveled 150 miles between refills [1].

Handbook of Fuels: Energy Sources for Transportation. Edited by Barbara Elvers

ISBN: 978-3-527-30740-1

4.2 Properties

Hydrogen (Greek: *hydro*=water and *genium*=to give birth) (symbol H), A_r 1.00797, atomic number 1, is the first element in the Periodic Table, having the electron configuration $1s^1$. Three isotopes with A_r 1, 2 and 3 are known; the isotope with A_r 3 is unstable. The differences in the relative isotopic masses are large so that physical and kinetic properties can differ considerably. The isotope with the relative mass 2 is named deuterium (symbol D), that with the mass 3 tritium (symbol T). Because of the expected increasing significance of hydrogen as a universal chemical and as an energy carrier, its physical and thermodynamic properties have been investigated extensively. The compilation of information given here is only an excerpt; comprehensive data can be found in various summaries and other literature sources, e.g., in [2–8].

4.2.1 Physical Properties

***Ortho-* and *Para*-Hydrogen** The atomic nuclei of hydrogen and its isotopes possess a spin, the value of which is 1/2 for H and T, and 1 for D. In the diatomic molecules both atomic nuclei can either rotate in the same or in opposing directions, i.e., parallel or anti-parallel. The molecules H_2, D_2 and T_2 can therefore exist in the form of two isomers with different nuclear spins, namely in the *ortho*-form (parallel nuclear spin) and in the *para*-form (anti-parallel nuclear spin); the two forms are interconvertible and in equilibrium with each other. The following forms of H_2 can be distinguished:

- o-H_2 — *ortho*-hydrogen
- p-H_2 — *para*-hydrogen
- e-H_2 — equilibrium hydrogen
- n-H_2 — normal hydrogen

e-H_2, e-D_2 and e-T_2 all correspond to mixtures of the equilibrium concentration at definite temperatures. n-H_2 (n-D_2, n-T_2) is a mixture of the equilibrium concentration at ambient temperature, i.e., 25 mol % p-H_2 or p-T_2 and 75 % o-H_2 or o-T_2. n-D_2 contains 33.33 % p-D_2 and 66.67 % o-D_2.

The conversions o-H_2 to p-H_2 and o-T_2 to p-T_2 are exothermic; both equilibrium composition and heat of conversion are a function of temperature, but independent of pressure, at low temperatures the *para*-form is prevalent.

Thermodynamic Properties The differences in the masses of the isotopes are relatively large, so that thermodynamic properties differ considerably. Small differences are also found among the spin – spin isomers so that for certain purposes attention must be paid to the actual form present.

Hydrogen gas is colorless, non-poisonous, odorless and tasteless. At 0 °C and 101.3 kPa, 1 m^3 of H_2 has a mass of 0.08988 kg. The most important properties of n-H_2 and p-H_2 are given in Tables 4.1 and 4.2. Contrary to most other gases, the inversion temperature of hydrogen lies below the ambient temperature. Liquid hydrogen is a colorless, very mobile liquid with low viscosity and surface tension. Some properties are also given in Tables 4.1 and 4.2.

Table 4.1 Selection of physical properties of n-hydrogen and p-hydrogen at fixed points

	p-Hydrogen	n-Hydrogen
Triple point		
Temperature, K	13.803	13.957
Pressure, kPa	7.04	7.2
Density (solid), kg/m^3	86.48	86.71
Density (liquid), kg/m^3	77.03	77.21
Density (vapor), kg/m^3	0.126	0.130
Boiling point (101.3 kPa), K	20.268	20.39
Heat of vaporization, J/mol	898.30	899.1
Liquid phase		
Density, kg/m^3	70.78	70.96
Specific heat capacity C_p, J/mol/ K	19.70	19.7
Enthalpy,* J/mol	–516.6	548.3
Entropy, J/mol/K	16.08	34.92
Viscosity, mPa s	13.2×10^{-3}	13.3×10^{-3}
Thermal conductivity, W/m/ K	98.92×10^{-3}	100×10^{-3}
Vapor phase		
Density, kg/m^3	1.338	1.331
Specific heat capacity C_p, J/mol/ K	24.49	24.60
Enthalpy,[a] J/mol	381.61	1447.4
Entropy, J/mol/ K	60.41	78.94
Thermal conductivity, W/m/K	16.94×10^{-3}	16.5×10^{-3}
Critical point		
Temperature, K	32.976	33.19
Pressure, MPa	1.29	1.325
Density, kg/m^3	31.43	30.12
Properties at STP (273.15 K, 101.3 kPa)		
Density, kg/m^3	0.0899	0.0899
Specific heat capacity C_p, J/mol/ K	30.35	28.59
Viscosity, mPa s	8.34×10^{-3}	8.34×10^{-3}
Thermal conductivity, W/m/K	182.6×10^{-3}	173.9×10^{-3}
Compressibility factor	1.0005	1.00042

[a] The reference state for enthalpy is zero for the ideal gas at 0 K.

Table 4.2 Vapor pressure and density of p-hydrogen at low temperatures

Temperature, K	Vapor pressure, kPa	Density, kg/m^3		
		• S	• L	• G
1	11.0×10^{-37}	89.024		
5	4.76×10^{-3}	88.965		
10	2.556×10^{2}	88.136		0.006
12	1.837×10^{3}	87.532		0.037
13.803[a]	7.0	86.503	77.019	0.126
20	93.5		71.086	1.247
20.268[b]	101.3		70.779	1.338
30	822.5		53.930	10.887
32.976[c]	1.293×10^{3}			31.43

[a] Triple point.
[b] 101.3 kPa.
[c] Critical point.

Solubility of Hydrogen in Metals Metals, with inner electron levels that are not completely occupied, have a particularly high solubility potential for hydrogen. Hydrogen is dissolved in metals not in the molecular but in the atomic form, whereby non-stoichiometric (alloy-type) or stoichiometric (partly hydride-type) compounds may be formed.

The solubility of hydrogen in metals at low hydrogen concentrations is proportional to the root of the hydrogen partial pressure (Sievert's law). The hydrogen solubility decreases with increasing temperature. Hydrides can be decomposed thermally.

Palladium and palladium – silver alloys with a silver content < 40 wt % have a particularly high solubility potential for hydrogen. The hydrogen atoms give their electrons to the electron gas of the palladium metal and occupy interstitial sites of the metal crystal structure. Palladium can absorb up to 2800 times its own volume of hydrogen.

Impurities in the metal and in the hydrogen can lower the solubility or cancel it completely. In particular, carbon monoxide, hydrogen sulfide and oxygen are poisonous in this respect.

The absorption of hydrogen in metals and metal alloys has become important for the storage of hydrogen in hydride storage systems. An overview is given in [9], [10].

4.2.2 Chemical Properties

Oxygen reacts with hydrogen according to

$$H_2 + {}^1/_2O_2 \rightarrow H_2O \qquad \Delta H^0_{298} = -286.02 \text{ kJ mol}^{-1} \qquad (1)$$

Above 550 °C the reaction occurs with flame propagation, explosion or detonation (oxyhydrogen reaction). The flame temperature is limited by the thermal dissociation of water vapor and attains a maximum of 2700 °C.

Atomic hydrogen is much more reactive than the molecular form. The dissociation reaction

$$H_2 \rightarrow 2H \qquad \Delta H^0_{298} = 432.2 \text{ kJ/mol} \qquad (2)$$

is highly endothermic. Atomic hydrogen can be absorbed in the metallic structure of some elements of Groups 8 – 10 (Fe, Co, Ni) of the Periodic system. The suitability of many catalysts for reactions involving hydrogen is based on the dissociation and solubility of hydrogen in the atomic form.

Further chemical properties can be found in [2].

4.2.3 Safety Aspects

When given sufficient thermal activation (ignition energy), hydrogen reacts more or less violently with oxidizing agents such as oxygen (air), fluorine or chlorine and N_2O. Depending on the conditions, combustion, deflagration or detonation may occur.

Table 4.3 Combustion and explosion properties of hydrogen, methane, propane and gasoline

Property	Hydrogen	Methane	Propane	Gasoline
Density of gas at standard conditions, kg/m^3 (STP)	0.084	0.65	2.42	4.4 [a]
Heat of vaporization, J/g	445.6	509.9		250–400
Lower heating value, kJ/g	119.93	50.02	46.35	44.5
Higher heating value, kJ/g	141.8	55.3	50.41	48
Thermal conductivity of gas at standard conditions, mW/cm/ K	1.897	0.33	0.18	0.112
Diffusion coefficient in air at standard conditions, cm^2/s	0.61	0.16	0.12	0.05
Flammability limits in air, vol %	4.0–75	5.3–15	2.1–9.5	1–7.6
Detonability limits in air, vol %	18.3–59	6.3–13.5	3.4–35	1.1–3.3
Limiting oxygen index, vol %	5	12.1		11.6 [b]
Stoichiometric composition in air, vol %	29.53	9.48	4.03	1.76
Minimum energy for ignition in air, mJ	0.02	0.29	0.26	0.24
Autoignition temperature, K	858	813	760	500–744
Flame temperature in air, K	2318	2148	2385	2470
Maximum burning velocity in air at standard conditions, m/s	3.46	0.45	0.47	1.76
Detonation velocity in air at standard conditions, km/s	1.48–2.15	1.4–1.64	1.85	1.4–1.7 [c]
Energy[d] of explosion, mass related, gTNT/g	24	11	10	10
Energy [d] of explosion, volume related, $gTNT/m^3$ (STP)	2.02	7.03	20.5	44.2

[a] 100kPa and 15.5 °C.
[b] Average value for a mixture of $C_1 - C_4$ and higher hydrocarbons including benzene.
[c] Based on the properties of *n*-pentane and benzene.
[d] Theoretical explosive yields.

Ignition and detonation properties of hydrogen – air mixtures are particularly important from the safety aspect [11]. The flammability limits (i.e., the difference between the rich and lean limit concentration) are exceptionally wide for hydrogen. A comparison of safety-relevant thermo-physical and combustion properties of hydrogen with those of methane, propane and gasoline is shown in Table 4.3 [12]. The flammability limits are affected by temperature as shown in Fig. 4.1, so that a preheated mixture has considerably wider limits for coherent flames [13]. An increase in pressure has only a small effect. Water vapor has a strongly inhibiting influence on the oxyhydrogen reaction. The flammability limits of hydrogen in pure oxygen are 4.65–93.9 vol % hydrogen.

Hydrogen flames are almost invisible in daylight. Smoke inhalation, one of the major causes of injury at fire hazards and, therefore, a main parameter of fire damage, is considered less serious in the case of hydrogen because the sole combustion product is water vapor. When considering the possible use of hydrogen in

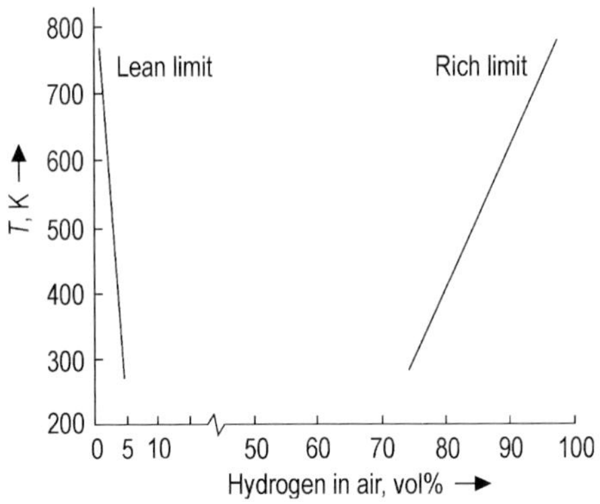

Fig. 4.1 Effect of temperature on flammability limits of hydrogen in air (pressure 100 kPa).

a future hydrogen energy concept, no major safety problems are anticipated. Hydrogen has been handled successfully for decades in the process industries. Handling as a liquid is also routine. In comparison with other inflammable gases or liquids, the safety risk of hydrogen as a fuel is not significantly different.

Ignition and detonation performance based on ignition range, ignition energy, flames propagation and transition from deflagration to detonation is discussed in detail in [2, Section 7.7].

4.2.4 Toxicology

Hydrogen does not inflict any physiological effects. It is nonpoisonous. Inhalation of the gas leads to sleepiness and a high-pitched voice. A danger of asphyxiation exists if the oxygen content sinks below 18 vol % because of hydrogen accumulation in the air. Direct skin contact with cold gaseous or liquid hydrogen leads to numbness and a whitish coloring of the skin and to frostbite. The risk is higher than, for example, with liquid nitrogen because of the greater temperature differences and the higher thermal conductivity of hydrogen.

4.3 Occurrence

Hydrogen is the most abundant element in the universe; 70 – 80 wt % of the observable universe consists of hydrogen, the remainder is mostly helium. More than 50 wt % of the sun consists of hydrogen. Of the elements in the earth's crust, hydrogen takes the ninth place with 0.88 wt %. Based on the atomic concentration of 15.4 atom % it is the third most common element after oxygen and silicon. In the free state, hydrogen can be found in some volcanic gases (up to ca. 30 vol %), in natural gas (from traces up to 40 vol %), small amounts are trapped in some minerals, rocks and meteorites. In the upper atmosphere, the troposphere and the stratosphere, hydrogen is present only in trace amounts of ca. 0.5 ppm (mL/m^3). Because of photolytic water cleavage the hydrogen concentration increases with increasing height. The exosphere (2000 – 20 000 km) contains atomic hydrogen.

Chemically bound hydrogen is present all over the earth: as a part of the earth's water mass (11.2 wt %), as part of organic matter (including fossil substances, such as natural gas oil and coal). About 10 wt % of the human body consists of hydrogen.

4.4 Production

Hydrocarbons are the main source for the production of hydrogen on an industrial scale. An even larger source, however, is water. When hydrogen is used as fuel, water is produced again. Thus, a cycle comparable to the carbon cycle (CO_2 – biomass – CO_2), but much simpler, is involved, so that water can be regarded as an unlimited source of hydrogen. With the help of an external energy source, hydrogen can be produced from either of these sources alone or from a mixture of both.

Figure 4.2 shows the production of hydrogen using various sources of energy. The many production routes for hydrogen, however, are by no means of equal economic importance. Most of the hydrogen for industrial uses is produced from natural gas and oil, either as a main product or as a byproduct, a process involving a chemical conversion. A smaller percentage is produced electrolytically or occurs as a byproduct of electrochemical processes. Here the energy applied is in the form of electricity, which itself may be produced using hydroelectric, nuclear or, even more predominantly, the energy stored in fossil fuels.

Thermal, thermochemical, biochemical and photochemical processes have so far not found industrial applications. The preference for producing hydrogen from hydrocarbons is due to the fact that the energy demand using hydrocarbons is much lower than that for the electrolysis of water (see Table 4.4).

Considering the strong correlation between energy prices and the prices for hydrocarbons, it is unlikely that the high percentage of the hydrocarbon processes for producing industrial hydrogen will change substantially in the near future. This last statement does not take into consideration environmental pollution by emission of carbon dioxide, sulfur compounds, nitrogen oxides and heat, so that this (environmental) aspect may one day reverse the situation.

4.4.1 Production from Coal and Hydrocarbons

The yield of hydrogen from hydrocarbons is summarized in Fig. 4.3. Hydrocarbons and coal are characterized by their H:C molar ratio. Chemically pure compounds such as methane and its homologues can be found on line (A) and give maximum hydrogen yield. Raw materials such as hard coal (a), lignite (b) and natural gas (h)

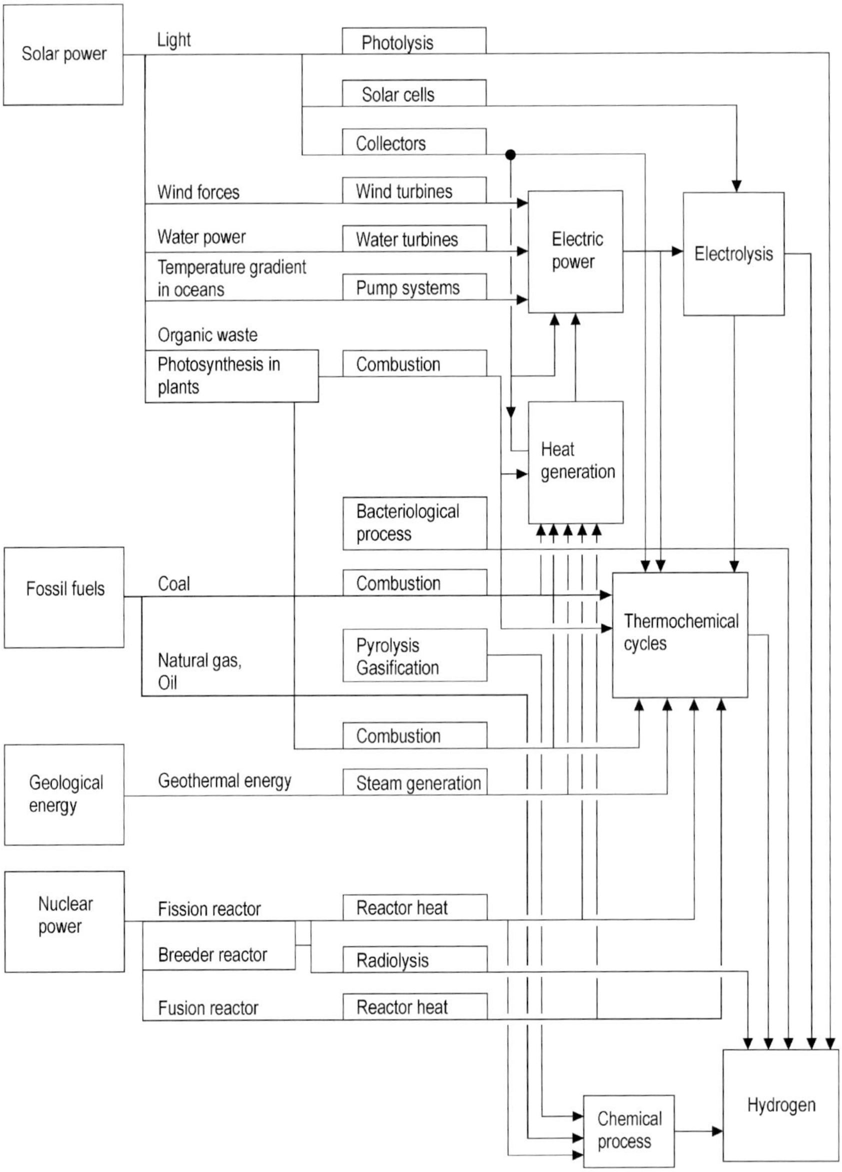

Fig. 4.2 Primary energy, raw materials, production methods, paths of production, transformation, and generation of and for hydrogen, adapted from [14].

give a lower hydrogen yield because they contain inert components (sulfur, nitrogen, oxygen or minerals, etc.). Other important sources for the production of hydrogen such as vacuum residue (c), naphtha (d) or hydrogen-rich off-gases, e.g., coke oven gas (j), are also shown in Fig. 4.3.

Table 4.4 Theoretical energy consumption for the production of hydrogen from various hydrocarbons, coal and water

Parameter	Natural gas as CH_4	LPG as $CH_{2.6}$	Naphtha as $CH_{2.2}$	Heavy oil as $CH_{1.4}$	Coal as $CH_{0.7}$	Water H_2O
Higher heating value [a]						
kJ/kg	50 050	46 130	44 300	41 300	38 100	
kJ/mol	890	2480				
Byproducts	CO_2	CO_2	CO_2	CO_2/S	CO_2/S	O_2
Specific production of byproducts, kmol/kmol H_2	0.25	0.30	0.32	0.37/0.003–0.015	0.43/0.002–0.01	0.5
Theoretical energy consumption,[a] kJ/kmol H_2	41 280	37 500	38 350	50 300	57 150	242 000

[a] Calculated from the heat of formation, oxidizing carbon completely to carbon dioxide (at 298 K and 0.1 MPa) [15], [16].

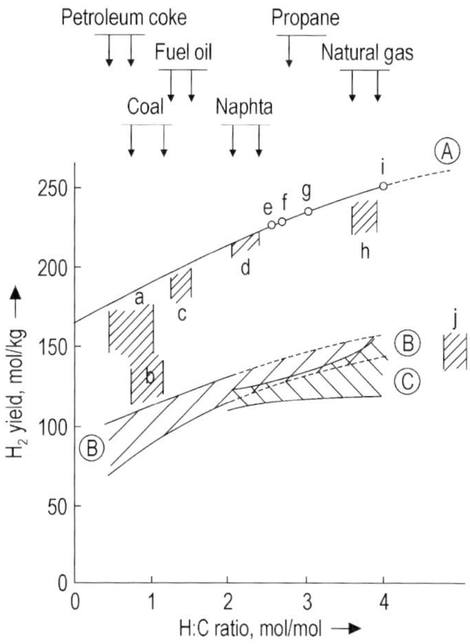

Fig. 4.3 Hydrogen production from hydrocarbons and coal A) Maximum theoretical yield from (a) hard coal, (b) bituminous coal, (c) vacuum residue, (d) naphtha, (e) butane, (f) propane, (g) ethane, (h) natural gas, (i) methane, (j) coke oven gas B) Commercial production by steam-reforming, including fuel, no credit for export steam. Lower curve: small plants, ca. 500 m^3(STP)/h Upper curve: large plants C) Commercial production by partial oxidation large plants only. Upper curve: without oxygen production.

Partial Oxidation (PO) A brief description of the partial oxidation of hydrocarbons in general is given below. The Texaco and Shell partial oxidation processes are applied commercially for the production of hydrogen [17–20].

Heavy residues from petrochemical processes are the preferred feedstocks for the production of hydrogen and carbon monoxide rich gases. Oil fractions with high sulfur and/or high heavy metal contents are difficult to handle in upgrading processes such as hydrogenation or coking. For environmental reasons they should not be used as fuels. In principle, PO processes from all hydrocarbons or even coal can produce hydrogen.

Gaseous hydrocarbons are also occasionally converted into hydrogen and carbon monoxide in PO processes. This takes place if the gases cannot be processed by catalytic steam reforming (see below) because of high concentrations of certain impurities such as olefins and H_2S or if the process must handle various feedstocks from natural gas up to fuel oils. The overall reactions during partial oxidation are as follows (heat of reaction, see Table 4.5):

Oxidation reactions

$$C_mH_nS_o + m/2O_2 \rightarrow mCO + (n/2 - o)\ H_2 + oH_2S \text{ (exothermic)} \tag{3}$$

$$2C + O_2 \rightarrow 2CO \text{ (exothermic)} \tag{4}$$

Crack reaction

$$C_mH_nS_o \rightarrow (n/4 - o/2)CH_4 + (m - n/4 + o/2)C + oH_2S \text{ (endothermic)} \tag{5}$$

Water gas reaction

$$C + H_2O \rightarrow CO + H_2 \text{ (endothermic)} \tag{6}$$

Shift reaction

$$CO + H_2O \rightarrow H_2 + CO_2 \text{ (exothermic)} \tag{7}$$

Methane reforming

$$CH_4 + H_2O \rightleftharpoons CO + 3H_2 \text{ (endothermic)} \tag{8}$$

The reactions of the hydrocarbon feedstock with sub-stoichiometric amounts of pure oxygen (oxygen content >95 %) follow Eq. 3. Other reactions, which take place simultaneously or immediately following, are the water gas reaction (Eq. 6), the CO shift (Eq. 7) and the methane formation or decomposition reaction (Eq. 8).

The reactions take place primarily in the flame and in the empty brick-lined reactor. The mean residence time is usually <5 s, the temperature 1250–1500 °C.

Table 4.5 Equations and reaction enthalpies for the most frequent reactions occurring during catalytic steam reforming

	ΔH at 101.3 kPa, 298 K, kJ/mol	Technical reaction conditions			
		Start of reaction at t, °C	End of reaction at t, °C	Reaction, pressure MPa	Heat of reaction ΔH, kJ/mol
Reforming reactions					
$CH_4 + H_2O \rightarrow CO + 3H_2$	206.36	500	870	2.0	234.66
$CH_4 + 2H_2O \rightarrow CO_2 + 4H_2$	165.13				
$C_2H_6 + H_2O \rightarrow CO + CH_4 + 2H_2$	141.22	500	870	2.0	166.56
$C_2H_6 + 2H_2O \rightarrow 2CO + 5H_2$	347.50	500	870	2.0	131.68
$C_3H_8 + H_2O \rightarrow C_2H_6 + CO + 2H_2$	150.68	500	870	2.0	179.24
$C_3H_8 + 2H_2O \rightarrow CH_4 + 2CO + 4H_2$	291.99	500	870	2.0	335.10
$C_3H_8 + 3H_2O \rightarrow 3CO + 7H_2$	498.36	500	870	2.0	563.64
Shift reaction					
$CO + H_2O \rightarrow CO_2 + H_2$	−41.16	220	250	2.0	−37.19
		350	420	2.0	−36.48
Oxidation reactions					
$C + O_2 \rightarrow CO_2$	−196.89				
$2CO + O_2 \rightarrow 2CO_2$	−566.37	800	1000	2.0	−542.58
$2H_2 + O_2 \rightarrow 2H_2O$	−484.17	800	1000	2.0	−480.06
$CH_4 + 2O_2 \rightarrow 2H_2O + CO_2$	−803.04	300	1400	0.1	−646.83
$2C_2H_6 + 7O_2 \rightarrow 6H_2O + 4CO_2$	−2857.6	300	1400	0.1	−2329.2
$C_3H_8 + 5O_2 \rightarrow 4H_2O + 3CO_2$	−2045.4	300	1400	0.1	−1673.9
Soot formation reactions					
$CH_4 \rightarrow C + 2H_2$	74.9				
$2CO \rightarrow C + CO_2$	−172.54				

Pressure is often adjusted to accommodate subsequent processes, 3 – 12 MPa being the usual range.

The subsequent formation of hydrogen by the water – gas shift reaction (Eq. 7), is of equal importance for hydrogen production as is the immediate formation of hydrogen (Eq. 3). Carbon monoxide is thereby shifted to hydrogen to the highest degree possible. The shift reaction occurs simultaneously with the oxidation – gasification reactions at temperatures >1200 °C without the presence of catalysts. Higher hydrogen yields, however, can be achieved by attaining equilibrium at 200 – 500 °C in the presence of steam and suitable catalysts.

Catalytic Reforming In catalytic reforming, hydrocarbons react with steam in an endothermic reaction to form carbon monoxide and hydrogen (Eq. 9). The most important feedstock for the catalytic steam reforming process is natural gas. Other feedstocks are associated gas, propane, butane, liquefied petroleum gas and some naphtha fractions. The choice is usually made on the availability and the price of the raw material. Preference is given to lighter feedstock, if the price on a heating value basis is comparable. The reason for this is the better operability of plants using lighter feedstocks as well as the easier pretreatment operations, and thus the lower investment costs for comparable plants.

The chemical process is described by

$$C_nH_m + nH_2O \rightarrow nCO + (n + m/2)H_2 \qquad (9)$$

The shift reaction, where carbon monoxide, formed according to Eq. 9, reacts with excess water, occurs concurrently to this reaction.

Indirect heating provides the heat of reaction of the overall endothermic process. In *autothermal reactors* (e.g., secondary reformers) the oxidation reaction (see Table 4.5) supplies the necessary energy.

The equilibrium of the hydrocarbon steam reaction (Eq. 9) and the water – gas shift reaction (Eq. 7) dictates the conditions that must be met to achieve maximum hydrogen yields:

- high temperature at the end of the reactor
- high excess of steam
- low pressure

These conditions can best be realized if the endothermic reforming reaction occurs in a fired tubular reactor at high temperature (800–900 °C) and low pressure (1.5–3.0 MPa). The amount of steam added to the hydrocarbon feedstock is quantified by the molar steam-to-carbon ratio (S/C=ratio of S moles of H_2O to C moles of carbon contained in the hydrocarbon feedstock). The S/C ratio has values from 2.5 to 6, depending on the feedstock and is subject to process optimization considerations.

The next step is the shift reaction at ca. 340–450 °C (high-temperature shift, HTS). In certain cases this is followed by a second shift reaction at 200–250 °C (low-temperature shift, LTS).

In commercial plants supported nickel catalysts catalyze steam reforming. The catalyst contains 15–25 wt % nickel oxides on a mineral carrier. Carrier materials are α-alumina, aluminosilicates, cement and magnesia. Before start-up, nickel oxide must be reduced to metallic nickel. This is done preferably with hydrogen but also with natural gas or even with the feed gas itself.

The catalysts are usually in the form of rings (e.g., outer diameter 16 mm, height 16 mm and inner diameter 8 mm). Other forms, such as saddles, stars and spoked wheels are also commercially available.

The main catalyst poison in steam reforming plants is sulfur, which is present in most feedstocks. Concentrations as low as 0.1 ppm form a deactivating layer on the catalyst [21]. To some extent activity loss of a poisoned catalyst can be offset by raising the reaction temperature. Additional catalyst poisons are chlorine, other halogen compounds and heavy metals, especially lead, arsenic and vanadium.

Carbon Deposits Formation of carbon deposits occurs predominantly because of the decomposition of methane, according to the data in Table 4.5.

$$CH_4 \rightarrow C + 2H_2 \qquad (10)$$

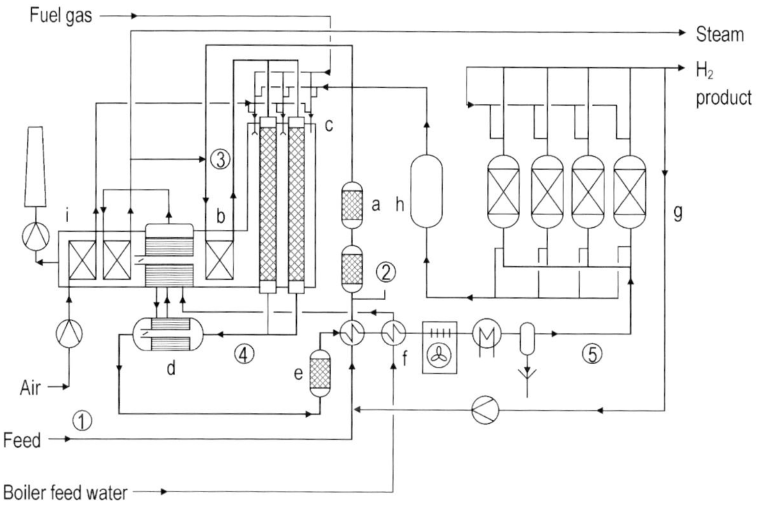

Fig. 4.4 Hydrogen production by steam reforming, gas purification by pressure swing adsorption (design: German Linde) a) Desulfurization; b) Feed preheater/superheater; c) Reformer; d) Waste-heat boiler; e) CO shift reactor (HT shift); f) Cooling of raw gas; g) Pressure swing adsorption; h) Off-gas puffer for fuel; i) Convection zone with steam production, steam superheating, and air preheating.

Other reactions leading to soot formation or carbon-coke deposits are cracking of higher hydrocarbons – especially aromatic compounds and olefins – and the Boudouard reaction in Table 4.5.

$$2CO \rightarrow CO_2 + C \qquad (11)$$

The Boudouard reaction is favored by low temperature and high pressure, typically found at the inlet portion of radiant tubes in a steam reformer.

Steam Reforming Plant Fig. 4.4 shows a simplified process flow diagram of a typical hydrogen plant. As in the majority of such plants, the feedstock is natural gas. For the production of pure hydrogen, the following process stages are necessary.

Feed Preheating and Desulfurization. Usually, natural gas contains only small amounts of sulfur compounds, predominantly in the form of hydrogen sulfide. Because of the high sulfur sensitivity of the catalyst in the reformer (and in the LT shift reactor, if installed) hydrogen sulfide and other sulfur compounds must be removed on a zinc oxide bed. If higher organic sulfur compounds (mercaptans, thiophene) or carbonyl sulfide in concentrations in the ppm-range are present, the zinc oxide bed alone is not sufficient and the sulfur compounds are converted into hydrogen sulfide in a hydrogenation stage. For hydrogenation and desulfurization the inlet gas is preheated to 350–400 °C. The gas leaves the purification section with a sulfur content of <1 ppm at essentially the same temperature.

Steam Addition, Reforming and Cooling. Steam is added to the inlet feed to give a ratio of 2.8–3.5 moles of steam per mole of carbon. This mixture can be further preheated in the convection section to 520–550 °C without cracking occurring. Reforming takes place in the reformer tube, the outlet temperature being 850 °C. After the reaction the gas is cooled quickly, thereby producing steam in the process gas boiler.

High-Temperature Shift Conversion. The gas cooled down to 330 °C in the process gas boiler still contains a significant amount of carbon monoxide, which can be converted into hydrogen by the water – gas shift reaction (Eq. 7, Table 4.5). An iron oxide based catalyst is used for this. The shift reaction is usually carried out in a one-stage adiabatic reactor. As the reaction is exothermic, the process gas temperature rises to 400 – 450 °C, so the conversion is incomplete.

Pressure-Swing Adsorption (PSA). After high-temperature shift conversion, the gas is cooled to ambient temperature. Water is condensed and removed before the gas is fed to at least three (and up to 12) parallel adsorbers, which run in a predetermined cyclic operation. The various purge and decompression gases of the adsorbers are led into a large buffer vessel to compensate for variations in pressure and gas concentrations and are subsequently fired in the reformer along with a small amount of fuel gas for trimming purposes.

4.4.2
Production with Electrolysis

Electrolysis has been used for approximately 100 years for the production of hydrogen. The first large installation was by Norsk Hydro in 1927 in Norway. Further plants were erected by Cominco in Trail, British Columbia, Canada in 1940 and, from 1945 onwards, other plants with capacities up to 33 000 m^3(STP)/h of hydrogen. The erection of large electrolysis plants depends on the availability of cheap electricity from hydropower stations. Plants for smaller amounts of hydrogen [50 – 500 m^3(STP)/h] are often used by the industry because they are simple to operate. Moderate power costs are an additional incentive. Annually 1 – 4 % of hydrogen is produced by electrolysis [22], [23].

Electrochemical Principles If a potential is applied to the electrodes of an electrolysis cell filled with a suitable aqueous electrolyte, the following reaction occurs at the electrodes:

Cathode:

$$2H_2O\ (l) + 2e^- \rightarrow H_2\ (g) + 2OH^-\ (aq) \tag{12}$$

Anode:

$$2OH^-\ (aq) \rightarrow {}^1/{}_2O_2\ (g) + H_2O\ (l) + 2e^- \tag{13}$$

Cell reaction:

$$H_2O\ (l) \rightarrow H_2\ (g) + {}^1/{}_2O_2\ (g) \tag{14}$$

where l denotes liquid, g gaseous and aq dissolved in water.

Pure water is not suitable as an electrolyte because of its very low conductivity. Therefore, aqueous solutions of potassium or sodium hydroxide, sodium chloride, hydrochloric acid, etc. are used. Depending on the electrolyte and the material of the electrodes the reaction at the anode may lead to other products, in particular to sodium hydroxide and chlorine.

The change in enthalpy for the electrochemical decomposition of water is

$$\Delta H = \Delta G + T\Delta S = -nFE + nFT \cdot \left(\frac{\partial E}{\partial T}\right)_p \tag{15}$$

where H is the enthalpy, G the Gibbs free energy, S the entropy, T the temperature, n the number of electrons transferred per formula conversion, F the Faraday constant and E the electrochemical potential.

The thermodynamic standard data at $p = 101.3$ kPa and $T = 298.15$ K are:

- reaction enthalpy: $\Delta H^0{}_{298}$ K = 286 kJ/mol
- reaction entropy: $\Delta S^0{}_{298}$ K = 163 J/mol/ K
- Gibbs free energy of reaction: $\Delta G^0{}_{298}$ K = 237 kJ/mol

The value of the Faraday constant is:

$$F = 96.485 \text{ A s/kmol}$$

Using these data the required electrical energy becomes:

$$W = \Delta G = F \times n \times E_{rev} \tag{16}$$

where E_{rev} is the ideal (reversible) decomposition potential. Under standard conditions this amounts to:

$$E_{rev} = \frac{\Delta G}{n \cdot F} = 1.228 \text{ V} \tag{17}$$

Because of irreversible processes on the electrodes and the resistance of the electrolyte, the actual decomposition potential is always higher than the ideal value.

According to Eq. 15 the free energy, ΔG, is smaller than the reaction enthalpy, ΔH, by the amount $T\Delta S$. In the ideal case, operation of the cell requires an amount of electrical energy, $W = \Delta G$, and the addition of heat, $Q = T\Delta S$.

If both amounts of energy are to be supplied in the electrical form, the potential under standard conditions is increased by the thermal potential ΔE_Q.

$$\Delta E_Q = \frac{T\Delta S}{n \cdot F} = 0.252 \text{ V} \tag{18}$$

Thus, the theoretical minimum decomposition potential under standard conditions is:

$$E_{th} = 1.480 \text{ V}$$

The potential actually required is the sum of the following contributions:

$$U = E_1 + E_2 + E_3 + E_4 \tag{19}$$

where U is the cell voltage (V), E_1 the theoretical cell voltage (E_{rev}) (V), E_2, E_3 the anodic and cathodic overvoltages (V) at the phase boundary electrolyte – electrode, respectively, and E_4 is the voltage drop (V), due to electrical resistance of the electrolyte system.

The overvoltages E_2 and E_3 can be influenced by suitable choice of the electrode materials and their surface conditions. The voltage drop in the electrolyte, E_4, depends on: (1) the conductivity of the electrolyte, (2) the permeability of the cell diaphragms, (3) the distance of the electrodes from one another and (4) the current density.

The efficiency of the electrolysis is related to the minimum voltage according to Eq. 15:

$$\eta = \frac{E + \frac{T\Delta S}{nF}}{U}$$

Under standard conditions:

$$\eta = \frac{1.48}{U} \frac{\text{V}}{\text{V}}$$

Principle Construction of a Water-Electrolysis Cell Figure 4.5 shows the principle construction of a cell. A diaphragm separates the anode and cathode regions. The cell is filled with electrolyte. Oxygen is formed during the electrolysis at the anode, and hydrogen at the cathode. Depending on the arrangement of the electrodes and the diaphragm, the cells may be designated as unipolar or bipolar cells, see Fig. 5.6. *Bipolar cells* are characterized by the fact that the metallic separator between two cells connected in series serves as the cathode in one of the cells and as the anode in the next cell.

Electrolysis Plants (see Fig. 5.7) Apart from the electrolysis unit, an electrolysis plant consists of a number of different installations, such as an electricity supply,

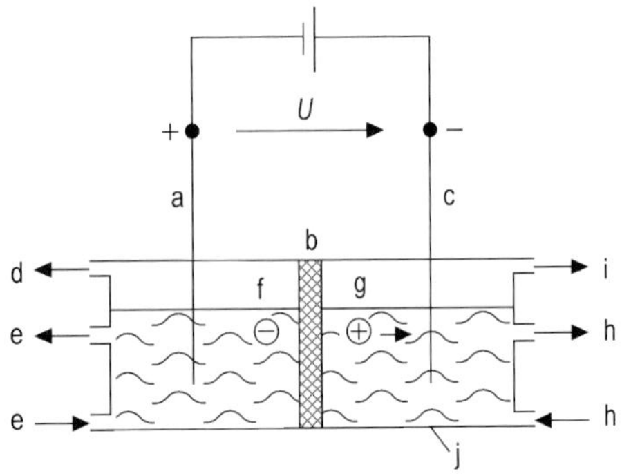

Fig. 4.5 Schematic of a water electrolysis cell a) Anode; b) Diaphragm; c) Cathode; d) Oxygen outlet; e) Anodic electrolyte cycle; f) Anion; g) Cation; h) Cathodic electrolyte cycle; i) Hydrogen outlet; j) Cell wall.

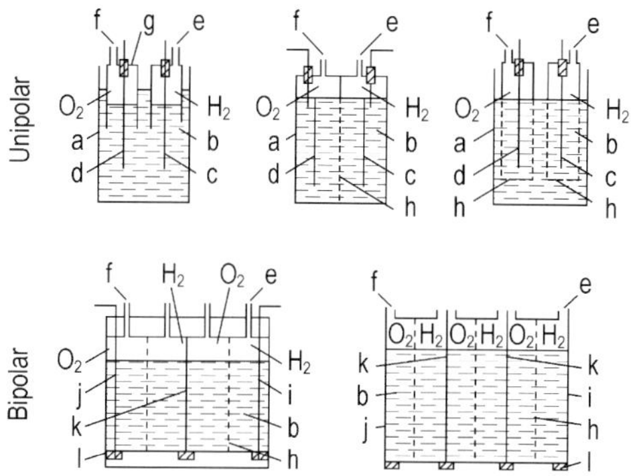

Fig. 4.6 Different modes of water electrolysis cell according to [24] a) Cell walls; b) Electrolyte; c) Cathode; d) Anode; e) Hydrogen outlet; f) Oxygen outlet; g) Gas collector; h) Diaphragm; i) Outer cathode; j) Outer anode; k) Bipolar electrode; l) Insulation.

water treatment, electrolyte circulation, the knockout system (phase separator) to separate electrolyte from the product gases, gas purification, and the instrument and control system.

The water consumed during electrolysis has to be continuously replaced; the consumption is 0.805 L per cubic meter of hydrogen plus evaporation losses. The water make-up must meet stringent purity requirements to avoid a build-up of substances in the electrolyte solution, which poison the electrodes or promote corrosion. The required water purity (conductivity 1 μS/cm) is obtained with the aid of ion exchangers.

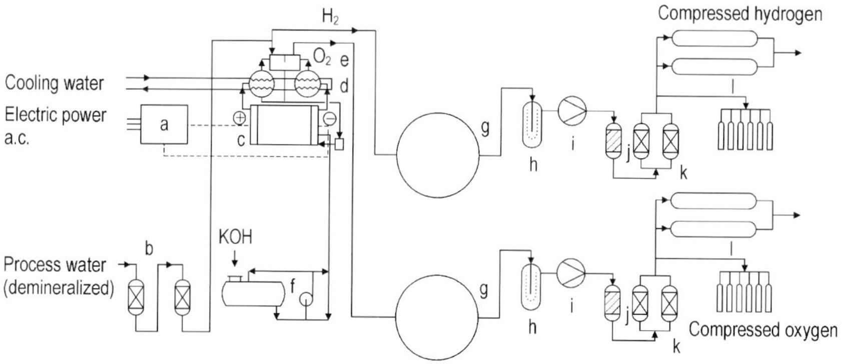

Fig. 4.7 Process diagram of an electrolysis unit with the most important components a) Rectifier unit; b) Process water demineralizer ion exchange unit; c) Electrolytic cell's water electrolyzer; d) Gas separator and cooler; e) Gas scrubber; f) Electrolyte tank and transfer pump; g) Gas holder; h) Filter; i) Compressor; j) Gas purifier; k) Drying; l) High-pressure storage tank and cylinder filling station.

Considerable amounts of potassium hydroxide are only necessary for the first filling. As the energy losses in the electrolyte are converted into heat, a large amount of cooling water is necessary to operate the plant. This is also true for the gas purification unit.

The *gas purification* depends on the purity requirements placed on the gas and may be a simple condensation to remove any moisture that is present. For higher purity, the catalytic conversion of the oxygen present may be required, followed by an adsorptive drying stage.

4.4.3 Other Production Processes

Hydrogen from Ammonia Hydrogen can also be produced from ammonia according to the reaction:

$$2NH_3 \rightarrow N_2 + 3H_2 \quad \Delta H_{298} = +\ 91.86\ \text{kJ/mol} \tag{20}$$

The endothermic reaction is favored by high temperature and low pressure and is accelerated by the presence of nickel or iron catalysts.

The possibility of hydrogen generation from ammonia leads to consideration of the use of anhydrous liquid ammonia as a medium for hydrogen distribution and storage. Liquid ammonia contains 17.8 wt % hydrogen, which is 50 % more hydrogen per volume than liquid hydrogen [25].

Hydrogen from Methanol Hydrogen and carbon monoxide can be obtained from the decomposition of methanol by reversal of the synthesis reaction. The endothermic decomposition occurs at temperatures >700 °C without a catalyst, or at 300–450 °C on catalysts based on copper – nickel or zinc – chromium alloys. Higher hydrogen yields are obtained by catalytic steam reforming of methanol. The process proceeds via initial methanol decomposition (Eq. 21) followed by a water-gas shift reaction (Eq. 22) as shown in the overall equation (Eq. 23).

$$CH_3OH \rightarrow CO + 2H_2 \qquad \Delta H_{298} = +91.7\ \text{kJ/mol} \tag{21}$$

$$CO + H_2O \rightarrow CO_2 + H_2 \qquad \Delta H_{298} = -41.0\ \text{kJ/mol} \tag{22}$$

$$CH_3OH + H_2O \rightarrow CO_2 + 3H_2 \qquad \Delta H_{298} = +50.7\ \text{kJ/mol} \tag{23}$$

Information on thermodynamic and kinetic investigations, suitable catalysts and a summary of the literature are given in [26–28]. The reaction is carried out in practice at pressures of 0.7 – 3 MPa on CuO – ZnO or CuO – Cr_2O_5 catalysts. The molar ratio of water to methanol will vary from 0.67 to 1.5; excess steam lowers the risk of carbon formation.

Hydrogen production from methanol is used in small to medium-sized plants (some of which are mobile) for the production of 1 – 2000 m^3(STP)/h. Purification of the product gas is performed by carbon dioxide scrubbing, pressure-swing adsorption or (in particular for small units and high purity), by diffusion through a

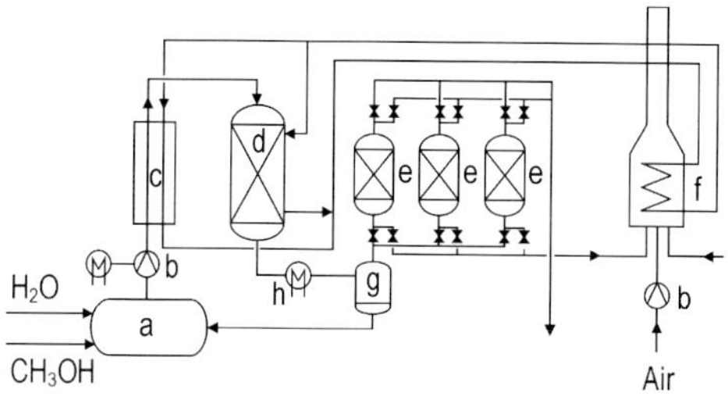

Fig. 4.8 Methanol reformer with pressure swing adsorption purification for the production of 50 m³(STP) hydrogen per hour, purity 99.9999 vol % (German Linde) a) Methanol/water vessel; b) Pump; c) Preheater; d) Reformer; e) Adsorber; f) Heater for heat transfer medium; g) Condensate feedback; h) Cooling water.

palladium – silver membrane. Figure 5.8 shows the schematic of a methanol cracker for 50 m^3(STP)/h of hydrogen at 1.4 MPa using a PSA purification unit to give 99.9999 % purity; 1.5 m^3(STP) of hydrogen is produced per kilogram of methanol.

4.4.4
Purification of Hydrogen

Hydrogen from synthesis gas may contain up to 50 % carbon monoxide (partial oxidation). Carbon monoxide can be recovered for CO production by a methane scrubber and used for synthesis, or it can be reacted with water to produce hydrogen and carbon dioxide (water – gas reaction, shift conversion):

$$CO + H_2O \rightarrow H_2 + CO_2 \tag{24}$$

The reaction is exothermal. Because of the equilibrium conditions, the carbon monoxide conversion is incomplete in adiabatic reactors. For increased carbon monoxide conversion a greater amount of water ($c_{H_2O}/c_{CO} > 3$) is necessary. At high carbon monoxide concentrations, a series of adiabatic reactors with intercooling are used. Isothermal reactors have also been suggested [29].

Pressure Swing Adsorption (PSA) Pressure-swing adsorption operates at almost constant temperature using partial-pressure differences. However, the small temperature gradients occurring with ad- and desorption contribute considerably to the mechanism of the process. The components to be removed are adsorbed at high system pressure and high partial pressure and are desorbed at low system pressure in accordance with the appropriate equilibrium isotherms of the various impurities.

When the adsorber capacity is exhausted, the adsorbent must be regenerated. Depressurization and hydrogen purging reduce the carbon monoxide loading on the absorbent. The PSA process is cyclical; typical cycle times are in the range of 3 to 10 min. Thus, removal of impurities present in high concentrations can be

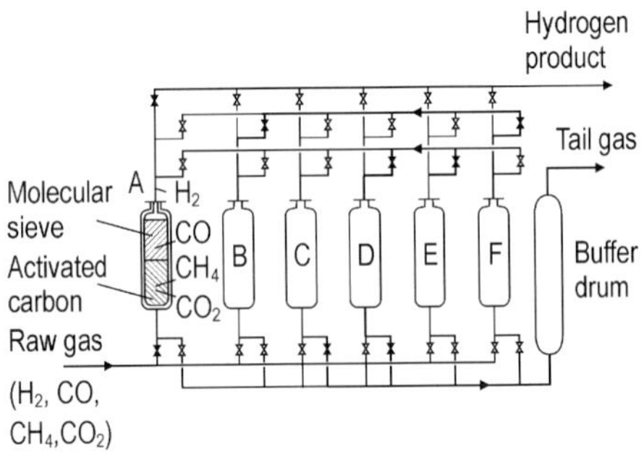

Fig. 4.9 Six-bed PSA unit Main flows for adsorption, repressurization, counterflow expansion (dump), counterflow purge, etc. for H2 production by adsorber A. After termination of this adsorption step, next on-stream adsorber is B, etc.

accomplished using moderately sized adsorber vessels. The essential steps carried out in a PSA unit are shown in Fig. 5.9 [30]. The chronological sequence of all steps constitutes the cycle of one adsorber:

1) Adsorption at the pressure of the raw gas (highest system pressure), hydrogen production (adsorber A),
2) Cocurrent depressurization, provision of gas for pressure equalization and purging (adsorbers F and E),
3) Countercurrent depressurization, production of part of the tail gas (adsorber C),
4) Countercurrent purging (lowest system pressure), production of another part of the tail gas (adsorber D),
5) Countercurrent repressurization (adsorber B).

4.5 Handling

4.5.1 Quality

The hydrogen content of technical hydrogen produced from hydrocarbons is between 97 and 99.5 vol %. The impurities are mainly methane and nitrogen; traces of oxygen, carbon monoxide, carbon dioxide and, depending on the production and purification stages, hydrogen sulfide or ammonia may be present. Hydrogen produced electrolytically is typically >99.5 vol % pure. Impurities in this case are mainly nitrogen, oxygen and water. The mercury content must be limited to a maximum of 0.15×10^{-3} mg/m^3 (for protection of the aluminum material in the downstream purification units).

Hydrogen is available in high and very high purities. The impurities are removed by means of catalytic combustion (oxygen), drying systems (water) and by adsorption and diffusion systems.

Particular requirements are placed on the purity of the hydrogen to be liquefied, because at liquid hydrogen temperatures all impurities except helium become solid and this would lead to a blockage of the refrigeration unit.

4.5.2
Compression

Hydrogen is often stored at high and very high pressures of up to 70 MPa. Because of its low molecular mass, however, it is difficult to compress in conventional machines. Particular types of compressors have therefore been developed.

For very large amounts of hydrogen (>50 000 m^3) radial-type high-pressure turbo compressors have become standard. Because the pressure ratio attainable with hydrogen is relatively low, multistage compressors are used. Leakage problems can be handled by appropriate construction.

For smaller amounts of gas and high-end pressures, multistage piston compressors are chosen.

Smaller amounts of low-pressure gas that require moderate compression ratios can be handled with screw compressors. This type of compressor is used especially in hydrogen liquefaction plants.

Apart from these possibilities, compression can also be achieved by pumping liquid hydrogen to the high pressure desired and then evaporating the liquid. This is used for the LH_2 injection in some hydrogen-fueled vehicle engines.

4.5.3
Liquefaction

The boiling point of hydrogen is 20.39 K at 101.3 kPa (see Table 4.1) for n-H_2 and 20.26 K for p-H_2. As for other gases, hydrogen is liquefied because storage and transportation are easier in the condensed state. A wide application for liquid hydrogen is predicted in the future hydrogen energy concept.

For the liquefaction of gases energy must be removed until the condensation point is reached. This is achieved either by cooling down to low temperatures through external means, such as liquid nitrogen or the magneto-caloric effect, or internally by irreversible or isentropic decompression.

The decompression of hydrogen in a throttle valve can only lead to a temperature decrease in a certain range of the Joule – Thomson curve. The range that is used in practice is from 77 K, the boiling point of liquid nitrogen at 0.1 MPa, down to 25 K. In this range the isenthalpic lines (H=constant) in the $T-S$ diagram are particularly steep. The decompression in the throttle leads to a decrease in temperature or, at the boiling point, to a high degree of liquefaction at a constant temperature [31].

Isentropic decompression in turbines or other machines is, generally speaking, possible over all temperature ranges. However, for technical reasons this process is only used for hydrogen above 25 K.

Hydrogen liquefaction is technically achieved in practice through a series of cyclic processes. The theoretical lowest energy consumption (minimum work) in a reversible Carnot process for the production of liquid p-H_2 (0.1 MPa) from gaseous n-H_2 (25 °C, 0.1 MPa) amounts to 14.23 kJ/g=3.95 kWh/kg.

In natural hydrogen (n-H_2) at 25 °C the ratio of the nuclear spin isomers o-H_2 and p-H_2 is ca. 0.75/0.25. At the boiling point of hydrogen the equilibrium lies almost completely on the side of p-H_2 (see Section 4.2). The reaction enthalpy ΔH of the exothermal o – p conversion (–1.448 kJ/mol at 25 K) is in the same range as the heat of evaporation for n-H_2 at 25 K (0.828 kJ/mol). Thus, if the reaction rate of the o – p conversion is large in comparison with the retention time in storage and in the transport vessels, a major portion of stored liquid n-hydrogen evaporates, even after a few days [32]. In addition, heat losses through the insulation occur. For this reason, practical hydrogen liquefaction units have o – p conversion reactors incorporated into the process. The o – p conversion takes place on iron oxide gel, iron hydroxide and chromium oxide on alumina catalysts.

The heat of reaction of the o – p conversion must be compensated for by additional refrigeration capacity of the liquefaction unit.

4.5.4
Transportation and Distribution

The consumers of hydrogen in the chemical industry usually produce hydrogen on site. Small- and medium-sized consumers can also be supplied with gaseous or

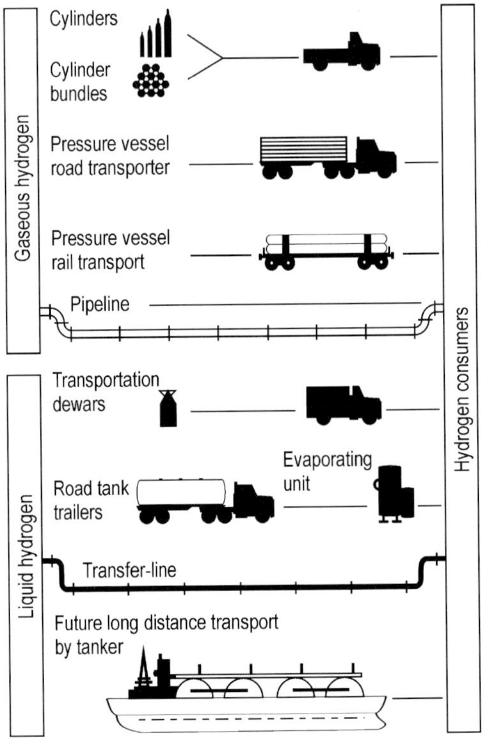

Fig. 4.10 Possibilities for hydrogen transport.

liquid hydrogen. The principle possibilities for hydrogen transport within the structure of regional hydrogen logistics are shown in Fig. 5.10 [33]. In any future hydrogen energy concept, transportation of large amounts of hydrogen between regions will be necessary.

High-Pressure Hydrogen Transport Small amounts of hydrogen can be transported in aluminum pressure cans (content ca. 5 L H_2) and in steel cylinders of various sizes (up to 70 L with H_2 at 20 MPa). Greater amounts of hydrogen are transported in bundles of steel or composite cylinders at 20 MPa by road or rail. Modern high-pressure transports have a capacity of 6.000 Nm^3 of hydrogen. The supply-pressure can not normally be utilized.

Pipelines play a particular role in the transport of large amounts of hydrogen. However, in contrast to hydrogen carried by road the area-wide pipeline transport requires a new infrastructure. Only a few regional hydrogen pipeline nets are run worldwide, e.g., in the USA by Air Products, and in France and in the Federal Republic of Germany by L'Air Liquide. The last-named system was built in 1938. It has a length of 210 km and can transport 250×10^6 m^3(STP)/a; four suppliers and 14 users are tied into the system. The experience gained with this pipeline has been used to provide specifications for planning, constructing, operating and maintaining such a system [34].

LH_2 Transport For laboratory uses liquid hydrogen is supplied in transport dewars. Larger amounts are transported in tank trailers that usually have a capacity of about 4.5×10^4 L. In the USA, liquid hydrogen is also transported by rail (10^5 L containers). Overseas containers, containing up to 2×10^4 L of liquid hydrogen are used to supply the European space project, Ariane, in French Guyana. During the Apollo program, barges with dewars containing up to 900 m^3 of liquid hydrogen were used. To keep evaporation losses small, the transport vessels have either perlite vacuum insulation or superinsulation. Transport dewars for use over large distances are equipped with a liquid nitrogen shield.

Transportation Costs The total costs of the transport chain from the hydrogen production to the user site have often been assessed, e.g., in [35]. Owing to the low load capacity of below 1.5 % in weight, the cost of transportation in high-pressure bundles is comparatively high. Over a distance of 100 km the cost of the latter is three times higher than that of transporting liquid hydrogen and actually escalates the higher the transport distance. Thus transport in high-pressure bundles is suitable for short distances and low demands. The load capacity of cryo-containers for the transportation of liquid hydrogen is about 8.75 % in weight.

The assumption is that high utilization rate pipelines are by far the most cost-effective means of transporting gaseous hydrogen over both short and long distances. To transport the same effective energy through transfer lines, hydrogen requires 3.5 – 4 times the amount of compression energy compared with natural gas; the total costs are higher by a factor of 1.3–1.6. The transport of liquid hydrogen via rail or road is ca. five times more expensive, the use of pressure gas cylinders is by far the most expensive (transportation costs are ca. 50 times higher, including the costs of idle equipment, such as gas cylinders and vehicles).

4.5.5 Storage

Conventional methods for storing hydrogen are well proven and in service. Depending on the end use various systems can be defined: (1) stationary, large-size storage systems, (2) stationary, small-size storage systems and (3) mobile storage systems. The last may be used for transport, for distribution, or as a fuel reservoir for prime movers. In these storage systems, hydrogen can be handled in the gaseous (GH_2) or liquid (LH_2) state or it can be chemically bound in metal hydrides. Newer developments with carbon nanotubes are described, e.g., in Ref. [36].

Compressed Hydrogen Storage Pressure storage systems of various sizes and pressure ranges are considered state-of-the-art. For hydrogen, steel cylinders (up to 70 L at 20 MPa), composite material cylinders (25 MPa), stationary high-pressure vessels (20 m^3 at 20 MPa) or low-pressure spherical vessels (15 000 m^3 at 1.2–1.6 MPa) are available. Low-pressure and high-pressure systems are suitable for the storage of daily or weekly supplies.

Underground storage in caverns or aquifers is expected to be possible; the storage pressure is up to 16 MPa. Gas losses of up to 1–3 % of the storage capacity are foreseen [37]. Specific problems due to hydrogen are not anticipated, however, the availability of suitable salt and aquifer formations is limited.

Liquid Hydrogen Storage The energy consumption to liquefy hydrogen amounts to ca. 29 % of its lower heating value. The technique for liquid hydrogen storage is well developed; it is used for the storage, transport and distribution. The vessels are insulated with double-walls (dewars); the space between the inside and outside wall is evacuated and filled with perlite (large vessels) or super-insulation. The rate of loss due to evaporation diminishes with an increasing size of the vessel. The losses are defined by the amount of hydrogen evaporated daily. Laboratory vessels have 2–3 %, mobile LH_2 transport tanks 1%, perlite-insulated large-volume vessels <0.1 % losses. For additional reduction of heat leakage the refrigeration available in the boil-off gas can be utilized to cool the insulation. Liquid hydrogen vessels are attainable from 100 L cans up to large tanks with volumes up to 5000 m^3. Tanks have a high specific storage capacity (ca. 100 kWh/kg) and are suitable as long-term storage systems.

Super-Insulation (Multi Layer Insulation, MLI) [38] A large number of layers of insulation foil are wound around the inside vessel (10–60 layers per centimeter). A layer consists of a plastic film, both sides of which have been sputtered with metal, or it is produced from a very fine aluminum foil, which is held apart by spacers of a fabric made from glass filaments or similar material. After evacuating the space between the inner and outer vessels, the energy transport by radiation and conduction of the insulation material is more or less completely interrupted. Values of 5×10^{-6} per Wm/K for the heat conductivity can be attained.

On-Board Liquid Hydrogen Storage A major concern in on-board liquid hydrogen storage is minimizing hydrogen losses from liquid boil-off. Because liquid hydrogen is stored as a cryogenic liquid that is at its boiling point, any heat transfer to the liquid causes some hydrogen to evaporate. Therefore cryogenic containers are designed to minimize conductive, convective and radiant heat from the outer container wall to the liquid. Three factors influence the boil-off:

- insulation
- shape
- *ortho* to *para* ratio

Minimizing heat leaks by thermal conduction is very challenging; special cryogenic know-how concerning layout, design, material selection (i.e., for the internal suspension, filling and extraction lines, etc.) is required. Newer developments allow storage of liquid hydrogen for up to two weeks without any losses by boil-off [39].

Metal Hydride Storage The capability of metals or metal alloys to reversibly store hydrogen in the form of hydrides has lead to a large number of investigations with respect to practical applications. The main areas covered have been the storage of hydrogen but also the separation of hydrogen from gas mixtures, the purification of gases and storage and separation of hydrogen isotopes. An overview is given in [40–43].

Physical Fundamentals Ionic and metallic hydrides are formed (exothermic process) and decomposed (endothermic process) according to:

$$y\mathrm{Me} + x\mathrm{H}_2 \rightleftharpoons \mathrm{Me}_y\mathrm{H}_{2x}$$

The hydrogen absorption and hydride formation behavior can be described by pressure – composition isotherms (see Fig. 5.11 A). After catalytic dissociation of the hydrogen molecule, some of the atoms lose their electron and occupy an interstitial position in the crystal structure of the host metal (solid solution, α-phase). As the hydrogen concentration is increased, local H–H interaction causes nucleation, which induces growth of the hydride phase (β-phase). The two phases (α + β) coexist until, through further hydrogen uptake, the stoichiometric limit is reached. Here, the isotherms reach a plateau level (constant pressure), which is termed the "dissociation plateau" or "plateau pressure". After the pure β-phase is reached, any further increase in the hydrogen to metal ratio requires a steep increase in hydrogen pressure. At higher hydrogen pressure, further plateaus may be found because of the existence of other hydride phases (γ, δ, . . .).

On reducing the hydrogen pressure, the process is run through in the reverse direction (desorption isotherm). Along the desorption plateau, the hydride phase decomposes into hydrogen-saturated metal and hydrogen gas. The decomposition process is endothermic. Absorption and desorption isotherms show a hysteresis effect. This is shown in Fig. 5.11 B and C for $LaNi_5$ and FeTi [44, 45].

The plateau pressure p, which varies strongly with temperature, is connected with the formation or decomposition enthalpy by the van't Hoff equation. The entropy term, ΔS^0, consists mainly of the standard entropy of gaseous hydrogen

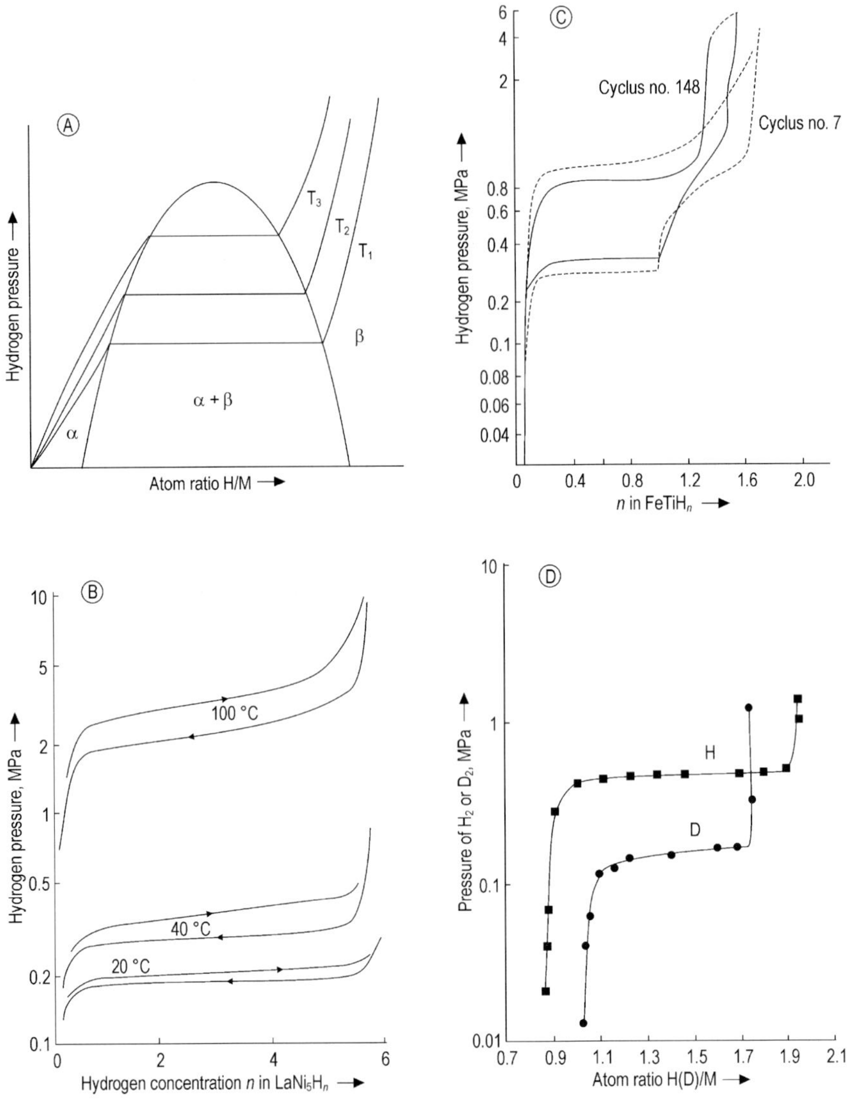

Fig. 4.11 A) Ideal pressure-composition isotherms for a hydrogen-metal system B) Absorption and dissociation isotherms of LaNi5 hydride with hysteresis C) Hysteresis loops of test cycles of FeTi at 25 °C. The plateau pressure of the monohydride stabilizes after ca. 20 cycles, whereas that of the dihydride formation is still rising D) Pressure-composition isotherms of hydrogen and deuterium in zone-refined vanadium at 40 °C.

($S^0[H_2^g]$=130.6 J/mol/K), because the entropy of the hydrogen dissolved in the metal can be neglected.

$$-RT \ln p_{H_2} = \Delta H - T\Delta S$$

$$\log p_{H_2} = \frac{\Delta H}{19.14\,(T/1000)} + 6.82$$

The heat evolved by the hydrogen – metal reaction depends on the type of hydride and can amount to ca. 30 % of the heat of combustion of the absorbed hydrogen. To liberate hydrogen, at least the same amount of energy is required (see Table 4.6).

Intrinsic Properties of the Hydride-Forming Materials To analyze the utilization possibilities of hydride-forming metals, additional properties, some of which are crucial, must be known. In particular, surfaces of intermetallic compounds containing transition metals accelerate the dissociation reaction, whereas thin oxide layers on the surfaces of elemental metals often block this. Easy hydrogen sorption is facilitated by surface segregation with continuous formation of reactive subsurfaces (self-activation). Addition of small amounts of transition elements, e.g., nickel, will always favor activation. Impurities in the hydrogen, such as carbon monoxide, hydrogen sulfide and sulfur dioxide slow down the sorption rate because of surface deactivation, whereas oxidizing impurities, such as oxygen and water, gradually decrease the capacity by reaction with the host compound. Such "poisoning" of the hydride material can in many cases be reversed by hydrogenation – dehydrogenation cycles with pure hydrogen at higher temperature.

Incorporation of the hydrogen into the lattice and the a/b-phase transition causes a deformation of the metal structure. The accompanying increase in volume can amount to 25 % and may lead to undesired distortion of the containing vessel.

Detailed analyses of the kinetics of hydride formation of $LaNi_5$ and MgNi are available [46]. Examples for hydride storage systems and systems for purification and recovery of hydrogen and for the use of metal hydrides for thermochemical machines and for the compression of hydrogen are described in [2].

New Developments for On-Board Metal Hydride Storage Novel light-metal nanocrystalline Mg hydrides allow capacities up to 2.5 kWh/kg and show fast absorption and desorption kinetics, for example at 300 °C within a few minutes [47]. The improved kinetics are explained by the high fraction of grain boundaries, which provide fast diffusion paths for hydrogen and heterogeneous nucleation sites for the hydride and the metallic phase as well as to using additional catalysts and a new processing route (high-energy milling). The goal of further developments is a desorption temperature of 200 °C and a refueling time $<$10 min [48].

Cost Comparison of Large-Volume Hydrogen Storage The energy stored in a certain volume of hydrogen, relative to that stored in the same volume of natural gas, is determined by the ratio of the volumetric heating values. Therefore, for otherwise equal systems, the hydrogen storage is more expensive by a factor of ca. 3

Table 4.6 Comparison of the hydrogen absorption data of some metal – metal hydride systems

Hydride system	Heat of reaction, kJ per mole H_2	Plateau region [a]		Hydrogen content		Energy density, [c] kJ/g
		pH_2, MPa	T, °C	wt %	g/mL [b]	
Elemental metals						
$Mg - MgH_2$	77.4	0.1	286	7.7	0.107	9.9
$Ti - TiH_2$	124	0.1	650	4.0	0.15	5.7
$V - VH_2$		0.4	40	2.1	0.19	2.95
Alloys						
$Mg_2Ni - Mg_2NiH_4$	64.5	1.0	350	3.16	0.081	4.5
$FeTi - FeTiH_{1.95}$	31.4	0.5	30	1.72	0.096	2.5
$LaNi_5 - LaNi_2H_{5.9}$	30.2	0.5	60	1.37	0.089	1.95
$Ti_{0.98}Zr_{0.02}V_{0.43}Fe_{0.09}Cr_{0.05}Mn_{1.5}$		1.0	24	1.75	0.09	2.5
Liquid hydrogen for comparison						
LH_2				100	0.077	142

[a] Under selected conditions, dissociation pressure at 50 % of H/Me ratio.
[b] Refers to the volume of compact hydride material without voids.
[c] Refers to HHV of absorbed hydrogen.

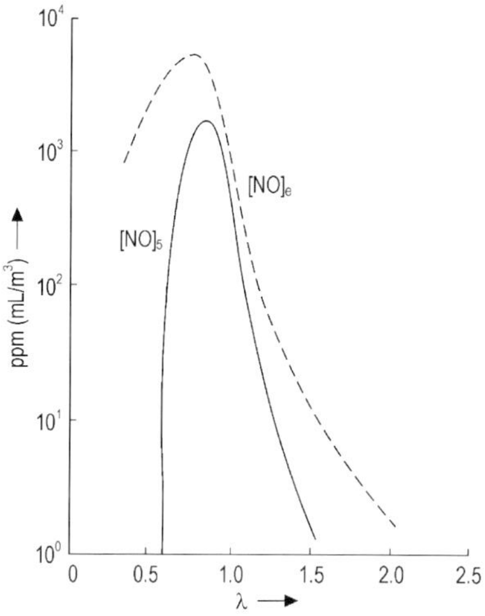

Fig. 4.12 NO concentration resulting from hydrogen – air flames as a function of excess air value ($[NO]_e$=equilibrium concentration; $[NO]_5$=concentration behind flame front within 5 ms; p=3.2 MPa).

than natural gas storage. Studies of the cost situation for various hydrogen storage systems allow the following conclusions to be drawn [49]:

1. Underground hydrogen storage is the most economic method for all areas of application.
2. Storage in pressure vessels is the most economic above-ground storage system when $t < 30$ h. It is suitable for all short-term storage systems.
3. LH_2 storage is the most economic storage system above ground for $t > 30$ h.
4. It is not clear whether the recent concepts for hydrogen storage (metal hydrides or cryo-adsorber systems) are technically and economically more favorable options.

4.5.6
Combustion and Heating

The energy contained in the hydrogen can be released most simply by combustion with oxygen or with air, see Section 4.2.

Emissions As only water is obtained as a product of the combustion, emissions, such as carbon monoxide, sulfur dioxide, incompletely combusted hydrocarbons, particles, soot, ash, which are obtained by combustion of fossil fuels, are avoided. If hydrogen is burned in the presence of air, the high temperatures favor the formation of nitrogen oxides. The NO_x formation depends on the flame temperature and can be influenced by the air – to fuel ratio, λ. Figure 5.12 shows typical

NO concentrations for a hydrogen – air flame. Maximum NO formation occurs in the range of the stoichiometric hydrogen – oxygen ratio, whereas under lean conditions (excess air) low nitric oxide concentrations may be attained [50].

Hydrogen Burners Differences in combustion properties, i.e., flame speed and flame stability in addition to density and heating value, make it necessary to use a different burner geometry when hydrogen or hydrogen-rich gases are to be burned. The flame speed of hydrogen – air mixtures is seven times higher than that of methane; the particular problem is therefore the flash-back over a wide range of hydrogen – primary air mixtures.

Catalytic Combustion The thermal energy of hydrogen can also be released by means of flameless catalytic combustion. Suitable catalysts, apart from platinum and palladium that initiate the reaction at room temperatures, are metal oxides, such as MnO_2 – CuO – Ag_2O (hopcalite) or Co_3O_4. A suitable carrier material in the form of porous plates or honeycombs can enlarge the surface accordingly. Catalytic burners operate either on the diffusion principle or by premixing hydrogen with air. Catalytic heaters react very flexibly; the temperature can be adjusted and limited by means of the throughput; the heat is produced in the infrared range. The combustion is free from emissions.

4.6 Hydrogen Fuel Systems

In all sectors of transportation (space and aircraft propulsion, earthbound traffic systems including individual transport) the fuel systems hydrogen – air or hydrogen – oxygen can be used as the prime mover.

4.6.1 Hydrogen for Aerospace

Hydrogen Propulsion Systems for Space Because of the large specific enthalpy of combustion, LH_2, combined with liquid oxygen (LOX), produces (apart from the system LH_2 – LF_2) the greatest specific impulse. Jet velocities up to 4500 m/s can be reached using combustion chamber pressures of 5 to 22 MPa.

The hydrogen – oxygen motor NASA Centaur (Pratt & Whitney) was used for the third stage of the Atlas Centaur and the fourth stage of the Titan III rocket. For the Apollo project, Rocketdyne motors were used for the second and third stage of the Saturn V [51]. The ultimate in the US development is the Space Shuttle main engine with a maximum thrust level of 1860 kN (vacuum: 2280 kN) [52].

The European Ariane 4 carrier system uses LH_2 – LOX in its third stage. Using 10 t of this fuel combination, the HM 7 motor develops a thrust of 60 kN over a period of 700 s [53]. The cryogenic main stage of the Ariane 5 (Vulcain HM 60) burns 25.8 t of hydrogen together with 132 t of oxygen in 570 s. It has a thrust of about 800 kN (vacuum: up to 1140 kN) [54].

Hydrogen as Aviation Fuel Liquid hydrogen has already been tested in 1957 in a B 57 Canberra by the NACA Lewis Research Center (Cleveland, OH, USA) and its suitability as an aviation fuel was proven. Since 1973, especially in the USA, investigation into the use of LH_2 as an aviation fuel has been followed up (for an overview see [55–57]). In 1988, test flights were made in the Soviet Union with a civil aircraft using LH_2 as the fuel.

Hydrogen is suitable for turbines in the sub- and supersonic regions. Engine designs for turbines converted for use with hydrogen fuel have certain advantages: They are smaller, easier to control and have a longer lifetime than normal fuel turbines. The cryogenic fuel is used to cool engine oil and turbine cooling air, whereas the turbine core exhaust heats up the fuel before injection into the combustor. The savings in the specific fuel consumption amount to 8 %. Ecological benefits are the lower noise level, smaller emission of NO_x, and the absence of CO, CO_2 and hydrocarbons.

The energy content related to volume and mass of LH_2 is the determining factor for the construction of a hydrogen aircraft. The four times greater fuel volume as compared with hydrocarbons for the same energy content and the cryogenic properties require a completely new concept for the aircraft fuselage and tanks.

The 2.8 times higher energy content of LH_2 compared with kerosene (related to mass) makes a much smaller lift-off weight possible for the same range and capacity. A comparison of design data and various performance characteristics of subsonic aircrafts for the fuels listed is shown in Table 4.7.

Similar studies have been presented for LH_2-fueled supersonic aircraft. Here the immense fuel requirements are even more in favor of LH_2. A comparison shows a reduction in the lift-off weight by ca. 50 %.

Table 4.7 Comparison of design and performance data of alternate fueled subsonic aircraft (basis: 400 passengers, 44 000 kg payload, 5500 NMi range, 0.85 Mach cruising speed)

Parameter	Liquid hydrogen (LH_2)	Liquid methane (LCH_4)	Synjet
Gross mass, kg	168 830	225 570	232 060
Block fuel, kg	21 620	58 980	72 530
Operating empty mass, kg	103 300	116 170	107 370
Wing area, m^2	296.8	385	380.3
Span, m	51.8	58.9	58.5
Fuselage length, m	65.7	61.4	60
Gross mass, kg	156 130		
Block fuel weight [a], kg	15 270		
Specific fuel consumption (cruise), kg/h/N	0.206	0.504	0.615
Lift/drag (cruise)	17.4	19.21	19.13
Thrust per engine, N	135 000	177 000	185 000
Energy utilization, kJ/seat/km	636	723	759

[a] With respect to laminar flow control.

Laminar Flow Control (LFC) The cryo-fuel LH_2 makes it possible to use the so-called cryogenic wall cooling technology. By cooling major aerodynamic surfaces of the aircraft to ca. 150 K, laminar flow in boundary layers is stabilized with a consequent reduction of skin-friction drag. The necessary refrigeration can be taken from the evaporating LH_2 directly or via a secondary refrigeration system (e.g., LN_2). Overall drag reduction in the order of 20 % leads to a fuel saving of about the same amount at cruising speeds [58].

Airport requirements for LH_2 as a fuel have been investigated in different studies. The facilities include the liquefaction plant, LH_2 storage capacity and LH_2 fuel handling systems [59, 60]. Because of planning and development issues, the use of LH_2 as an alternative fuel for aviation will not occur before the next two aircraft generations, i.e., about 2030.

4.6.2 Hydrogen for Automotive Vehicles

The influence of environmental legislation in all areas, including road traffic is increasing. In this respect hydrogen is an alternative fuel for road vehicles. Current driving forces behind the hydrogen technology are the Californian zero-emission-vehicle legislations, the reduction of CO_2 emissions according to the Kyoto Protocol, and in the longer term, the limitation of oil resources.

One party on its own cannot implement a hydrogen fuel infrastructure; therefore there is co-operation between legislators, vehicle manufacturers and gas and energy companies to implement hydrogen as an available economically viable vehicle fuel.

The Californian Zero Emission Vehicle Program Owing to its geographical position and the large amount of road traffic the so-called "California air basin" suffers from decreased air qualities.

California experiences high levels of smog, especially in the summer time. Smog, a combination of the words smoke and fog, is a mix of pollutants, mainly ozone (O_3) and particulate matter (PM). Ozone is formed from nitrogen oxides (NO_x) and volatile organic compounds (VOCs) in the presence of sunlight and hot stagnant air [61].

In 1990 the State of California embarked on a plan to reduce vehicle emissions to zero through the gradual introduction of zero emission vehicles (ZEVs).

The California Environmental Protection Agency, Air Resources Board adopted a requirement for each vehicle manufacturer that "at least 10 % of the new cars produced and delivered for sale in California in 2003 and subsequent model years must be ZEVs" [62]. The Californian market is very important for vehicle manufacturers worldwide: firstly because of its size, and secondly, the Californian environmental legislation for vehicles is considered forward-looking and is used by governments worldwide as a benchmark for road vehicle emission legislation.

Hydrogen is the only practical, technically feasible, carbon-free fuel. For the drive train in the automotive vehicles sector two different systems are used. Most manufacturers, such as DaimlerChrysler, GM, Toyota, etc., focus on fuel cells for an electric drive train (see Chap. 5). Others, such as BMW and MAN prefer internal

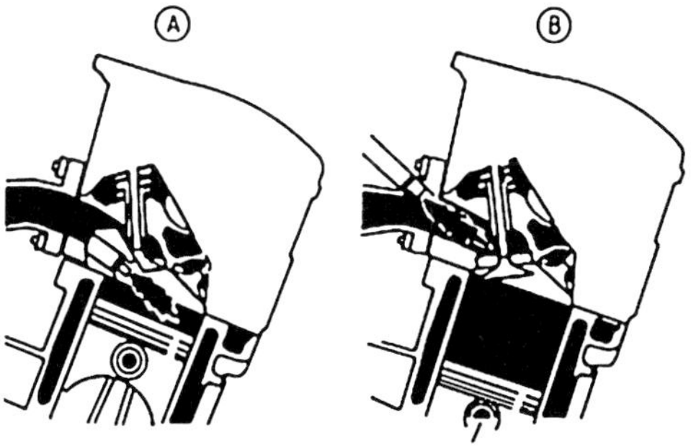

Fig. 13 Internal mixture formation (A) using cryogenic hydrogen (direct injection) and external mixture formation (B) using ambient-temperature hydrogen.

combustion engines (ICE). Ford is developing both techniques. An overview of the hydrogen vehicles built from 1967 to 2001 can be found in [63].

Internal Combustion Engines Appropriately modified, a hydrogen-powered car with an ICE does not differ from a conventional gasoline or diesel car in its driving and power properties or in its acoustics.

Some properties of hydrogen make engine modifications necessary. Low ignition energy and high flame speeds can cause self-ignition during the mixture preparation or flame flashback. The octane number of hydrogen is much lower than that of gasoline. However, the wide ignition range allows burning of lean mixtures and gives a large control range.

Various techniques are used to prepare the feed mixture for the combustion chamber. With external mixing a homogeneous hydrogen – air mixture is introduced. Uncontrolled pre-ignition or flash-back into the intake manifold is prevented by adding ballast, preferentially water, and by timed individual port injection of hydrogen close to the cylinder intake. Internal mixture formation is achieved by direct injection into the combustion chamber. The necessary supply pressure must be at least 1.5 MPa, or up to, depending on the time of injection, more than 10 MPa. This is only possible by using piston pumps for liquid hydrogen. Figure 5.13 shows the principle layout of internal and external mixture formation.

Attainable specific charge heating values (energy content per cylinder loading) are compared in Table 4.8, under various operating conditions. Revolutions per minute (rpm) and effective efficiencies are comparable to the gasoline motor. Because of the lower amounts of air possible and the lower heating value of the mixture, external mixture formation with water injection gives power ratings comparable to that of diesel motors, whereas with the internal mixture ca. 40 kW/L cylinder volumes were attained [64].

Hydrogen-powered engines emit only water, nitrous oxides (NO_x), and due to combustion of small amounts of engine lubricants minor traces of hydrocarbons (C_xH_y) and carbon monoxide (CO). Numerous investigations concerning the

Table 4.8 Specific charge heating values of hydrogen – air mixtures in combustion engines compared with gasoline

Operating conditions	Maximum of spec. charge HV, $H_{spec.max}$		Maximum of efficiency, η_{emax}		Maximum of excess air, φ_{min}	
	φ	kJ/L	φ	kJ/L	φ	kJ/L
Gasoline	1	37.7	1.1	34.4	0.83	31.6
Hydrogen (external mixture formation)	1	31.9	0.4	15.5	0.2	8.3
Hydrogen (internal mixture formation)	1	45.3	0.4	18.1	0.2	9

calculation and measurement of the nitric oxide emission have been carried out, e.g., [65], [66]. In Fig. 5.14 typical NO_x values for various modes of operation with hydrogen and gasoline are shown as a function of the equivalent ratio ($\varphi=1/\lambda$). Especially in lean modes of operation, i.e., with excess air, hydrogen motors operate more or less emission-free.

The motor concept chosen for a vehicle depends on the method of storing the hydrogen on board. Low-pressure gaseous hydrogen allows only external mixture formation, whereas liquid hydrogen, because of its cryogenic properties, is more flexible (liquid compression, cooling, injection).

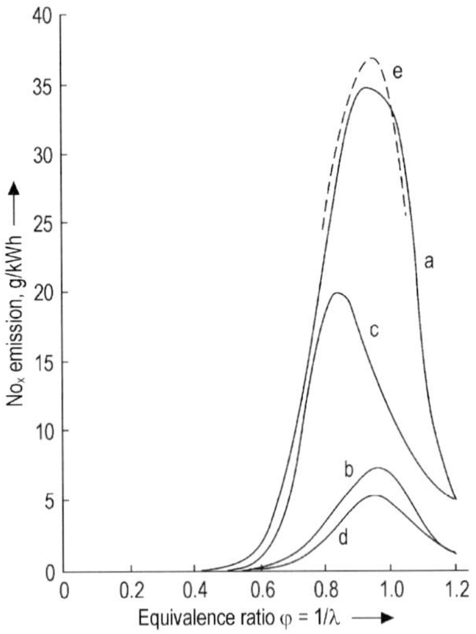

Fig. 4.14 Typical NO_x values for various modes of operation of hydrogen (a – d) and gasoline engines a) External mixture formation, load governing; b) External mixture formation, quality governing; c) External mixture formation, optimum attainable values; d) Internal mixture formation; e) Typical values for Otto and Diesel engines.

Fig. 4.15 Year 2001 BMW 745h with 4.4-L 8-cylinder-V-engine for liquid hydrogen or gasoline [67].

State of the Art The development project of BMW in the Federal Republic of Germany is described, as this is representative of the many investigations being carried out in several countries.

The near-series BMW 745h, built in 2001, has a bivalent 4.4-L 8-cylinder-V-engine for liquid hydrogen or gasoline. It has an output of 135 kW and a top speed of 215 km/h (135 mph) at a range of approximately 300 km (180 miles) in hydrogen drive plus 650 km (400 miles) in gasoline mode [67].

Whereas the basic engine is series-produced, for hydrogen drive certain modifications in the mixture process are necessary. For this purpose, an electronic fuel-mixture system exactly regulates the hydrogen intake and the charge cycle. The combustion process generally involves an excess of air. This absorbs heat in the combustion chamber and keeps flame temperature below the critical limit at which uncontrolled ignition can occur. At the same time, the substantial surplus amount of air and concomitantly reduced combustion temperature avoids the risk of generating oxides of nitrogen (NO_x).

The liquid hydrogen is kept in a highly-insulated140-L tank in the luggage compartment. The insulation system allows loss-free storage of hydrogen for about three days and can be refilled in around 3 min either by a fully automated robot, or simply by hand as usual.

Figure 5.15 shows the described BMW 745h.

Fuel Cell Drive See Chap. 5 for a detailed discussion of fuel cells.

Some Projects to Promote Drive In April 1999 the "California Fuel Cell Partnership" — a cooperation of car manufacturers, public institutions, mineral oil and energy suppliers — was founded. Starting from the year 2000 the California Fuel Cell

Partnership wants to launch innovative automobiles on the streets of California. The hydrogen and methanol automobiles, cars, and busses have to stand the test on real everyday life terms. With this fleet test the cooperation partners want to demonstrate the advantages of hydrogen drive as well as solve the question about the suitable fuel and the fuel infrastructure. In early 2007 there were 23 hydrogen fueling stations and 167 fuel cell vehicles in operation all over California [62].

The Clean Energy Partnership (CEP) is an international association of the following companies: Aral, BMW, Berliner Verkehrsbetriebe (BVG), DaimlerChrysler, Ford, GM/Opel, Hydro, Linde, TOTAL, Vattenfall Europe and Volkswagen AG. Its goal is to tap the technological potential of hydrogen as a source of energy, conducting tests with a view to suitability for routine use and system capability. The Clean Energy Partnership is supported by the German Federal Government as part of its national sustainability strategy and initially runs until December 2007 [68].

Other important projects to promote hydrogen drive are e.g.:
Icelandic New Energy – Promoting Hydrogen in Iceland [69],
Wasserstoff-lnitiative Bayern WIBA [70],
Clean Urban Transport for Europe CUTE [71].

4.7 References

1. A. S. Brown, *CEP*, Feb. 2002, **12–14**.
2. *Ullmann's*, 5th ed., **A 13**, pp. 299ff.
3. *Gmelin* System no. 2, 8.
4. R. D. McCarthy: *Hydrogen Technology Survey – Thermophysical Properties,* NASA SP-3089, Washington D.C., 1975.
5. JANAF: *Thermochemical Tables,* NSRD NBS 37, Washington D.C., 1975.
6. H. M. Roder in R. D. McCarthy, J. Hord (eds.), *Selected Properties of Hydrogen,* Dept. of Commerce NBS Report, Boulder, 1981.
7. K. E. Cox, K. D. Williamson Jr. (eds.), *Hydrogen: Its Technology and Implications,* 2nd vol. 1–5, CRC Press, Boca Raton, 1979.
8. L'Air Liquide : *Encyclopedie des Gaz,* Elsevier, Amsterdam, 1976.
9. G. Alefeld, J. Völkl (eds.): *Hydrogen in Metals,* vols. 1, 2, Springer, Berlin, Heidelberg, New York, 1978.
10. H. J. Hoge, R. D. Arnold, *J. Res. Natl. Bur. Stand. (U.S.)* **47** (1951) 63.
11. K. M. Kalyanam, D. R. Hay, *Safety Guide for Hydrogen,* National Research Council Hydrogen Safety Committee, Ottawa, 1987.
12. J. Hord, *Int. J. Hydrogen Energy* 3 (1978) 157.
13. I. L. Drell, F. E. Belles, *Survey of Hydrogen Combustion Properties,* NACA Report 1383, Washington D.C., 1952.
14. N. Getoff, *Wasserstoff als Energieträger,* Springer, Wien, New York, 1977, p. 5. NASA/ASEE Systems Design Study "Hydrogen Energy Carrier," I, II, 1973.
15. *Catalyst Handbook* ICI, Wolfe Scientific Books, London, 1972.
16. D. R. Stull, E. F. Westrum, G. C. Sinke: *The Chemical Thermodynamics of Organic Compounds,* Wiley, New York, 1969, pp. 631ff. W. D. Seider, R. Gautam, C. W. White, "Computation of Phase and Chemical Equilibrium. A Review," *ACS-Symp. Ser.* **124** (1980) 115–134.
17. H. Jungfer, R. Lohmüller, *Erdöl Kohle Erdgas Petrochem.* **39** (1986) no. 4, 188–1892.
18. C. P. Marion, J. R. Muenger, *AiChE,* (Houston, Texas, USA, 1981) 4.
19. H.-J. Madsack, *Hydrocarbon Process.* **61** (1982) no. 7, 169–172.

20. Y. Shimizu, *CEER Chem. Econ. Eng. Rev.* **10** (1982) no. 7, 9–14.
21. J. Rostrup-Nielsen, *Chem. Eng. Prog.* **73** (1977) no. 9, 87–92.
22. C.E.G. Padro: Survey of the economics of hydrogen technologies, *National Renewable Energy Laboratory (NREL)*, Sep. 1999.
23. http://www.hyweb.de.
24. M. S. Casper, *Chemical Technology Review* no. 102, Noyes Data Corp. (Park Ridge, NJ, 1978).
25. G. Strickland, *Int. J. Hydrogen Energy* 9 (1984) 759.
26. T. Yamamoto, *CEER Chem. Econ. Eng. Rev.* **13** (1981) no. 1–2, 11.
27. J. C. Amphlett et al., *Can. J. Chem. Eng.* **59** (1981) 720.
28. J. C. Amphlett, M. J. Evans, R. F. Mann, R. D. Weir, *Can. J. Chem. Eng.* **63** (1985) 605.
29. F. J. Brykowski (ed.): *Ammonia and Synthesis Gas, Recent Energy-Saving Processes,* Noyes Data Corp., Park Ridge, NJ, 1981, pp. 130–147.
30. F. G. Wiebner, *Gas Separation and Purification* **2** (1980) no. 3, 115–119.
31. H. W. Wooley, R. B. Scott, F. G. Brickwedde, *J. Res. NBS* 41 (1948) 379, RP 1932.
32. H. Blechschmidt, *Kälte* **9** (1976) 357.
33. K. Baumgärtner, R. Ewald, *Chem. Ing. Tech.* **59** (1987) 296.
34. G. Kaske, H. G. Fasold, P. Schmidt in D. Behrens (ed.): *Transport- und Verteilungsleitungen in Wasserstofftechnologie,* DECHEMA, Frankfurt/Main 1986, p. 462.
35. J. Reijerkerk, B. Valentin: *Wirtschaftlichkeit einer zukünftigen Tankstellenversorgung mit Wasserstoff, Deutscher Wasserstoff Energietag,* Essen, Nov. 2002.
36. http://physicsweb.org/article/world/11/1/9.
37. L. Paterson, *Int. J. Hydrogen Energy* **8** (1983) 53.
38. W. Peschka (ed.): *Flüssiger Wasserstoff als Energieträger,* Springer, Wien, New York 1984.
39. J. Wolf: *Wasserstoffspeicherung, Transport und Logistik,* Sommerseminar Fachhochschule Lübeck, Sep. 2001.
40. G. Alefeld, J. Völkl (eds.): *Hydrogen in Metals,* vols. I and vol. II. *Topics and Applied Physics,* vols. 28, 29, Springer Verlag, Berlin, Heidelberg, New York 1978.
41. G. G. Libowitz, G. D. Sandrock (eds.): *Proceedings and Applications of Metal Hydrides,* vols. 1 and 2, Elsevier, Amsterdam 1980.
42. H. Wenzl, *Int. Met. Rev.* **27** (1982) 140.
43. G. D. Sandrock, S. Suda, Applications in L. Schlapbach (ed.): *Hydrogen in Intermetallic Compounds,* Springer Verlag, Berlin, Heidelberg, New York 1986.
44. F. A. Kuijpers, H. H. VanMal, *J. Less-Common Met.* **23** (1971) 395.
45. P. D. Goodell, G. D. Sandrock, E. L. Huston, *J. Less-Common Met.* **73** (1980) 135.
46. M. J. Song, M. Pezat, B. Darrier, P. Hagenmüller, *J. Solid State Chem.* **56** (1985) 191.
47. T. Klassen, W. Oelerich, R. Bormann: *Nanocrystalline Mg-based Hydrides: Hydrogen Storage for the Zero-Emission Vehicle, Energietechnik mit Wasserstoff und Brennstoffzellen,* Sommerseminar Fachhochschule Lübeck, Sep. 2001.
48. http://www.gkss.de/Themen/W/WTP/wasserstoff/index.html.
49. C. Carpetis in C. J. Winter, J. Nitsch (eds.): *Wasserstoff als Energieträger,* "Speicherung, Transport und Verteilung von Wasserstoff," Springer Verlag, Berlin, Heidelberg, New York, Tokyo 1986, p. 245.
50. K. Hoyermann, H. Janaer, H. G. Wagner, *Ber. Bunsenges. Phys. Chem.* 79 (1975) 858.
51. J. L. Sloop, *Liquid Hydrogen as a Propulsion Fuel,* The NASA History Series, BASA-SP-4404, TL 785 S 58, 1978.
52. http://www.boeing.com/defense-space/space/returntoflight/vehicleupgrades/ssme.html.
53. H. Dederra, E. Kirner, MBB: *Vulcain HM60-Thrust Chamber Critical*

Development Items, 15th ISTS-Congress, Tokyo, May 1986.
54. http://www.arianespace.com/site/launcher_sub_index.html
55. D. Brewer, *Astronaut. Aeronaut.* **12** (1974) 40.
56. P. Hofmann *The Forever Fuel. Hydrogen as Aircraft Fuel*, Westview Press, Boulder, CO 1982, p. 139.
57. International Symposium (DGLR and DFVLR), *Hydrogen in Air Transportation*, Sept. 11th to 14th, 1979, Stuttgart.
58. E. Reshoto, *J. Aircr.* **16** (1979) 584.
59. G. D. Brewer (ed.): NASA Contractor Report CR 2700, Washington, DC 1976.
60. H. P. Alder, *Int. J. Hydrogen Energy* **12** (1987) 571.
61. Transboundary Air Issues Branch (TAIB): *Phase II Federal Smog Management Plan*, Government Environment Canada, March 1998.
62. http://www.cafcp.org.
63. http://www.h2cars.de.
64. BMFT Studie, *Neuen Kraftstoffen auf der Spur*, Bonn 1974.
65. L. A. Sinclair, J. S. Wallace, *Int. J. Hydrogen Energy* **12** (1984) 123.
66. H. S. Homan, P. C. T. De Boer, W. J. McLean, *Int. J. Hydrogen Energy* **8** (1983) 131.
67. http://www.bmw.com/cleanenergy.
68. http://www.cep-berlin.de.
69. http://www.newenergy.is/.
70. http://www.wiba.de.
71. http://www.fuel-cell-bus-club.com/

5
Fuel Cells

Matthias Duisberg, Marco Lopez, Katharina Seitz, Gerhard Sext

5.1
The Function of Fuel Cells

In future energy supply scenarios fuel cells will play an important role. Fuel cell systems are more environmentally friendly than conventional systems and their size can be adapted to the market requirements, ranging from fuel cells used to drive cars to those for electricity generation in power plants.

Fuel cells directly convert chemical into electrical energy – without noise emission or open combustion. They use hydrogen or hydrogen-containing gases as the fuel and pure oxygen or air as the oxidant. In contrast to the detonating oxyhydrogen reaction, which releases thermal energy by way of explosion, the electrochemical reactions in a fuel cell release energy in the form of electrical current and only minor amounts as heat. In an ideal case only pure water is formed as the reaction product. To control the generation of energy, the gases are converted separately on catalyst layers, which function as anodes and cathodes. Hydrogen (H_2) is chemically oxidized at the anode. The electrons released in this process flow through the external circuit to perform electrical work before they reduce oxygen molecules to oxonium anions (O_2^-) at the cathode. Hydronium ions (protons, H^+), which are produced by the oxidation of hydrogen, move through the gas-impermeable electrolyte and recombine with oxonium anions to form water. Figure 5.1 shows the structure of a polymer electrolyte membrane fuel cell (PEMFC), the fuel cell with the greatest prospects for use in car engines.

Oxidation reaction at the anode:

$$2H_2 \rightarrow 4H^+ + 4e^-$$

Reduction reaction at the cathode:

$$O_2 + 4H^+ + 4e^- \rightarrow 2H_2O$$

Handbook of Fuels: Energy Sources for Transportation. Edited by Barbara Elvers

ISBN: 978-3-527-30740-1

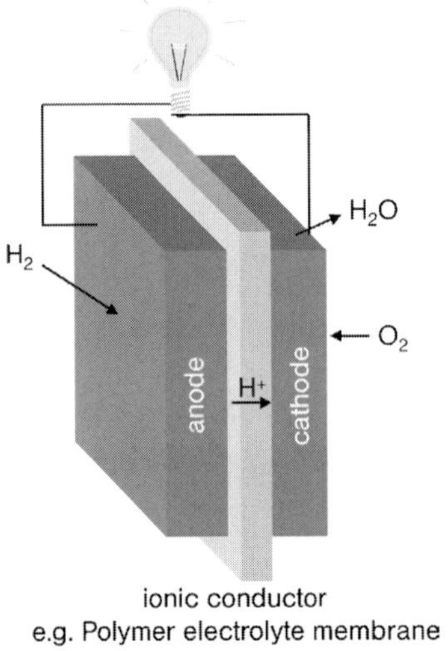

Fig. 5.1 Schematic of the structure of a PEMFC.

5.2 History

The working principle of the fuel cell has been known since the 19th century: in 1839 the physicist Sir William Robert Grove discovered the reversibility of the electrolysis of water at a platinum wire surface. In the apparatus designed by Grove (Fig. 5.2), hydrogen and oxygen are converted in a nitric acid solution, so that a current flows between the two electrodes. Grove named his apparatus a gas battery.

Later the chemists Ludwig Mond and Charles Langer tried to develop a device that used air and coal gas as fuels. These two chemists also introduced the term fuel cells. At the end of the 19th century Wilhelm Ostwald pointed out the

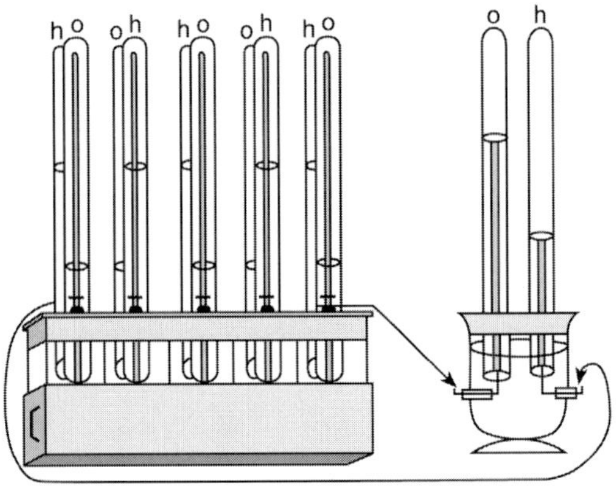

Fig. 5.2 Schematic of Grove's original gas battery apparatus [1].

advantages of the fuel cell, which enabled the direct production of electrical energy [2]. This principle found its first technical use in 1939. FRANCIS BACON, a British engineer, constructed a fuel cell that used hot caustic potash solution at 200 °C as the electrolyte, which had to be used under pressure, with nickel nets as the electrodes. This fuel cell developed by BACON was utilized in submarines during the Second World War.

As a result of the high costs, shortly after the war fuel cells only found application in space travel applications, as a replacement for batteries. Based on BACON's cell, the NASA Apollo program of the early 1960s used alkaline fuel cells (AFCs) as an energy source. Owing to their high energy efficiency AFCs were also used in later space shuttle programs.

In parallel with the AFC, the PEMFC was developed. Similar to the AFC, the first applications were in space travel and in the military. A PEMFC was used, for

Table 5.1 Development steps of fuel-cell technology

Year	Development
1800	JOHANNES V. RITTER produces hydrogen and oxygen through the electrolysis of a sulfuric acid solution
1839	Sir W. GROVE recognizes the reversibility of RITTER's experiments and constructs the first prototype of a fuel cell
1889	LUDWIG MOND and CHARLES LANGER construct a fuel cell using air and coal gas, on the basis of GROVE's experiments
1894	WILLIAM OSTWALD points out the relevance of GROVE's discovery in a discourse
1939	FRANCIS T. BACON constructs the first fuel cell with an alkaline electrolyte (200 °C hot caustic potash solution and pressurized hydrogen)
1950s	Siemens and Varta investigate AFC technology
1960s	The PEMFC is tested as an energy source in NASAs Gemini space travel program
1960s	Use of alkaline fuel cells in the Apollo space travel program
1982	Submarines with PEMFCs
1994	The first cogeneration power plant based on phosphoric acid fuel cells (PAFC) comes into service in Hamburg
1997	Daimler–Benz introduces a public-service bus (new electric bus, NEBUS) with fuel cells
1997	Start of the NECAR (new car) model ranges as an experimental platform for automobile applications
2000	Introduction of a go-cart based on direct methanol fuel cell (DMFC)
2002	Crossing the USA with a NECAR 5 from DaimlerChrysler
2003	DaimlerChrysler, Ford, Honda, Nissan and Toyota start first demonstration fleets

example, at the beginning of the Gemini space travel program. However, it was not applied in the Apollo program because of technical problems.

Growing environmental awareness in the late 20th century and the large increase in environmental pollution due to emission of waste gases of all types resulted in rapidly increasing interest in the low-emission fuel-cell technology. Efforts to improve air quality led to the introduction of the Clean Air Act in California in 1988. This law demanded a set proportion of emission-free automobiles (zero emission vehicles) to be in operation from 2004 onward. This put the automobile industry under strong pressure to develop clean driving techniques. Various prototypes of fuel-cell vehicles have been developed since the mid-1990s [3].

An overview of the most important steps of the development of fuel-cell technology is presented in Table 5.1 [4].

5.3
Fuel Cells and their Applications

Fuel-cell systems and their components have to conform to various different requirements. While portable fuel cells, for example for computers or mobile phones, correspond to the batteries used at present with regard to size and power (from 0.1 to a few 10 or 100 W), stationary (standby) systems require about 1 kW of power and systems in use require about 30 kW of power. In recent decades various types of fuel cells have been developed, which differ mainly in the components employed and the working temperatures (Fig. 5.3).

All fuel cells consist of a gas-impermeable ion conductor as the electrolyte, which is catalytically coated on both sides or which has catalytic characteristics itself. This ion conductor separates the anodic and cathodic spaces of the electrochemical cell from each other. Depending on the working temperature, which can range from room temperature to 1000 °C depending on the cell type, hydrogen, alcohols or hydrocarbons (natural gas, gasoline, diesel fuel) are used as the fuel and air or oxygen as the oxidant. Table 5.2 summarizes the various operating conditions, the catalysts used and application areas.

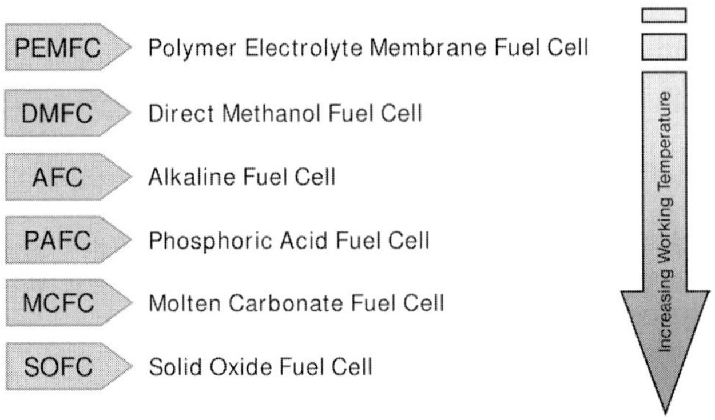

Fig. 5.3 Major types of fuel cells.

Table 5.2 Operating conditions of major types of fuel cells

	AFC	PEMFC	PAFC	MCFC	SOFC
Operating temperature (°C)	60–90	80–110	160–200	600–800	800–1000 (new developments: 500–600)
Fuel	H_2 (high-purity)	H_2, methanol	H_2, natural gas	natural gas	natural gas
Oxidant	O_2 (high-purity)	O_2, air	air	air	air
Anode catalyst (examples)	Raney Ni, Pt/C	Pt/C, Pt-Ru/C	Pt/C,	Ni/Al Ni/Cr	Ni
Electrolyte	aq. KOH	sulfonated PTFE polymer	phosphoric acid	Li/Na carbonate	yttrium doped zirconium oxide
Cathode catalyst	Raney Ni	Pt/C	Pt/C, Pt-Co-Cr/C	NiO	La-Sr-MnO_3
Preferred application	space travel	mobile, portable, stationary	stationary (block-type thermal power units	stationary (block-type thermal power units	stationary (block-type thermal power units

In practice, many separate cells are connected in series in a stack, similar to the construction of a battery. Figure 5.4 shows the construction of a fuel-cell stack of a PEMFC with seven separate cells connected in series. The stack is fitted with electrical connections and inlets for the fuel and air. If the stacks exceed a certain power and size, they have to be cooled to prevent uncontrolled overheating due to the heat of the reaction. In such a case, further connections for a cooling media (air- or liquid-cooling) are present. The individual membrane–electrolyte assemblies (MEA) are separated by bipolar plates, which enable the collection of the current. With their coiled-type gas distributors, the bipolar plates also facilitate the entry of fuel and oxidant (dark-gray openings in Fig. 5.4). Often the plates contain channels for cooling media to remove the heat of reaction. Two end plates close the stack off, which makes the whole structure mechanically stable. Simultaneously the generated electrical energy is collected through the end plates.

Noble-metal containing catalysts are mainly used in the types of fuel cells that work at relatively low temperatures (100–200 °C); the PEMFC, AFC, PAFC and DMFC belong to this category. Transport of matter and charge occurs through protons in the electrolyte in these fuel-cell types.

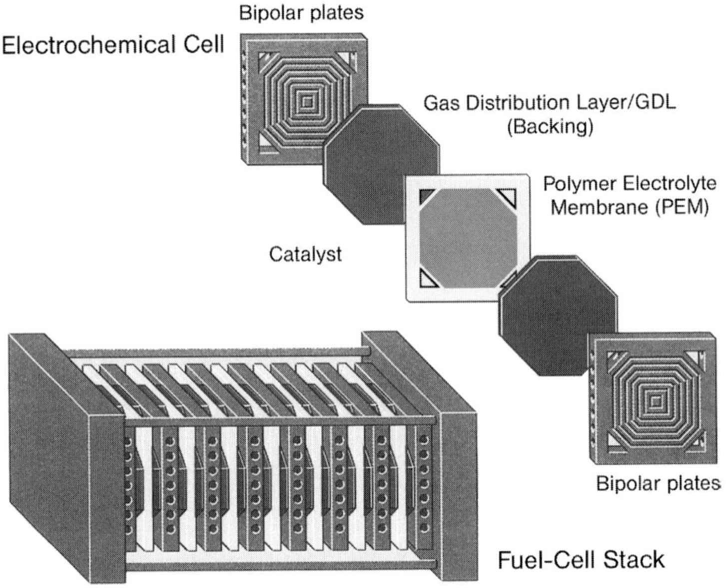

Fig. 5.4 Principles of the construction of a PEM fuel-cell stack.

High-temperature fuel cells, such as the molten carbonate fuel cell (MCFC) and the solid oxide fuel cell (SOFC) do not require a noble-metal catalyst, due to the high working temperatures at which they are used. In these cells the charge transport occurs through carbonate ions (MCFC), or O^{2-} ions of the crystal lattice (SOFC).

With MCFC systems it is possible to reform simple hydrocarbon-containing fuels with high methane content inside the cell at the anodic catalyst (nickel or nickel–chromium alloy). Nickel oxide serves as the cathodic material. Carbon monoxide, produced as a side product in the reforming process, does not interfere with the subsequent oxidation of hydrogen. A mixture of lithium–potassium or lithium–sodium carbonate is used as the electrolyte, which is stabilized in a matrix ($LiAlO_2$). Owing to the strong corrosive characteristics of the carbonate melt, the requirements on the resistance of the construction materials are particularly high. The electrical efficiency is approximately 50 %. When the energy of the hot fuel-cell off-gas is converted into electrical energy within a gas turbine (CHP, combined heat and power), the electrical efficiency increases to over 70 %.

The working temperature of SOFCs lies between 800 and 1000 °C (although 500–600 °C is also now possible) and, as in the MCFC, hydrocarbons can also be internally reformed. Nickel or nickel oxide is suitable for the anodic material; $LaSrMnO_3$ serves as the catalyst material for the cathode. A porous yttrium-doped zirconia functions as oxide ion conductor. When the generated electrical energy and the heat produced are considered, an efficiency of over 85 % is reached.

Concentrated phosphoric acid (ca. 85 %) is used as the electrolyte in PAFC systems; the electrolyte is contained in a porous silicon carbide (SiC) matrix. Hydrogen is generated from hydrocarbons in an external reformer, in which, in

Table 5.3 Thermodynamic and technically achievable electrical efficiencies of various types of fuel cell

Type of fuel cell	AFC	PEMFC	PAFC	MCFC	SOFC
Efficiency (theoretical), %	83	83	80	78	73
Efficiency (technical), %	60	60	55	55–65	60–65

addition to CO_2, CO is always produced. Owing to the relatively high working temperature of ca. 200 °C these systems tolerate a CO content of up to 1.5 vol % in the reformate. The electrical efficiency of the PAFC systems is approximately 50 %. If the waste heat is taken into the energy balance (CHP system) the efficiency can reach ca. 85 %.

The AFC has only found use in military projects and in space travel. This is because pure oxygen, or at least CO_2 free air, must be used in these fuel cells, otherwise the alkali hydroxide in the electrolyte would react with the carbon dioxide in the reformate or the air to form carbonate, which would result in blocking of the porous structure in the catalyst layer(s). In addition, the electrolytes would be slowly consumed, bringing about energy losses. Raney nickel and platinum catalysts on a carbon black carrier are used as catalysts. No other type of fuel cell can reach such a high electrical efficiency of ca. 83 % (Table 5.3 [5]).

The polymer electrolyte-based system PEMFC is the most important type of fuel cell at present. On the basis of the PEMFC system, alternative concepts for the supply of energy to homes, working systems and for powering automobiles are being developed.

5.4 Polymer Electrolyte Fuel Cells

5.4.1 Introduction

Figure 5.5 shows the construction of a PEMFC. The power range of the PEMFC system extends from a few watts for portable applications up to 100 kW for automobile applications.

The polymer electrolyte of the PEMFC separates the anode and cathode compartments of the cell. The polymer used is a copolymer of tetrafluoroethylene and perfluoro-2-(fluorosulfonylethoxy)propyl vinyl ether (Fig. 5.6), which is stable up to 100 °C [6]. This class of polymer was originally developed by DuPont as a cation-exchange resin (trade name Nafion), only to be used later as a proton-conducting electrolyte [7]. In the meantime proton-conducting polymers and films of similar composition have become available from other producers, e.g., from Gore, 3M, Asahi Glass Corp. and Asahi Kasei [8]. The polymers differ essentially in their

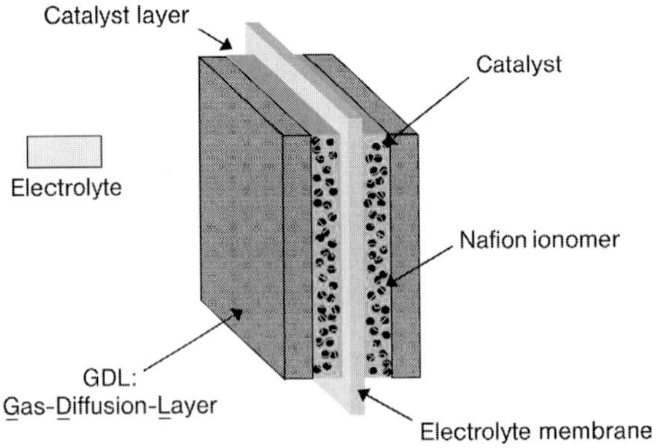

Fig. 5.5 Schematic of the PEMFC structure.

molar masses, three-dimensional structure and the equivalent mass of the monomer. The equivalent mass indicates how many grams of polymer correspond to a mole of the sulfonic acid group.

As with other types of low-temperature fuel cells, PEMFCs require catalysts based on noble metals in order to achieve sufficient activity at the working temperature of 100 °C. Both sides of the electrolyte membrane are coated with a catalyst layer, which is composed of the electrocatalyst and short-chain electrolyte ionomers, which are distributed in the layer. Platinum catalysts on an electrically conductive carbon black carrier or a pure noble metal catalyst are used. The catalyst layer has to fulfill a number of requirements: it must be porous and gas permeable in order to provide an uninterrupted diffusion of hydrogen to the catalyst particles of the anode and oxygen to the cathode, in addition to ensuring the transport of the product, water, from the cathode. The catalyst particles have to be embedded in a layer that is electrically conductive; in the case of supported catalysts this function is fulfilled by the carbon black carrier. At the same time the particles have to be in contact with the proton conductor. To this end, electrolyte ionomers are included in the catalyst layer, which form a porous three-dimensional network, due to their hydrophobic/hydrophilic structure. The sulfonic acid groups project into the pores of the network and enable the proton and ion conduction [6], [9]. For catalysts without a carrier ("full metal catalyst") the catalyst itself provides the porosity of the layers on the membrane. To this end a catalyst material must be produced with

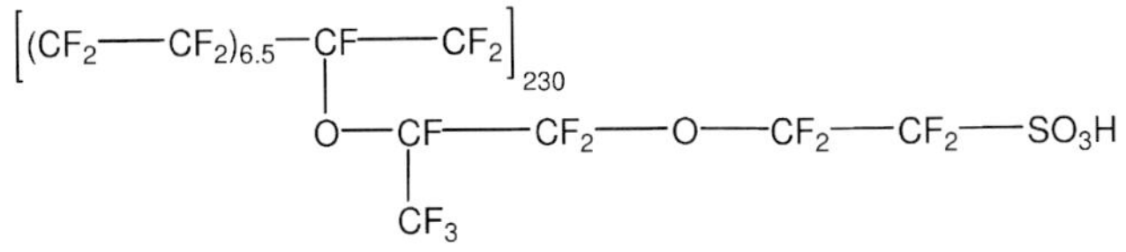

Fig. 5.6 Structure of Nafion.

a high porosity that has to be maintained during further processing. Thus, in supported catalysts the carbon black carrier, and in pure metal catalysts the metal itself, acts as the conductor.

To provide an even distribution of the gases in the anode and cathode compartments, the gas distribution layer (GDL) is deposited on the catalyst layer. Useable GDLs normally consist of carbon fiber paper or cloths, which are electrically conductive and provide the charge balance with the bipolar plates.

5.4.2 Efficiency of PEMFCs

A number of effects are responsible for the difference between the theoretical cell voltage U_{H0} and the actual achievable terminal voltage U_{eff}, leading to significant losses in energy (Fig. 5.7). One of the reasons for energy loss is the kinetic hindrance of oxidation and reduction reactions at both electrodes, giving rise to the overvoltage ΔU_{kin}, *and ohmic losses,* ΔU_{Ω}, due to the membrane and the cell construction. In high current density regions, over 800 mA/cm^2, mass-transport effects cause a larger or smaller loss in cell voltage. At this current density the diffusion of the gases to the electrodes and the diffusion of the reaction product, water, from the reaction site proceeds slower than the electrochemical conversion. The different contributions to the loss in cell voltage and the methods for reduction of these losses are discussed in the literature. Reference [10a] provides a good overview.

The types of PEMFC differ mainly in the type of fuel employed. In addition to pure hydrogen, natural gas, propane, gasoline or methanol can also be used.

5.4.3 PEMFC for Operation with Hydrogen

PEMFCs that operate with pure hydrogen use catalysts with carbon black as the carrier and platinum as the active component for hydrogen oxidation (anode) and oxygen reduction (cathode). The size of the platinum particles ideally lies in the nanometer region to allow maximum electrochemical usage of the noble metal [11]. Conductive carbon black with large surface areas and good electrical conductivity serves as a carrier [12].

When using air as the oxidant the greatest loss of voltage is caused by a cathodic overvoltage (Fig. 5.7). The high overvoltage is a result of the kinetically hindered oxygen reduction, which is the sum of the different rate-determining steps. To minimize these effects, use of platinum alloys has been proposed. Binary platinum alloys PtX with X = Mn, Fe, Co, Ni or Cr have been tested as alternatives; these allow acceleration of the rate-determining step in the oxygen reduction reaction [12b], [13].

There are several models that explain how metal alloys influence the oxygen reaction. For more details see [13b], [14].

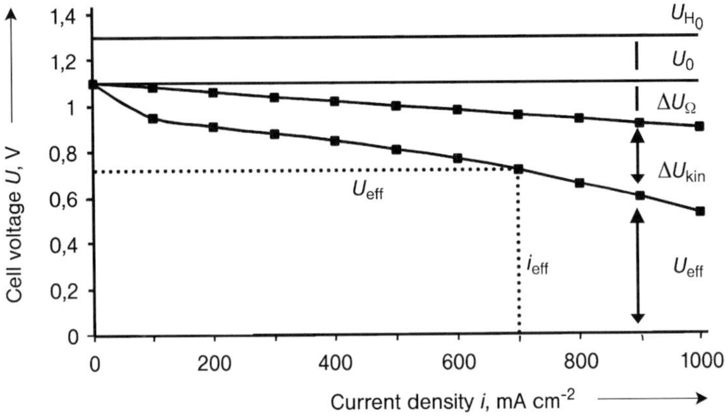

Fig. 5.7 Graph of current potential versus energy loss for a typical PEMFC.

5.4.4 PEMFC for Use with Hydrogen Produced in Reformers

Owing to the low working temperatures of PEMFCs, neither hydrocarbons nor alcohols can be used directly as fuel. They must first be transformed into a hydrogen-containing gas. The processes of transforming a raw material with chemically bonded hydrogen into a hydrogen-containing gas mixture (reforming), and subsequent purification of this gas from unwanted components (gas purification), are known collectively as fuel processing (see Section 5.5). In addition to hydrogen, reformed gas also contains a proportion of CO.

CO is a strong catalyst poison because it binds irreversibly to platinum. As a consequence, the active surface of the anode catalyst that is available for the adsorption of hydrogen molecules and their electrochemical oxidation is reduced. Figure 5.8 shows schematically the situation at a platinum surface if the reformed hydrogen gas contains CO: only those Pt particles that are free from CO can adsorb hydrogen molecules and split them into atoms. As a result, the turnover at the anode and the performance of the fuel cell are severely reduced.

Several strategies have been developed to produce a viable PEMFC that can function in spite of residual CO in the feed gas: some of the simplest measures include the operation of the fuel cell under pressure or the mixing of a small amount (a few vol %) of air with the hydrogen obtained by reforming. The amount of air added must be metered carefully so that it is clearly below the lower explosion limit

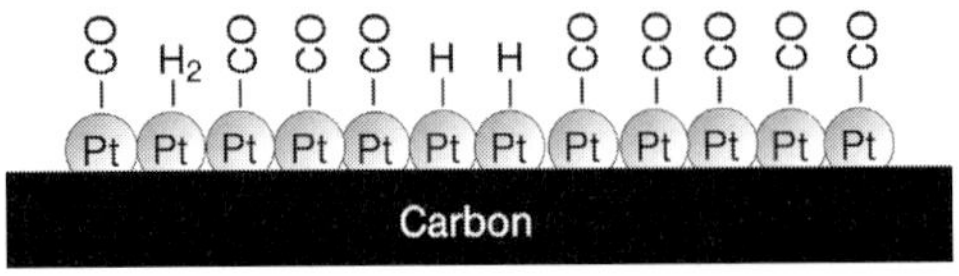

Fig. 5.8 Competing H_2 and CO_2 adsorption at a Pt surface.

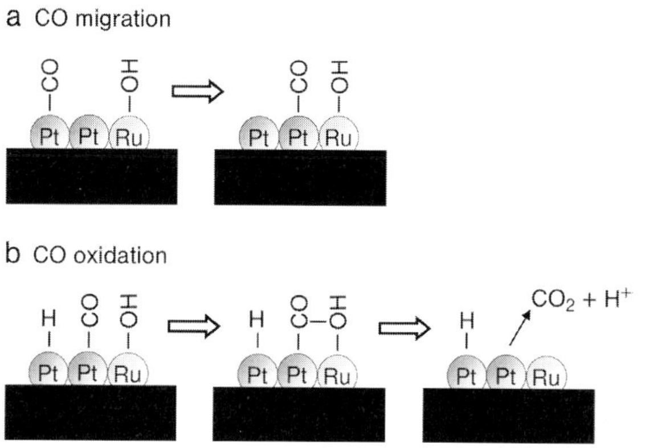

Fig. 5.9 Mechanism for CO oxidation on a PtRu catalyst.

of a hydrogen–air mixture. Oxygen in the air oxidizes the CO adsorbed at the surface of the catalyst to CO_2, which is subsequently desorbed. This procedure is referred to in the literature as an air bleed technique. It has the disadvantage that the added oxygen also reacts with hydrogen to form water, which reduces the efficiency of the fuel cell. Furthermore, this technique requires various peripheral aggregates (e.g., air compressors) and extensive safety features that act as "parasitic" consumers of energy, reducing the overall efficiency.

A higher CO tolerance is best achieved by the use of special anode catalysts. These involve bimetallic catalysts that can significantly improve the performance of PEMFCs, utilizing hydrogen obtained by reforming [10]. A platinum–ruthenium alloy has been shown to be particularly advantageous [15]. Accordingly, over the last ten years, PtRu catalysts have undergone constant development and improvement. Other suitable alloy metals are molybdenum [16], tungsten [17] or tin [18].

The effect of the second metal is attributed to its ability to positively influence the electronic structure of platinum. CO is adsorbed at the surface of the Pt catalyst in two forms, which differ in their binding strength. In the case of the bimetallic catalyst (e.g., PtRu or PtSn), the weaker-binding form accumulates, so that more CO can be oxidized on the platinum under the operating conditions. In addition, the metal alloy itself is able to catalyze the oxidation of the platinum-adsorbed CO to CO_2, thereby counteracting any poisoning of the catalyst. Thus, the number of free centers of adsorption at the platinum surface is increased further. Figure 5.9 shows a possible mechanism of CO oxidation at a metal surface with Pt and Ru clusters [19].

5.4.5
Direct Methanol Fuel Cell

The direct methanol fuel cell (DMFC) is a special type of PEMFC. Both PEMFCs and DMFCs are constructed according to similar principles; the anode and cathode catalysts and also the polymer electrolyte are similar. However, in the case of the DMFC, instead of hydrogen, it is methanol in an aqueous solution that is oxidized at the anode catalyst (a PtRu alloy) to form water and CO_2. The reaction releases six electrons per molecule of methanol (Eq. 1).

$$CH_3OH + H_2O \rightarrow CO_2 + 6H^+ + 6e^- \quad (1)$$

The anode catalyst of the DMFC must be tolerant to CO as the oxidation occurs stepwise with CO as an intermediate (Eqs. 2–8) [10].

$$Pt + CH_3OH \rightarrow Pt{-}(CH_3OH)_{ads} \quad (2)$$

$$Pt{-}(CH_3{-}OH)_{ads} \rightarrow Pt{-}(CH_3O)_{ads} + H^+ + e^- \quad (3)$$

$$Pt{-}(CH_3O)_{ads} \rightarrow Pt{-}(CH_2O)_{ads} + H^+ + e^- \quad (4)$$

$$Pt{-}(CH_2O)_{ads} \rightarrow Pt{-}(CHO)_{ads} + H^+ + e^- \quad (5)$$

$$Pt{-}(CHO)_{ads} \rightarrow Pt{-}(CO)_{ads} + H^+ + e^- \quad (6)$$

$$M + H_2O \rightarrow M{-}(H_2O)_{ads} \quad (7)$$

$$Pt{-}(CO)_{ads} + M{-}(H_2O)_{ads} \rightarrow Pt + M + CO_2 + 2H^+ + 2e^- \quad (8)$$

M = Ru, Mo, W, Sn, Os, etc.

The rate-limiting step in the decomposition of a methanol molecule at the surface of the anode catalyst is the oxidation of the bound CO (Eqs. 7 and 8) [21]. As with the reformate PEMFC, bimetallic catalysts are used. The alloying metal adsorbs either water or OH molecules, which then enable the oxidation of CO. Standard anode catalysts are carrier [22] or pure platinum–ruthenium catalysts [22b], [23]. In addition, ternary (e.g., PtRuW [24] or PtRuOs [25]) and quaternary [26] catalyst formulations are known.

A further aspect that must be considered in the design of a DMFC is the undesired cross-diffusion of methanol through the electrolyte membranes [27]. Methanol is then oxidized at the cathode, which causes a reduction of the cell voltage, because of the mixed potential at the cathode due to methanol oxidation and oxygen reduction. One possibility to prevent methanol cross-diffusion is the use of thicker electrolytic membranes, as in the cases of reformate and hydrogen cells, and the use of methanol-tolerant cathode catalysts. Unfortunately, the former method also results in an increase in the ohmic resistance of the single cell.

As a fuel, methanol has the advantage over hydrogen of having a higher energy-storage capacity and less stringent safety requirements regarding the machinery and its operation. Methanol fuel cells are, therefore, particularly suitable for small portable applications. However, the performance of the DMFC per surface area

(W/cm^2 MEA) is a factor of three to five lower than that of a PEMFC using reformate hydrogen. Furthermore, the DMFC requires the use of a large amount of noble metal in the catalyst layers.

5.5 Fuel Processing

At present, a widespread supply of hydrogen is not possible due to the inadequate infrastructure. In addition, problems in the handling and storage of hydrogen have not yet been solved satisfactorily. For example, the transport of compressed hydrogen by road consumes more energy than the hydrogen itself provides. Therefore, decentralized generation of hydrogen is an appropriate alternative. This can be either a decentralized production of stockpiles of hydrogen at, for example, fueling stations, or the transformation of available energy carriers into a hydrogen-rich gas (reforming) directly before its use in a fuel cell.

The reforming of hydrocarbons produces, in addition to hydrogen and carbon dioxide, up to 10 vol % carbon monoxide. Because of the high proportion of CO, reformed gas has a high calorific value. Through oxidation of the CO to CO_2 by the water–gas shift reaction (WGS), further hydrogen is formed (see also Section 4.4.1). The remaining CO content, maximum 1 vol %, must then be removed, as CO is a poison to the anode catalyst of the fuel cell. The aim of the purification step is to reduce the CO content to levels at which the losses in performance of the fuel cells are negligible. Therefore, with today's catalysts the CO content in the reformed gas should be $\leq$100 ppm. Consequently, several catalytic steps are required in the fuel-processing reaction chain to ensure that PEMFCs can be driven by hydrogen produced by reforming (see Fig. 5.10).

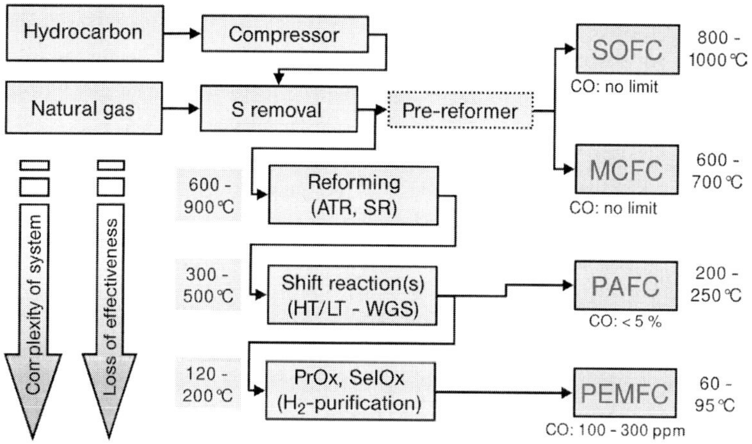

Fig. 5.10 "Fueling the Fuel Cells": production of fuel gas for various types of fuel cell, starting from liquid hydrocarbons or natural gas.

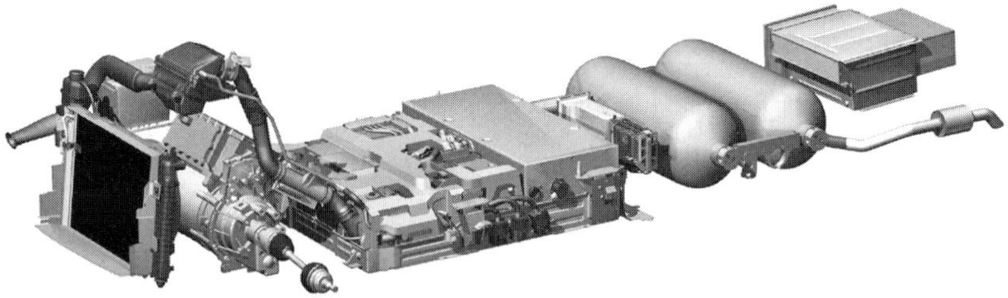

Fig. 5.11 Fuel cell system for vehicle propulsion (Courtesy of DaimlerChrysler).

5.6
Fuel Cells for Vehicle Propulsion

A schematic of a fuel cell system used for powering of vehicles is depicted in Fig. 5.11.

Two examples of vehicles developed by DaimlerChrysler and equipped with PEMFC stacks are shown in Fig. 5.12 and 5.13. Figure 5.12 depicts an A-class

Fig. 5.12 A-class passenger car with PEMFC stacks [29] (Courtesy of DaimlerChrysler)

Fig. 5.12 *(Continued)*

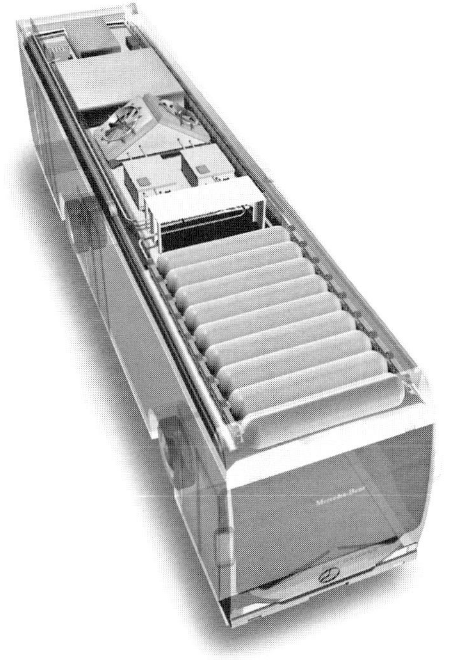

Fig. 5.13 Citaro bus with PEMFC stacks [30] (Courtesy of DaimlerChrysler).

passenger car, where the whole fuel cell system is accommodated in the sandwich floor construction of the car with long wheelbase. The car's fuel tanks contain hydrogen compressed under 350 bar pressure and give the car an operating range of ca 150 km. Hydrogen consumption is equivalent to 4.2 L of diesel per 100 km. The electrical energy gained drives an electric motor. The electric motor generates 65 kW, accelerating the car from 0 to 100 km/h in 16 s and on to a maximum speed of 140 km.

The 12 m long Mercedes-Benz Citaro low-floor bus with fuel cell drive system has a range of some 200 km and can accommodate more than 60 passengers, depending on the individual customer specification. The PEMFC stacks (Mark 900, produced by the Canadian company Ballard) with a power output of over 200 kW and the compressed gas cylinders containing hydrogen compressed at 350 bar are located on the roof of the Citaro bus (see Fig. 5.13). Maximum speed is up to 80 km/h. The electric motor, transmission, driveshaft and mechanical rear axle are mounted in the rear section of the bus.

In the frame of the EU sponsored project CUTE Citaro fuel cell powered buses went into service in Amsterdam, Barcelona, Madrid, Hamburg, London, Luxembourg, Porto, Stockholm, Stuttgart and Reykjavik in 2003 [30]. The project ended in 2005, but because of its great success was prolonged and extended under the new name HyFLEET:CUTE, the successful demonstration phase of HyFLEET:CUTE ending in February 2007 [31].

5.7 References

1. W. R. Grove: On a New Voltaic Combination of Gases by Platinum, *The London and Edinburgh Philosophical Magazine and Journal of Science, Philos. Mag.* **14** (1839) 127.
2. W. Ostwald, *Z. Elektrotech. Elektrochem.* **1** (1894) no. 4, 122.
3. A. E. Hammerschmidt, M. F. Waidhas, *Spektrum der Wissenschaft* (1999) no. 2, A44.
4. a) Internet page from Smithonian Institutes http://americanhistory.si.edu/csr/fuelcells/index.htm; b) http://dc2.uni-bielefeld.de/dc2/fc/folien/f-geschi.htm; c) http://www.fuelcelltoday.com.
5. H. Wendt, M. Götz, *Chem. unserer Zeit* **31** (1997) no. 6, 301.
6. H.-G. Haubold, Th. Vad, H. Jungbluth, P. Hiller, *Electrochim. Acta* **46** (2001) 1559.
7. a Q. Sun, M. A. Harmer, W. E. Farneth, *Ind. Eng. Chem. Res.* **36** (1997) no. 12, 5541; b) A. Heidekum, M. A. Harmer, W. F. Hoelderich, *J. Catal.* **188** (1999) no. 1, 230.
8. M. Wakizoe, O. A. Velev, S. Srinivasan, *Elektrochim. Acta* **40** (1995) 335.
9. K. Linsmeier, *Spektrum der Wissenschaft* (2001) no. 9, 66.
10. W. Vielstich, A. Lamm, H. A. Gasteiger (eds.): *Handbook of Fuel Cells,* John Wiley & Sons Ltd., Chichester (2003). a) Volume 1: Fundamentals and Survey of Systems; b) Volume 2: Electrocatalysis; c) Volume 3: Fuel Cell Technology and Applications Part I; d) Volume 4: Fuel Cell Technology and Applications Part II.
11. a) K. A. Starz, E. Auer, T. Lehmann, R. Zuber, *J. Power Sources* **84** (1999), 167; b) M. Watanabe, S. Saegusa, P. Stonehart, *J. Electroanal. Chem. Interf. Electrochem.* **271** (1989) 213.

12. a) Y-C. Liu, X.-P. Qiu, Y.-Q. Huang, W.-T. Zhu, *J. Power Sources* **111** (2002) 160; b) T. R. Ralph, M. P. Hogarth, *Platinum Metals Rev.* **46** (2002) no. 1, 3; c) T. R. Ralph, M. P. Hogarth, *Platinum Metals Rev.* **46** (2002) no. 3, 117.
13. a) T. Toda, H. Igarashi, M. Watanabe, *J. Electroanal. Chem.* **460** (1999) 258; b) T. Toda, H. Igarashi, H. Uchida, M. Watanabe, *J. Electrochem. Soc.*, **146** (1999) 3750; c) U. A. Paulus, A. Wokaun, G. G. Scherer, T. J. Schmidt, V. Stamenkovic, V. Radmilovic, N. M. Markovic, P. N. Ross, *J. Phys. Chem. B* **106** (2002) 4181.
14. F. A. Uribe, T. A. Zawodzinski Jr., *Electrochim. Acta* **47** (2002) 3799.
15. H. Binder, A. Koehling, G. Sandstede, in G. Sandstede (ed.): *From Electrocatalysis Fuel Cells*, Univ. Wash. Press, Seattle, Washington, (1972), 43.
16. a) H. A. Gasteiger, N. M. Markovic, P. N. Ross Jr., *J. Phys. Chem. B* **99** (1995) 8290; b) H. A. Gasteiger, N. M. Markovic, P. N. Ross Jr., *J. Phys. Chem. B* **99** (1995) 16757.
17. B. N. Grugur, N. M. Markovic, P. N. Ross Jr., *J. Phys. Chem. B* **102** (1998) 2494.
18. P. K. Shen, A. C. C. Tseung, *J. Electrochem. Soc.* **141** (1994) 3082; b) M. M. P. Jannsen, J. Moolhuysen, *Electrochim. Acta* **21** (1976) 861.
19. H.-F. Oetjen, V. M. Schmidt, U. Stimming, F. Trila, *J. Electrochem. Soc.* **143** (1996) 3838.
20. a) B. Beden, J.-M. Leger, C. Lamy, in J. O. M'Bockris, B. E. Conway R. E. White (eds.): *Modern Aspects of Electrochemistry*, Plenum Press, New York, (1992), *22*, 97; b) B. Beden, J.-M. Leger, C. Lamy, in J. O. M'Bockris, B. E. Conway R. E. White (eds.): *Modern Aspects of Electrochemistry*, Plenum Press, New York, (2001), *34*, 53.
21. J.-M. Léger, *J. Appl. Electrochem.* **31** (2001) 767.
22. a) V. Radmilovic, H. A. Gasteiger, P. N. Ross Jr., *J. Catal.* **154** (1995) 98; b) L. Liu, C. Pu, R. Viswanathan, Q. Fan, R. Liu, E. S. Smotkin, *Electrochim. Acta* **43** (1998) 3657.
23. M. Watanabe, M. Uchida, S. Motoo, *J. Electroanal. Chem.* **229** (1987) 395.
24. C. Roth, M. Goetz, H. Fuess, *J. Appl. Electrochem.* **31** (2001) 793.
25. K. L. Ley, R. Liu, C. Pu, Q. Fan, N. Leyarovska, C. Segre, E. S. Smotkin, *J. Electrochem. Soc.* **144** (1997) 1543.
26. a) S. Wasmus, A. Küver, *J. Electroanal. Chem.* **461** (1999) 14; b) E. Reddington, A. Sapienza, D. Gurau, R. Viswanathan, S. Sarangapani, E. S. Smotkin, T. E. Mallouk, *Science* **280** (1998) 1735.
27. X. Ren, T. A. Zawodzinski Jr., F. Uribe, H. Dai, S. Gottesfeld, *Proc. Electrochem. Soc.* **95** (1995) no. 23, 284.
28. a) C. Rice, S. Ha, R. I. Masel, A. Wieckowski, *J. Power Sources* **115** (2003) 229; b) C. Rice, S. Ha, R. I. Masel, P. Waszczuk, A. Wieckowski, T. Barnard, *J. Power Sources* **111** (2002) 83.
29. www.daimlerchrysler.com/dccom/0-5-7154-1-149991-1-0-0-149730-0-0-135-7145-0-0-0-0-0-0-0.html.
30. www.daimlerchrysler.com/dccom/0-5-7153-1-9433-1-0-0-0-0-0-8-7145-0-0-0-0-0-0-1.html.
31. www.hysolutions-hamburg.de/index.php?id=5.

6
Octane Enhancers

6.1
Methyl Tert-Butyl Ether and Ethyl Tert-Butyl Ether

Udo Peters, Franz Nierlich, Ekkehard Schulte-Körne, Michael Sakuth, Rula Deeb, Maryline Laugier, Martti Suominen, Michael Kavanaugh

6.1.1
Introduction

Methyl *tert*-butyl ether, 2-methoxy-2-methylpropane (MTBE) [1634-04-4], M_r 88.15, was first synthesized (by the classical Williamson ether synthesis) and characterized in 1904 [1].

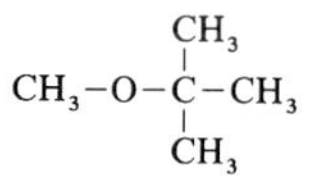

Methyl *tert*-butyl ether (MTBE)

Extensive studies in the United States during World War II demonstrated the outstanding qualities of MTBE as a high-octane fuel component [2]. Even so, it was not until 1973 that the first commercial plant went on stream in Italy.

Reduction of the lead content in gasoline in the mid-1970s led to a drastic increase in demand for octane enhancers, with MTBE being used increasingly in this role. Political decisions about gasoline quality (e.g., lower aromatics content, lower vapor pressure and a defined oxygen content) especially in the United States had led to a significant increase in MTBE consumption in the 1990s. After years with two-digit growth rates (1990–1995) MTBE consumption reached some 21×10^6 t/a in 2001/2002 and decreased sharply in 2005/2006. At the end of 2006 the remaining worldwide MTBE capacity was approximately 13×10^6 t/a.

There were two main reasons for the decline of interest in MTBE:

1. Negligently maintained underground storage tanks and pipelines as well as careless handing of MTBE containing gasoline caused a contamination of various drinking water supplies (ground water and lakes) with MTBE. As a consequence

Handbook of Fuels: Energy Sources for Transportation. Edited by Barbara Elvers

ISBN: 978-3-527-30740-1

Milwaukee (1998), California (2003), New York State and Connecticut (2004) decided to phase out MTBE in gasoline. In summer 2005 the Energy Bill passed the US Congress and became law in August 2005. This bill did not include a provision that limited the liability of the hydrocarbon processing industry from water contamination lawsuits, and resulted in an indirect ban of MTBE all over the USA as of May 2006 [3]. The concept of how to deal with the loss of nearly 4% of the total gasoline pool and maintaining the goals of the Clean Air Act [4] is described in [5].

2. In 2003 the European Commission adopted the EU Directive 2003/30/EC [6]: The aim of this Directive was the stepwise increase of utilization of biofuels for transportation. The target share of biofuels in transportation fuels (based on energy content) was 2.0% in 2005, rising up to 5.75% in 2010. Tax incentives and other governmental actions led to an increased blending of bioethanol and ETBE (ethyl *tert*-butyl ether) in gasoline. In order to balance supply and demand of "bio-containing" gasoline additives most of MTBE plants were retrofitted to ETBE plants. By the end of 2006 more than 60% of the total ether production capacities in Europe (approx 5×10^6 t/a) have been converted into ETBE production facilities [7].

In 2007 the world of ether octane enhances is divided into three regions: The United States, which phased out the use of all ethers for environmental reasons, Europe which have moved from MTBE to ETBE for ecological reasons, and the rest of world, which is still using MTBE and other ethers.

Because ETBE became the most important octane enhance in Europe during the last two years, some data for ETBE have been included into Tables7. 1 and 7.2 for comparison.

6.1.2 Physical and Chemical Properties

Physical Properties Methyl *tert*-butyl ether and ethyl *tert*-butyl ether are colorless, readily mobile liquid with a characteristic terpene-like odor. Their most important physical properties are listed in Table 7.1.

Vapor pressure, density, and solubility in water, as well as the composition and boiling points of azeotropes with water and methanol/ethano are given in Tables 2 and 3 [8]. Both ethers have unlimited miscibility with all ordinary organic solvents and all hydrocarbons.

Chemical Properties Both ethers are stable under alkaline, neutral, and weakly acidic conditions. In the presence of strong acids, they are cleaved to the corresponding

Table 6.1 Selected physical properties of MTBE and ETBE

	MTBE	ETBE
mp, °C	−108.6	−94
bp, °C	55.3	70
n_D^{20}	1.3692	1.3756
Dielectric constant (20 °C)	4.5	
Viscosity, mPa·s	0.36	0.366
Surface tension (20 °C), mN m^{-1}	20.0	19.8
Specific heat (20 °C), kJ kg^{-1} K^{-1}	2.18	2.11
Heat of vaporization (at *bp*), kJ kg^{-1}	337	322
Heat of formation (25 °C) kJ mol^{-1}	−314	−313
Heat of combustion MJ kg^{-1}	−34.88	
Flash point (Abel–Pensky) °C	−28	−19
Ignition temperature (DIN 51 794) °C	460	310
Explosion limits in air, vol %	1.65–8.4	1–10
Critical data		
t_{cr} °C	224	241
p_{cr} MPa	3.43	3.04

alcohol and isobutene. Depending on reaction conditions the latter can form isobutene oligomers.

6.1.3
Resources and Raw Materials

MTBE and ETBE are produced by the reaction of isobutene, contained in C_4-fractions, methanol and ethanol respectively. At present, *isobutene* from the following sources is used as feedstock for ether production [9]:

Table 6.2 Vapor pressure and water solubility of MTBE and ETBE

	Vapor pressure, kPa*		Water solubility, wt %	
Temp°C	MTBE	ETBE	MTBE	ETBE
0	10.9	4.1		
10	17.6	7.4		
20	27.2	12.6		
25			2.7	1.2
30	40.9	20.4		
40	59.7	31.3		

* Antoine Parameters for $\log p = A - B / (C + \vartheta)$; $[\vartheta] = °C$

	A	B	C
MTBE	5.029	694.833	173.515
ETBE	5.011	697.181	158.338

Table 6.3 Binary azeotropes of MTBE and ETBE

Azeotrope	MTBE		ETBE	
	bp, °C	conc., wt %	*bp*, °C	conc., wt %*
Ether – water	52.6	96	65	94.0
Ether– alcohol	51.6	86	65	78.1
MTBE – methanol (1.0 MPa)	130	68		
MTBE – methanol (25 MPa)	175	54		

* Ether concentration

1) isobutene in raffinate 1, which is formed as a coproduct of butadiene production from steam cracker C_4 fractions (see Table 4 for typical composition)
2) pseudoraffinate 1, which is obtained by selective hydrogenation of butadiene in mixed C_4 fractions from steam crackers (composition, see Table 4)
3) isobutene contained in the C_4 fraction of fluid catalytic crackers (FCC – C_4; for typical composition, see Table 4).
4) isobutene from dehydrogenation of isobutane, which is obtained both from refineries and from field butanes after isomerization; and
5) isobutene by dehydration of *tert*-butanol, a coproduct of propylene oxide synthesis

Raffinate 1 and pseudoraffinate 1 provide isobutene for approximately 21 % of the world's total MTBE production [10].

Raffinate 1, pseudo raffinate 1, and FCC – C_4 can be utilized directly in ether synthesis. By contrast, isobutane must be dehydrogenated. The same holds true for field butanes after isomerization of the *n*-butane fraction to isobutane. A number of industrial processes have been established during the last decades [11], [12].

Table 6.4 Typical composition of C_4 hydrocarbon streams from steam crackers (raffinate 1) and fluid catalytic crackers (FCC – C_4)

Compound	Raffinate 1, wt %	Pseudoraffinate 1, wt %	FCC – C_4, wt %
Isobutane	4	2	36
n-Butane	12	7.5	13
Isobutene	44	24	15
Butene-1	24	39	12
cis-Butene-2	6	8	9
trans-Butene-2	9	19	14
Butadiene-1,3	0.5	0.0	0.3
Balance	0.5	0.5	0.7

For the primary butane isomerization, the Butamer process is most commonly used. For the isobutane dehydrogenation, the commercially available processes Oleflex (UOP), Catofin (ABB Lummus global), STAR and uhde GmbH are at present of industrial importance.

To produce MTBE from *tert*-butanol [13], isobutene must first be obtained by elimination of water from the alcohol before the olefin can be used for ether production.

Methanol the second reactant in MTBE synthesis, is produced at a typical purity of $> 99.9\%$ and is used directly for ether synthesis without further purification. Some 20% of the present methanol output of 32×10^6 t/a is still consumed by MTBE [14].

Bioethanol is produced by fermentation of sugarcane, maize, grain and sugar beets (see Section 3.8.4). Additional plants have been built in order to satisfy the fast growing market. Less than 10% of the fuel ethanol production is used as feedstock for ETBE manufacturing, the much bigger part is blended directly into the gasoline pool.

6.1.4
Production

Methyl and ethyl *tert*-butyl ether can be obtained by the acid-catalyzed addition of methanol to isobutene [15]. Suitable catalysts are solid acids such as bentonites [16], zeolites [17], [18], [19] and – commonly used in industrial world scale MTBE-production units – macroporous acidic ion-exchange resins [20]. The reactions are weakly exothermic with a heat of reaction of -37.7 kJ/mol.

The kinetics of MTBE formation have been investigated intensively [20], [21], [22]. It has been shown, that the observed reaction rates can be described sufficiently both by a kinetic model according to a Langmuir–Hinshelwood mechanism and by an Eley–Rideal (ER) approach [23]. However, the ER-model seems to be the most probable one. The commonly used kinetic model has been developed by Rehfinger et al. [24]. Because of equilibrium limitations only 92% conversion can be achieved with equimolar amounts of isobutene and methanol at 333 K. An excess of methanol not only increases the conversion of isobutene but also suppresses its dimerization and oligomerization. Dimerization of isobutene is the most important side reaction of MTBE synthesis [25], [26]. At a molar methanol excess as low as 10%, the selectivity for MTBE is practically 100%.

In addition to the Snamprogetti [27] and Hüls (now Oxeno) [28], [29], [30] esterification processes had been developed by Arco [31], IFP [32], and CDTECH (ABB Lummus Crest and Chemical Research Licensing) had been established [33].

Other industrial processes have been developed by DEA (formerly Deutsche Texaco) [34], Shell (Netherlands) [35], Phillips Petroleum [32], and Sumitomo [36].

MTBE plants can be retrofitted in order to produce ETBE. The key aspects of adaptation work are redesign of various heat exchangers (taking into account the higher boiling points of ETBE, ethanol and changed mass flows) and the alcohol extraction unit (changed alcohol load and different partition coefficient of ethanol in water and C^4 hydrocarbons). Examples of recent retrofits have been published by PCK Schwedt [38] and Degussa AG/Oxeno GmbH [39].

The process description below is valid for both ethers in principal.

Process Description Most commercially available processes are comparable and consist of a reaction and a refining section.

Reaction Section. All processes have in common the reaction of isobutene with a certain molar excess of methanol on a macroporous acidic ion exchanger at 50–90 °C. In the Snamprogetti, Hüls, and Arco processes, the pressure in the reaction section is chosen so that the reactants over the catalyst are entirely in the liquid phase over the catalyst bed, i.e., between 1.0 and 1.5 MPa. This enhances catalyst life time by reducing the polymerization of isobutene at the catalyst's surface and promotes the selectivity for the MTBE formation. For example, only ca. 100–200 ppm by weight of isobutene is converted to diisobutene (2.2.4-trimethylpentene-1 and -2) in the "Hüls-MTBE-Process".

Design of the reaction section is largely determined by the heat of reaction, which must be removed. The heat of reaction depends on the isobutene concentration in the feedstock. In the three processes mentioned (Snamprogetti, Hüls, Arco) adiabatic reactors are always employed with an *FCC – C_4 feedstock.* If *raffinate 1* is used as feedstock, Snamprogetti and Oxeno prefer tubular reactors for reasons of process engineering, whereas Arco employs recycle reactors. In the adiabatic reactors employed by IFP (Institute Français du Pétrol) or CDTECH, which also requires a recycle at higher isobutene concentration, an isobutene conversion of approximately 85% is achieved. The heat of reaction is removed by partial evaporation of the C_4 hydrocarbons in the reactor as the reaction proceeds. This process design, however, gives rise to a certain loss in selectivity for MTBE. A final isobutene conversion of ca. 97% is achieved by a subsequent *catalytic distillation.*

In most industrial plants, isobutene conversion of 95–97% is sufficient. *Residual butenes* are mainly used for the manufacture of alkylate gasoline, recycled to the cracker, or simply burned. If they are to be utilized for other chemical purposes such as the production of polymer-grade butene-1, the degree of isobutene conversion must be significantly increased. To overcome the conversion limit of 95 to 97% posed by the chemical equilibrium, a higher excess of methanol in the feed stream can be used. Unfortunately, this excess methanol remains in the MTBE product. It can, nevertheless, be recovered by distillation, e.g., as a methanol-poor minimum boiling azeotrope with MTBE [37]. Not only does this give MTBE with purities >99.7% (balance mainly *tert*-butanol produced by a reaction of isobutene with water dissolved in the feedstock), it also results in isobutene conversions of ca. 98%. In the Oxeno-MTBE process isobutene conversions of >99.9% could be achieved in a two-stage process (see Fig. 6.16.). This conversion corresponds to ca. 300 ppm per weight of residual isobutene in the spent C_4 fraction. Usually, if the feedstock is a raffinate 1 stream, the spent C_4 fraction is called raffinate 2.

This high conversion can also be obtained by using highly sulfonated acidic resins in the reaction section followed by an additional catalytic distillation column in the refining section.

Refining Section. Catalytic distillation (CD) or reactive distillation (RD) refers to a process, where both catalytic reaction and distillation are carried out simultaneously

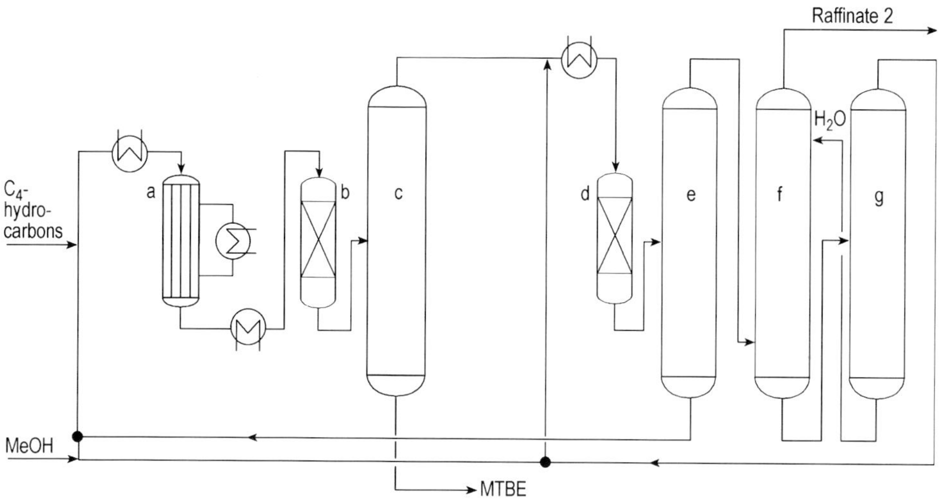

Fig. 6.1 Two-stage Oxeno-MTBE-Process
First stage: a) Multitubular reactor; b) Adiabatic reactor; c) First C_4 distillation tower (debutanizer)
Second stage: d) Secondary adiabatic reactor; e) Second C_4 distillation tower Methanol recovery: f) Methanol extraction; g) Methanol tower.

in the debutanizer column. From the viewpoint of reaction engineering, this column acts as a two-phase countercurrent flow, fixed-bed catalytic reactor. Usually CD is defined as a process in which a heterogeneous catalyst is located in a distillation column. A more common term for this operation is RD. This term does not distinguish between a homogeneous or heterogeneous catalyzed reaction in distillation columns.

The most important advantage of using CD for MTBE synthesis lies in the elimination of equilibrium limitation of isobutene conversion as a result of continuous removal of the reaction product MTBE from the reaction mixture.

The partly reacted mixture from the reaction section – which is usually in chemical equilibrium – enters the CD column below the catalyst packing zone to ensure the separation of the high-boiling component MTBE from the feed stream. The catalyst packing is installed in the upper mid-portion of the column with normal distillation sections above and below.

The CD-based MTBE process was first developed by CR & L using a patented catalyst support system (commercialized as CD MTBE or CD ETHEROL) [40], [41]. The catalyst used in the CD columns is essentially the same as that employed in the reaction section. It is an acidic ion-exchange resin, which is supported by a fiberglass cloth reinforced with stainless steel wire mesh. These fiberglass cloths are rolled into catalyst bales and stacked on sieve trays in the column.

A CD-process has been developed also by UOP and Hüls in 1992, utilizing Koch Engineering's KataMax-technology [42] (commercialized as ETHERMAX process).

In this kind of process technology the catalytic section of the CD-column uses a conventionally structured, wire-cloth distillation packing, which contain pellets of strong acidic ion-exchange resin [43]. The benefit in using this type of packing is a very good distribution of liquid and vapor phase at a low pressure drop, an efficient contact of reactants with catalyst pellets, and instantaneous distillative removal of reactants. Therefore, the packing exhibits excellent separation performance combined with an efficient mass and heat transfer for chemical reaction. An analogous catalytic packing system have been developed by Sulzer (commercialized as KataPak) [44] and Montz (commercialized as MultiPak) [45].

CD-technology has the important advantage over conventional unit design (reaction section followed by debutanizing in a distillation column) of lower capital investment costs, if a very high conversion of isobutene has to be achieved. Because in the CD-process both chemical reaction and distillation are carried out in the same device, at least one process step is eliminated. Beyond this, the exothermic heat of reaction can be used to vaporize the reaction mixture. Therefore, the maximum temperature in the reaction zone is limited to the boiling point of the mixture, so that an efficient and secure temperature control is possible.

One evident disadvantage of the CD-process technology is the fact, that since the debutanizer column must be taken out of operation during exchange of the catalytic packing, also the whole MTBE unit must be shut down. In contrast, in the conventional design, where the reaction section usually consists of two or more fixed bed reactors (e.g., in the two stage Hüls-MTBE-Process), a continued production can be performed at a lower feed rate, while the catalyst is exchanged. In the CD-technology it must also be ensured that a small excess of methanol exists along the reaction zone. In the absence of methanol the exothermic dimerization of butenes take place at high reaction rates. The ensuing sharp temperature rise ("hot spot") will cause irreversible catalyst deactivation and catalyst damage.

Nevertheless, since the mid-1980s most of the MTBE plants that have been built were equipped with a CD-column.

In the removal of nonconverted C_4 hydrocarbons from MTBE by distillation, excess *methanol* appears in the distillate, because of formation of a methanol-C_4 azeotrope. The methanol content in the azeotrope depends on pressure and C_4-composition, e.g., at 0.6 MPa the azeotrope contains ca. 2 to 4 wt% methanol. This methanol can completely be removed from the C_4-stream in an extraction column using a countercurrent water wash. The methanol can be recovered by distillation and is recycled to the reaction section.

If methanol losses of ca. 200 ppm by weight are acceptable, methanol recovery from the distillate with molecular sieves is also appropriate [46]. This 200 ppm residual methanol together with dimethyl ether, also formed in trace amounts, can be separated from the butene fraction in a subsequent molecular sieve adsorption unit [18].

Construction Materials. Because corrosive media are not used anywhere and do not result from catalyst disintegration, the entire plant is fabricated from standard carbon steel. Moreover, under normal operating conditions, residues, process wastewater, and gaseous emissions are not formed.

6.1.5 Environmental Protection

The small amounts of byproducts (see Section 6.1.6) in MTBE need not be removed if the product is to be used for gasoline. The catalyst is regenerable, and no environmental problems arise from its disposal.

Because MTBE is soluble in water, suitable precautions against groundwater contamination must be taken in loading and storage areas. For the environmental significance of MTBE, see Section 6.1.11. For emission control in the storage and handling of MTBE, see Section 6.1.8.

6.1.6 Quality Specifications

The usual purity of commercial MTBE is 98–99 wt%. The byproducts, *tert*-butanol and diisobutenes, as well as excess, residual methanol, do not have a detrimental effect on octane number improvement through the use of MTBE. Depending on the quality of the C_4 feedstock mixture, the MTBE product may also contain C_5 and C_6 hydrocarbons. A removal of these components by distillation is not necessary in most cases.

A typical MTBE composition adopted worldwide for use in the fuel sector is given below:

MTBE	98–99 wt%
Alcohols (methanol, *tert*-butanol)	0.5–1.5 wt%
Hydrocarbons (C_5, C_6-hydrocarbons, diisobutenes)	0.1–1.0 wt%
Water	50–1500 ppm by weight
Total sulfur	max. 10 ppm by weight
Residue on evaporation	max. 10 ppm by weight

For special applications (see Section 6.1.10.2), a high-purity MTBE (ether content > 99.98% wt) is marketed under the trade name Driveron S by Oxeno. High purity MTBE is also available from Shell and ARCO.

6.1.7 Chemical Analysis

Pure MTBE is analyzed by gas chromatography, preferably in capillary columns with a highly polar stationary phase, e.g., TCEP [1,2,3-tris-(2-cyanoethoxy)propane], Carbowax 20 M, or DX-1. For GC analysis of MTBE-containing fuels, an oxygen-specific detector (O-FID) [47] or a column combination technique [48], [49] can be employed. For other references on analysis of MTBE, see [50], [51], [52].

6.1.8 Storage and Transportation

Being noncorrosive, MTBE can be handled in the same way as fuel, and the existing distribution system can also be used for fuels containing MTBE. The ether has an indefinite storage life even in the presence of air. Because water is miscible with MTBE (though only slightly), dry storage is required. The product can be stored in nonpressurized containers. Carbon steel can be used as the container material, as well as aluminum, brass, copper, polyethylene, or polypropylene. Teflon, Buna-N, and other fuel-resistant plastics and rubbers can be used for seals; the use of vinylidene fluoride-hexafluoropropene copolymer (Viton) is not recommended.

Methyl *tert*-butyl ether has a vapor pressure of 61 kPa at 40 °C. Emission from storage facilities can be controlled or prevented by ordinary measures. The usual safety precautions for highly inflammable liquids must be employed.

Extinguishing agents for fire fighting are powder, carbon dioxide, and alcohol-resistant foam such as Tutogen L. With alcohol-resistant foams, a higher application rate is necessary for MTBE than for pure hydrocarbons.

Transport classifications are as follows:

ADR/RID II	class 3, no. 3 b; Name of substance: 33/2398 Methyl-*tert*-butyl ether
ADNR II	class 3, no. 3 b
GGVSee/IMDG	class 3.1
UN no.	2398
ICAO Code/IATA-DGR II	class 3
UN no.	2398

6.1.9 Legal Aspects

In the *European Community*, the Council Directive 85/536/EEC, December 5, 1985 [53], does not require special labeling at filling stations for fuels containing up to 15 vol% MTBE or ETBE.

In the *United States*, the use of oxygenates in gasoline is now restricted to ethanol blending in order to fulfil the requirements of the Clear Air Act [4].

6.1.10 Uses

6.1.10.1 As an Octane Enhancer (see Automotive Fuels, Chap. 3)

More than 95% of MTBE and nearly 100% of ETBE produced is used in the gasoline pool. In addition to many other favorable characteristics, the current importance of ethers is based primarily on its exceptionally good octane-enhancing properties when used as a gasoline blendstock [54], [55]. These antiknock properties are especially important because the use of cheap but toxic alkyllead compounds has been restricted or banned by law, both on environmental grounds and to permit the use of exhaust catalytic converters. Depending on the composition of the base gasoline, blend octane numbers of 115–125 (research octane number, RON) and 92–120 (motor octane number, MON) can be achieved. The highest blend values are obtained with saturated, paraffin-rich gasolines; the lowest, with olefin-rich

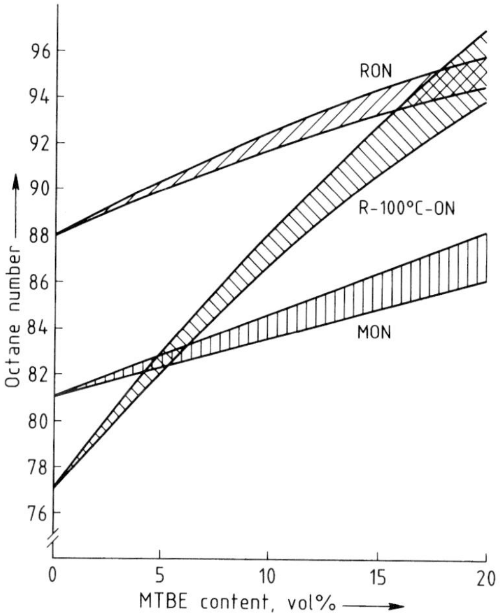

Fig. 6.2 Ranges of octane improvement by addition of MTBE to an unleaded gasoline with RON = 88, MON = 81, and R-100 °C-ON = 77.

gasolines. The lead and aromatics content also influences the blend octane numbers of MTBE (Fig. 6.2).

Besides increasing the octane number, addition of MTBE and ETBE to fuel has other positive effects. It does not necessitate any modifications to existing vehicles. The fuel vapor pressure (Reid vapor pressure, RVP) is decreased, so that vapor emissions during automobile fueling and operation are reduced. Addition of MTBE cuts down exhaust emissions, particularly carbon monoxide, unburned hydrocarbons, polycyclic aromatics, and particulate carbon. The blending of ETBE helps to reach the EC goals of increasing the use of biofuels.

6.1.10.2 **Other Uses**

Methyl *tert*-butyl ether is also used in the petrochemical industry. Production of isobutene by splitting MTBE is the only application that has been used on an industrial scale.

By reversal of its formation reaction, MTBE can be cracked to isobutene and methanol on acidic catalysts at >100 °C [56]. The methanol obtained as a coproduct is recycled to MTBE synthesis. For economic and ecological reasons, the MTBE-cracking route and that involving the splitting of *tert*-butanol obtained by direct hydration of isobutene [57] will continue to displace the conventional sulfuric acid extraction process for the isolation (recovery) of pure isobutene from C_4 streams [58]. At present, isobutene production with MTBE as feedstock is employed by Exxon Chemical and Sumitomo.

Methyl *tert*-butyl ether itself can be used in a number of chemical reactions, for example, the production of methacrolein and methycrylic acid [59] and of isoprene [60]. For other possible applications, see [61–66].

The lack of acidic hydrogen atoms makes MTBE a suitable *solvent* for chemical reactions such as Grignard reactions. The preferred grade for this purpose is high-purity MTBE with a residual alcohol content <500 ppm by weight, such as the product marketed by Oxeno under the trade name Driveron S. Because of its negligible tendency to form peroxides, high ignition temperature, and narrow explosion limits, MTBE is a good solvent for analytical use [66]. It is also used as an *extractant*, for example, in solvent dewaxing of hydrocarbon oils [66],[67].

6.1.11
Toxicology and Occupational Health

6.1.11.1 Toxicological Profile

As regards its widespread application and the amounts used, MTBE ranges within the top 2 % of all tested chemical substances. According to all of the toxicity data reviewed, in respect of European Union regulations MTBE has been classified as a skin irritant.

Acute Toxicity Tests with different animal species on oral, dermal, and inhalation application routes indicate a low acute toxicity for MTBE (oral LD_{50} well above 2000 mg per kilogram bodyweight, dermal $LD_{50} > 10\,000$ mg per kilogram bodyweight, inhalation LC_{50} values in the range from 85–142 mg/L for 4 h). At high concentrations anesthesia is the most immediate effect (CNS depression). At nonlethal concentrations local irritations, typical for organic solvents, have been observed besides transient behavioral effects.

MTBE was found to be only a slight irritant to the rabbit eye but causes reversible moderate to severe skin effects in rabbits. Animal studies showed no sensitizing potential for neat MTBE.

Chronic Toxicity The main effects observed in repeated dose toxicity studies via the oral and inhalation routes were CNS depression, irritation of the respiratory and gastrointestinal tract and enlargement of the liver. For the oral route the no observed adverse effect level (NOAEL) in a 90 day test is 300 mg/kg bodyweight per day while it is 800 mL/m^3 or 2880 mg/m^3 in 90 days inhalation tests.

Genotoxicity, Mutagenicity, Carcinogenicity Neither the structure of MTBE nor the comprehensive number of available in vitro and in vivo mutagenicity tests with MTBE and its metabolite *tert*-butanol provide any indication of possible genotoxic or mutagenic effects. The other metabolite, formaldehyde, has a proven potential of DNA damage, but will be rapidly detoxified when formed in the body. Therefore, available test data strongly suggest that with regard to MTBE itself and its metabolites, no genotoxic activity should be assumed by MTBE exposure. The absence of genotoxic potential is important when discussing possible carcinogenic effects. MTBE produces tumors in rats and mice at high concentrations. No observed adverse effect concentrations (NOAEC) of 400 mL/m^3 by inhalation and 250 mg/kg bodyweight at oral dose are derived from the tests results. The tumors found have been formed via nongenotoxic mechanisms, and are most probably not relevant to humans. Based on the available data several independent bodies, such as the International Association of Research on Cancer (IARC, WHO-agency), the European Union Working Group on Classification and Labelling of Dangerous Substances, and other bodies have recently come to the conclusion that MTBE is not a human carcinogen. Based on the available data MTBE is not classifiable with respect to its carcinogenic potential for humans.

Reproductive Toxicity Tests on reproductive toxicity and developmental development shows that MTBE is not significantly toxic to reproduction nor to fetal development.

6.1.11.2 Human Experience

As the odor threshold of MTBE in air (0.053 mL/m^3 = 0.19 mg/m^3) as well as odor and taste thresholds in water (15 μg/L and 40 μg/L, respectively) are very low, at low exposure levels effects on humans may be transient and subjectively caused by taste and odor. At 180 mg/m^3 no objective symptoms on the CNS and no signs of irritation of the respiratory tract were reported. Although it is not reported, repeated exposure on skin may result in de-fatting of the skin.

As indicated by all reported human experiences there is no evidence of any adverse effect caused by low exposure levels.

6.1.11.3 Health Risk Characterization, Occupational Health

Since the exposure concentrations for workers involved in production of MTBE, blending, distribution and handling of gasoline containing MTBE, service-station attendants and garage workers, and also for consumers filling MTBE containing gasoline in cars are significantly lower than the no effect levels in toxicity tests, MTBE production and use indicates no concern for human health.

However, compliance by workers with the short- and long-term occupational exposure limits and protection against skin contact is necessary to be protective against the described effects of MTBE. The lowest recommended occupational exposure limits are 270 mg/m^3 (75 mL/m^3) in the short-term and 90 mg/m^3 (25 mL/m^3) in the long-term value.

6.1.12 Ecotoxicology

6.1.12.1 Ecotoxicity Data

The result of a large amount of acute aquatic toxicity test data available for organisms at different trophic levels indicates low toxicity to aquatic organisms. All measured EC_{50} and LC_{50} values for microorganisms, algae, invertebrates and fish are well above 100 mg/L. Because of the fact that the measured concentrations of MTBE in the environment are significantly lower than the no effect levels in the mentioned tests and the calculated no effect levels for sediment organisms and terrestrial organisms like earth worms, it is verified that the risk for environmental organisms is low.

6.1.12.2 Behavior in the Environment

A calculation with McKay level 1 partitioning model estimates that in case of a spillage most of the MTBE (95%) will distribute in the atmosphere. There MTBE is readily degraded by photochemically produced hydroxyl radicals. The half-life for photodegradation in the air is about 3–7 days.

Aerobic and anaerobic degradation in water and groundwater depends strongly on the conditions. With inoculum of predominantly domestic sewage nearly no biodegradation occurs within 28 days in water in a standard test. However, with adapted bacteria MTBE is readily biodegradable under aerobic conditions.

As MTBE is relatively easily soluble in water (about 50 g/L) and will not adsorb on soil, it is easily washed into the groundwater. Biodegradation in contaminated soil and groundwater (remediation) is discussed in Section 6.1.13.

Due to its physical-chemical properties and metabolism in organisms MTBE is not expected to bioaccumulate and this has been shown to be correct by a bioaccumulation test.

6.1.13
Removal of MTBE from Impacted Groundwater

As a component of gasoline, MTBE can enter subsurface environments through a number of release scenarios including leaking underground fuel tanks (LUFT), dispensing equipment, accidental spills during tank refills, leaking fuel pipelines, minor on-going spill incidents, and larger volume surface spills. In many cases, gasoline retail stations and production terminals are sources of MTBE in groundwater. The concentration of MTBE in groundwater and the extent of groundwater contamination depend largely on the volume and duration of a gasoline release and whether remediation has been initiated at the site. As with any remedial effort for most contaminants, source removal is critical for the success of remediation systems at MTBE-impacted sites. A quick response following an accidental release of an MTBE-blended gasoline is of great importance, regardless of the selected remediation method, since it decreases the size of the area requiring characterization and the volume of groundwater needing treatment, thereby minimizing site characterization and remediation costs.

6.1.13.1 Overview of MTBE Remediation Technologies

The selection of a remedial strategy at an MTBE-impacted site depends on the volume of contaminated soil and groundwater, contaminant concentrations and mass, site-specific hydrogeological characteristics, and risk-based or regulatory-driven cleanup objectives. Of particular interest at MTBE-impacted sites is the behavior of MTBE relative to other gasoline components in subsurface environments.

The properties of MTBE greatly influence its fate in the environment, govern transport mechanisms, phase partitioning and biological interactions, and impact the success of remedial technologies. Key physiochemical properties of MTBE were presented in Tables 6.2 and 6.3. As compared with toluene, a common gasoline component and groundwater contaminant, MTBE has a lower density and boiling point, but a higher vapor pressure and much higher water solubility. The properties of MTBE suggest the following regarding its fate in subsurface environments: (1) MTBE does not readily sorb to soil particles in aquifers; (2) MTBE volatilizes readily from non-aqueous phase liquid (NAPL) sources; and (3) MTBE strongly partitions into the aqueous phase and preferentially remains in water. As a consequence, MTBE can be highly mobile in subsurface environments relative to other gasoline constituents.

Contrary to early reports, most of the technologies used historically at gasoline-contaminated sites are effective at removing MTBE from groundwater.

Successful technologies include groundwater extraction followed by air stripping, advanced oxidation or granular activated carbon systems. In-situ technology options include air sparging, multi-phase extraction, soil vapor extraction, bioremediation and phytoremediation. In addition to single technologies traditionally used at gasoline-contaminated sites, emerging technologies, modifications to existing technologies or the use of a treatment train involving more than one technology can greatly improve MTBE removal efficiency and/or reduce life cycle remediation costs.

6.1.13.2 Phase Transfer Technologies

Phase transfer technologies move MTBE from the water phase to either the gas phase (e.g., air stripping) or the solid phase (e.g., granular activated carbon) without chemically transforming it. Consequently, additional treatment steps to destroy MTBE (e.g., off-gas treatment from air strippers) or waste disposal (e.g., spent granular activated carbon) may be required.

6.1.13.2.1 Ex-Situ Phase Transfer Technologies

6.1.13.2.1.1 Air Stripping

Air stripping is a physical process for removing volatile organic compounds from water. In general, the effectiveness of air stripping increases with a compound's Henry's constant and the operating air-to-water ratio. MTBE's Henry's constant is much lower than that of common groundwater contaminants such as toluene (0.018–0.026 vs. 0.27, [68]). Thus, high air-to-water ratios (100 to 250) are needed to remove MTBE from water. Air stripping has been used to successfully remove MTBE at concentrations that are typically associated with groundwater contamination from LUFTs [69]. Evaluation of MTBE performance data from eight case studies of low profile and packed tower air strippers showed that air stripping has the potential to be more widely used in both drinking water and remediation applications [70]. For each of the studies evaluated, MTBE removal efficiencies exceeded 90%. Because state and/or local air quality regulations sometimes required stripper off-gas treatment (granular activated carbon, thermal and catalytic oxidation, advanced oxidation), process costs can increase significantly.

6.1.13.2.1.2 Granular Activated Carbon (GAC)

Although MTBE has a low affinity for sorption to solids, several types of activated carbon including coconut shell or coal-based GAC can be used to remove low concentrations of MTBE in water (200 μg/L) cost-effectively [69], [71]. GAC can be used to treat low flows of water from private wells with low MTBE concentrations. For treatment of higher concentrations of MTBE, GAC can be used as a polishing step in a treatment train following air stripping or chemical oxidation [71]. Background water quality and co-contaminant concentrations influence MTBE removal efficiencies. High concentrations of natural organic matter (NOM) and other gasoline constituents in influent water streams compete with MTBE for GAC sorption sites thereby increasing GAC usage rates and process costs.

6.1.13.2.1.3 **Other Ex-Situ Phase Transfer Technologies**

Results from laboratory and pilot scale studies suggest that synthetic resins have the potential for successful remediation of MTBE-impacted sites [67]. One of the advantages of using synthetic resins is the ease of regeneration relative to GAC. One of the limitations of resins is the experimentally observed reduction in MTBE sorption due to the presence of other gasoline constituents, such as benzene, toluene, ethylbenzene, and xylene (BTEX) compounds and *tert*-butanol (TBA, *tert*-butylalcohol), in contaminated water. This is analogous to the effects of NOM and BTEX on GAC sorption as discussed above. The use of synthetic resins is costly relative to GAC but resins can be designed to achieve a higher degree of selectivity and tend to have a longer lifetime than GAC.

In addition to synthetic resins, other sorbents such as high silica zeolites have been shown to be effective in removing MTBE from contaminated water [72].

Other technologies, such as membranes and solvent extraction, can be used to remove MTBE from contaminated groundwater [73], [74], [75] but only a few bench- and pilot-scale studies have been conducted.

6.1.13.2.2 **In-Situ Phase Transfer Technologies**

Demonstrated in-situ phase transfer technologies include soil vapor extraction (SVE), multi-phase extraction (MPE), in-situ air sparging (IAS) and phytroremediation. Emerging phase transfer technologies include steam injection/vacuum extraction and six-phase heating.

6.1.13.2.2.1 **Soil Vapor Extraction (SVE) and Multi-Phase Extraction (MPE)**

Soil vapor extraction (SVE) is commonly used for the remediation of vadose zones at gasoline-contaminated sites. In order to remove MTBE from both saturated and unsaturated zones, other remediation technologies can be used in combination with SVE. The success of SVE has been documented at MTBE-impacted sites when implemented either as the sole technology or in conjunction with other groundwater treatment technologies [76–78]. The combination of SVE and groundwater extraction, known as multi-phase extraction, is effective in removing MTBE from contaminated aquifers [73] and is routinely used at many sites. Effectiveness with MTBE can be enhanced if the remediation system is operated under high vacuum.

6.1.13.2.2.2 **In-Situ Air Sparging (IAS)**

In-situ air sparging involves the injection of air below the groundwater table to transfer contaminants from the aqueous phase into the gaseous phase. The injection of air has the additional advantage of introducing oxygen to subsurface environments thereby potentially enhancing aerobic contaminant biodegradation rates. Laboratory air sparging column studies in different soils showed that the rate of MTBE volatilization is very rapid in sandy soils [79]. Results from another laboratory study showed that 85% of MTBE in water equilibrated with gasoline was successfully removed using air sparging [80]. MTBE recovery rates were shown to increase with increased air injection rates and water saturation levels. Results from a number of field studies revealed that MTBE groundwater concentrations at gasoline-contaminated sites were reduced by 88 to 99% after two years of air sparging [81]. As a general rule, if the site hydrogeology is conducive to successful use of IAS, e.g., soil porosity is satisfactory, this technology could be very successful at removing MTBE from groundwater.

6.1.13.2.2.3 Phytoremediation

Phytoremediation involves the use of plants or trees to either remove or degrade groundwater contaminants. Phytoremediation occurs by plant uptake of the contaminants followed by translocation and accumulation into plant tissues, or transpiration through leaves. In some cases, the breakdown of contaminants can take place either within plant tissues or in the rhizosphere [82]. Phytoremediation is most effective for contaminants with log K_{ow} values in the range of 0.5 to 3.0 [83]. Since MTBE has a log K_{ow} of 1.2, phytoremediation is expected to be an effective technology at MTBE-impacted sites. Several plant and tree species have been recently reported to uptake MTBE including alfalfa plants [84], poplar trees [82], [85], [86] and Eucalyptus trees [86], [87]. In laboratory studies, poplar trees were shown to remove 30% of the MTBE in test solutions at concentrations of 300 to 1600 μg/L in a week [83]. Results of several laboratory studies suggest that phytoremediation has the potential to be effective for the removal of MTBE from groundwater, especially at sites where the contaminated groundwater is shallow.

6.1.13.3 Transformation or Destruction Technologies

The underlying principle of transformation or destruction technologies is the transformation of MTBE in water to another chemical form either partially (e.g., acetone) or completely (carbon dioxide). Transformation or destruction technologies can either be implemented ex-situ following groundwater extraction (e.g., chemical oxidation, biological treatment) or in-situ (e.g., in-situ chemical oxidation, in-situ bioremediation).

6.1.13.3.1 Ex-Situ Transformation Technologies

6.1.13.3.1.1 Advanced Oxidation Processes (AOPs)

Advanced oxidation processes involve the use of ozone – peroxide, ozone – UV, peroxide – UV, Fenton's reaction (hydrogen peroxide in combination with iron(II)salts), high-energy electron beam irradiation (E-beam) or sonication-hydrodynamic cavitation. Most of these processes are characterized by the production of hydroxyl radicals. AOP technologies have been tested for MTBE destruction in both laboratory and pilot-scale studies. In addition, some full-scale applications have been reported [88-93]. Common problematic byproducts include bromate due to the reaction of ozone with bromide, and TBA, *tert*-butyl ether (TBE), acetone and formaldehyde from the use of UV – H_2O_2 [93]. Some of these compounds are either more toxic than MTBE or more resistant to oxidation [89], [90].

AOPs are believed to be most efficient for groundwater with MTBE concentrations ranging between 0.1 and 80 mg/L [89]. Although the use of AOPs is effective at treating high concentrations of MTBE, AOP systems can be expensive and can lead to the formation and accumulation of toxic or undesirable byproducts. More research to evaluate the cost-effectiveness and performance of AOPs at field-scale levels is needed.

6.1.13.3.1.2 **Biological Treatment**

Alkyl ethers, such as MTBE, are relatively difficult to degrade because of the high energy required by microorganisms to cleave the ether bond and the resistance of the branched carbon structure to microbial attack [94]. Several early studies reported that MTBE is biologically recalcitrant under most environmental conditions [95–97]. Recent laboratory and field studies, however, have shown that a number of bacterial and fungal cultures from various environmental sources are capable of degrading MTBE under aerobic [98], [99] and anaerobic [100–104] conditions. Microorganisms can either utilize MTBE as a sole source of carbon and energy, or can degrade it cometabolically following growth on short-chain alkanes such as propane.

The biodegradation of MTBE has been demonstrated using both suspended growth and fixed film bioreactors [105–111]. In a growth bioreactor microorganisms were able to reduce MTBE concentrations from 2400 mg/L to 1.6 mg/L with an average removal rate of 96.2% over the course of the study.

Several types of ex-situ reactors can be used for MTBE treatment including fluidized bed, porous pot, membrane and trickling filter bioreactors. Membrane bioreactors sustain high densities of microorganisms and are therefore suited for treating high strength waste streams as well as waste streams containing slowly biodegraded compounds such as MTBE. In one study, the use of a membrane bioreactor reduced MTBE from high (2000 mg/L) to very low (40 g/L) concentrations during normal operation [110]. The potential for success of ex-situ bioreactors has increased over the last several years with the isolation of pure cultures capable of degrading MTBE and TBA at relatively rapid rates [98]. For example, two pure cultures designates PM1 and ENV735 are being used commercially in fluidized bed reactors (FBRs). Such systems show great promise for biologically treating MTBE contaminated water streams.

6.1.13.3.2 **In-Situ Transformation Technologies**

6.1.13.3.2.1 **In-Situ Chemical Oxidation**

In-situ chemical oxidation (ISCO) is based on the same principles of advanced oxidation. Most ISCO technologies involve the injection of oxidants into groundwater to produce hydroxyl radicals. Common oxidants include peroxide, permanganate, ozone, and Fenton's reagent. In a successful field scale demonstration, hydrogen peroxide was injected at a site in New Jersey. MTBE concentrations dropped from 6000 μg/L to below detection limits in less than a year [112]. Injection of Fenton's reagent into groundwater at a site in Texas reduced MTBE concentrations by 83.4% [113]. One of the advantages of ISCO is the potential to destroy MTBE in a relatively short time frame. Technology limitations include effective oxidant delivery and the potential of byproduct formation.

In addition to the injection of ozone as an oxidant, technologies involving ozone sparging are currently being tested. Fine bubbles with a high surface-to-volume ratio are injected into the saturated zone to extract dissolved MTBE from contaminated groundwater. Ozone contained within the bubble and thin film around the bubble is expected to react rapidly and destroy MTBE in the gas phase.

6.1.13.3.2.2 **In-Situ Bioremediation**

In-situ bioremediation involves the use of indigenous or exogenous microorganisms to either destroy or immobilize contaminants. In-situ bioremediation strategies could involve direct metabolism, cometabolism, bioaugmentation or some combination thereof. Results from pilot studies showed that bioremediation approaches involving oxygen injection and/or bioaugmentation have a strong potential for success at MTBE-impacted sites [114], [115]. In addition, a recent study has demonstrated the success of propane injection and bioaugmentation at an operating service station in New Jersey where a 93% reduction in MTBE concentration was achieved within two months of system operation [110].

Fortum Oil and Gas Oy and Oxeno Olefinchemie GmbH, two member companies of EFOA (The European Fuel Oxygenates Association – www.efoa.org) contributed to Sections 6.1.11, 6.1.12 and 6.1.13.

6.1.14 References

1. C. Henry, *Chem. Zentralbl.* (1904) 1065.
2. H. C. Barnett et al., *Natl. Advis. Comm. Aeronaut.*, Wartime Report ACR No. E 4 H 03 (1944).
3. Hydrocarbon Processing or ICIS Business News (wird nachgeliefert)
4. U.S. Environmental Protection Agency, www.epa.gov/oar/oaqps/peg_caa/pegcaain.html <24.03.2007>.
5. Energy Information Administration, www.eia.doe.gov/pub/oil_gas/petroleum/feature_articles/2006/mtbe2006.pdf <24.03.2007>.
6. Bundesministerium für Umwelt, Naturschutz und Reaktorsicherheit, Richtlinie 2003/30/EC, www.erneuerbare-energien.de/inhalt/4736/. <24.03.2007>
7. Oxeno's market research, unpublished data, Feb. 2007.
8. *Ullmann*, 4th ed., **17**, 58.
9. M. L. Morgan, DGMK Conference "C_4 Chemistry – Manufacture and Use of C_4-Hydrocarbons", Oct 6–8 (1997) Aachen.
10. K. D. Miller, 23rd Annual Dewitt Petrochemical Review, March 31–April 2 (1998) Houston.
11. K. C. Rock, R. O. Dunn: "Economics of Producing Butene Intermediate Feedstocks from Butanes", *Energ. Prog.* **8** (1988) no. 4, 191–195.
12. P. R. Sarathy, G. S. Suffride, *Hydrocarbon Process.* (1993) Feb, 43–51.
13. D. J. Miller: "Arco's Processes for Oxygenated Fuels – MTBE/TAME and ETBE", *Dewitt 1989 Petrochemical Review*, Houston, March 28–30, 1989.

14. http://www.methanol.org/pdfFrame.cfm?pdf=methanolsupplyDemand02.pdf.
15. F. Ancilotti et al., *J. Mol. Catal.* **4** (1978) 37–48.
16. A. Gicquel, B. Torck, *J. Catal.* **31** (1983) 27–31.
17. J. M. Adams et al., *Clays Clay Miner.* **34** (1986) no. 5, 597–609.
18. P. Chu, G. H. Kühl, *Ind. Eng. Chem. Res.* **26** (1987) 366.
19. A. Kogelbauer et al., *J. Catal.* **152** (1995) 122–129.
20. F. Ancilotti et al., *J. Catal.* **46** (1977) 49–57.
21. K. Sundmacher, U. Hoffmann, *Chem. Eng. Sci.* **49 (18)** (1994) 3077–3089.
22. T. Zhang, R. Datta, *Ind. Eng. Chem. Res.* **34** (1995) 730–740.
23. D. Parra et al., *Chem. Eng. Sci.* **49 (24A)** (1994) 4563–4578.
24. A. Rehfinger, U. Hoffmann, *Chem. Eng. Sci.* **45** (1990) no. 6, 1605–1617.
25. F. Obenaus, *Oil Gas J.* **77** (1979) 76–77.
26. A. Rehfinger, U. Hoffmann, *Chem. Eng. Technol.* **13** (1990) 150–156.
27. S. Rossini et al., DGMK-Conference "C_4 Chemistry – Manufacture and Use of C_4-Hydrocarbons", Oct. 6–8 (1997), Aachen.
28. F. Obenaus: "Wege zur besseren Nutzung der C_4-Kohlenwasserstoffe", *DGMK-Meeting Karlsruhe* , February 8, 1982.
29. *Hydrocarbon Process.* **72** (1996) Nov., 113.
30. W. Droste, *Tech. Mitt.* **80** (1987) 277–282.
31. Th. A. Beck, D. J. Miller: *Arco MTBE – The Maximum in Flexibility* , Dewitt 1988, Petrochemical Review, Houston, TX., March 23–25, 1988.
32. *Hydrocarbon Process.* **75** (1996) Nov., 114.
33. P. H. O. Dixon et al., *Catalytic Distillation Technology and MTBE-Production* , 1989 NPRA Annual Meeting, San Francisco, March 19–21, 1989.
34. DEA (formerly Deutsche Texaco), DE 27 06 465, 1987 (H. Humberg, H. Wegner).
35. Shell NL, DE-OS 29 11 077, 1979 (H. Groeneveld, A. Kwantes).
36. Sumitomo Chemical Company, EP-A 0 206 594, 1986 (S. Arakawa et al.).
37. Hüls AG, DE 2 629 769, 1976 (F. Obenaus et al.).
38. DGMK-Veröffentlichung PCK Schwedt (wird nachgeliefert).
39. A. Rix et al, *Chem. Ing. Tech.* **2006**, no. 9, 1172 ff.
40. CR & L, EP-A 0 008 860, 1979 (L. A. Smith Jr.).
41. L. A. Smith, M. N. Huddleston, *Hydrocarbon Process.* (1982) March, 121–123.
42. J. L. DeGarmo et al., *Chem. Eng. Prog.* (1992), March, 43–50.
43. Koch Engineering Company, EP-A 0 428 265, 1991 (A. P. Gelbein, M. Buchholz).
44. Gebr. Sulzer AG, EP 0 396 650, 1992 (R. Shelden, J.-P. Stringaro).
45. DE-OS 19 701 045, 1998 (A. Górak, L. Kreul).
46. E. Trubac: "Oxygenate Removal from MTBE Raffinate", lecture presented at the *AIChE Meeting,* Minneapolis, Minn., Aug. 16–19, 1987.
47. W. Schneider, J.-Ch. Frohne, H. Bruderreck, *J. Chromatogr.* **245** (1982) 71–83.
48. J. Winskowski, *Erdöl Kohle Erdgas Petrochem.* **41** (1988) no. 1, 30–36.
49. J. M. Levy, J. A. Yancey, *J. High Resolut. Chromatogr. Commun. HRC CC J. High Resolut. Chromatogr.* **9** (1986) no. 7, 383–387.
50. AFNOR (Association Française de Normalisation) M 07-084, Dec. 1984, Norme experimentale.
51. DIN 51 413, Parts 4 and 5.
52. BSI (British Standards Institution) proposed method JP-STG-6 C (The Institute of Petroleum).
53. EEC Council Directive 85/536/EEC, Dec. 5, 1985 (Column 3).
54. G. H. Unzelmann, *Oil Gas J.* (1989, April 10) 33–37.
55. W. J. Piel, *Energ. Prog.* **8** (1988) no. 4, 201–204.
56. Fattore et al., *Hydrocarbon Process.* **60** (1981) Aug., 101. August (1981) 101.
57. *Chem. Eng. N.Y.* **90** (1983) no. 25, 60.
58. H. Kröper, K. Schlömer, H. M. Weitz, *Hydrocarbon Process* **48** (1969) no. 9, 195–198.

59. Nippon Shokubai Kagaku Kogyo Co., EP-A 0 304 867, 1988 (K. Kinumi et al.).
60. Kuraray Co, JP-Kokai 60 193 932 <85 193 932>, 1985 (Y. Ninagawa et al.). Sumitomo Chem., US 3 574 780, 1971 (Y. Watanabe et al.).
61. Degussa AG, EP 0 050 869, 1981 (B. Lehmann).
62. Anic S.p.A., DE-OS 3 130 428, 1982 (F. Ancillotti et al.). BASF AG, DE-OS 27 40 590, 1979 (F. Merger et al.).
63. BASF AG, DE-OS 29 11 466, 1980 (H. Naarmann et al.).
64. BASF AG, DE-OS 29 28 098, 1981 (W. Hoffmann).
65. Bayer AG, DE-OS 30 02 203, 1981 (G. Bonse et al.).
66. H. Grangette, C. Bernasconi, S. Randriamahefa, R. Gallo, *Inf. Chim.* **246** (1984) Jan./Feb., 135–140.
67. Exxon Research and Engineering Co., EP-A 00 88 603, 1983 (T. H. West).
68. J. Zogorski, A. Morduchowitz, A. Baehr, B. Baumann, D. Conrad, R. Drew, N. Korte, W. Lapham, J. Pankow, E. Washington: Fuel Oxygenates and Water Quality. Coordinated by Interagency Oxygenation Fuel Assessment, Office of Science and Technology Policy, Executive Office of the President, Washington, DC, 1997.
69. California MTBE Research Partnership: Treatment technologies for removal of methyl *tert*-butyl ether from drinking water. National Water Research Institute, Fountain Valley, CA 2000.
70. R. Deeb, S. Sue, D. Spiers, M. Kavanaugh, Proceedings of the Air and Water Management Association's 94th Annual Conference and Exhibition, Orlando, FL 2001.
71. California MTBE Research Partnership: Treating MTBE-impacted drinking water using granular activated carbon. National Water Research Institute, Fountain Valley, CA (2002).
72. M. Anderson, *Environ. Sci. Technol.* **34** (2000) no. 4, 725–727.
73. L. Vane, F. Alvarez, B. Mullis, *Environ. Sci. Technol.* **35** (2001) no. 2, 391–397.
74. A. Keller, B. Bierwagen, *Environ. Sci. Technol.* **35** (2001) no. 9, 1875–1879.
75. J. Choi, I. Song, W. Lee, *J. Membr. Sci.* **166** (2000) 159–175.
76. D. Creek, J. Davidson, Proceedings of the 1998 Petroleum Hydrocarbons and Organic Chemicals in Ground Water: Prevention, Detection, and Remediation, National Ground Water Association, Westerville, OH 1998, pp. 560–568.
77. P. Johnson, R. Ettinger, in C. Ward, J. Cherry, M. Scalf (eds.): *Subsurface Restoration* , Ann Arbor Press, Inc., Chelsea, MI 1997.
78. United States Environmental Protection Agency: MTBE fact sheet #2: Remediation of MTBE contaminated soil and groundwater. EPA 510-F-98–002, 1998.
79. A. Mortensen, K. Jensen, T. Sonnenborg, E. Arvin, *Groundwater Monit. Rem.* (2000) 87–95.
80. C. Bruce, C. Gilbert, R. Johnson, P. Johnson, in G. Wickramanayake, R. Hinchee (eds.): *Physical, Chemical, and Thermal Technologies: Remediation of Chlorinated and Recalcitrant Compounds* , Battelle Press, Columbus, OH 1998, pp. 293–298.
81. California MTBE Research Partnership. Overview of MTBE remediation technologies. National Water Research Institute, Fountain Valley, CA 2002.
82. E. Rubin, A. Ramaswami, *Water Res.* **35** (2001) no. 5, 1348–1353.
83. J. Burken, J. Schnoor, *Environ. Sci. Technol.* **32** (1998) no. 21, 3379–3385.
84. Q. Zhang, L. Davis, L. Erickson, *Environ. Sci. Technol.* **35** (2001) no. 4, 725–731.
85. M. Hong et al., *Environ. Sci. Technol.* **35** (2001) no. 6, 1231–1239.
86. L. Newman et al., *Prepr. Extend. Abstr. Am. Chem. Soc.* **40** (2000) no. 1, 288–289.
87. L. Newman et al., *J. Soil. Contam.* **7** (1998) no. 4, 531–542.
88. J. Acreo et al., *Environ. Sci. Technol.* **35** (2001) no. 21, 4252–4259.
89. S. Cater, M. Stefan, J. Bolton, A. Safarzadeh-Amiri, *Environ. Sci. Technol.* **34** (2000) no. 4, 659–662.

90. P. Chang, T. Young, *Water Res.* **34** (2000) no. 8, 2233–2240.

91. J. Kang, H. Hung, A. Lin, M. Hoffman, *Environ. Sci. Technol.* **33** (1999) no. 18, 3199–3205.

92. A. Safarzadeh-Amiri, *Water Res.* **35** (2001) no. 15, 3706–3714.

93. M. Stefan, J. Mack, J. Bolton, *Environ. Sci. Technol.* **34** (2000) no. 4, 650–658.

94. G. White, N. Russell, E. Tidswell, *Microbiol. Rev.* **60** (1996) 216–232.

95. Y. Fujiwara, T. Kinoshita, H. Sato, I. Kojima, *Yugagaku* **33** (1984) 111–114.

96. H. Jensen, E. Arvin, in F. Arendt, M. Hinsenveld, W. van den Brink (eds.): *Contaminated Soil '90* , Kluwer Academic Publishers, Dordrecht 1990, 445–448.

97. J. Suflita, M. Mormile, *Environ. Sci. Technol.* **27** (1993) 976–978.

98. R. Deeb, K. Scow, L. Alvarez-Cohen, *Biodegradation* **11** (2000) no. 2–3, 171–186. C. Yeh, J. Novak, *Water Environ. Res.* **67** (1995) 828–834.

99. A. Stocking et al., *Biodegradation* **11** (2000) no. 2–3, 187–201.

100. P. Bradley, F. Chapelle, J. Landmeyer, *Appl. Environ. Microbiol.* **67** (2001) no. 4, 1975–1978.

101. P. Bradley, F. Chapelle, J. Landmeyer, *Environ. Sci. Technol.* **35** (2001) no. 23, 4643–4647.

102. K. Finneran, D. Lovley, *Environ. Sci. Technol.* **35** (2001) no. 9, 1785–1790.

103. J. Landmeyer, F. Chapelle, H. Herlong, P. Bradley, *Environ. Sci. Technol.* **35** (2001) no. 6, 1118–1126.

104. United States Environmental Protection Agency: Natural attenuation of MTBE in the subsurface under methanogenic conditions. EPA/600/R-00/006, 2000.

105. K. Acuna-Askar, A. Englande, C. Hu, G. Jin, *Water Sci. Technol.* **42** (2000) no. 5–6, 153–161.

106. D. Dupasquier, S. Revah, A. Auria, *Environ. Sci. Technol.* **36** (2002) no. 2, 247–253.

107. N. Fortin, M. Deshusses, *Environ. Sci. Technol.* **33** (1999) no. 17, 2980–2986.

108. N. Fortin, M. Deshusses, *Environ. Sci. Technol.* **33** (1999) no. 17, 2987–2991.

109. P. Hatzinger, R. Steffan, S. Drew, *Contaminated Soil, Sediment and Water Journal, AEHS* , 2001, 81–84.

110. R. Steffan, P. Hatzinger, Y. Farhan, S. Drew, *Proceedings of the Petroleum Hydrocarbons and Organic Chemicals in Ground Water.* November 14–16, Houston, Texas 2001, pp. 252–263.

111. S. Vainberg, A. Togna, P. Sutton, R. Steffan, Submitted to *J. Env. Eng.* 128 (2002) 842–851.

112. United States Environmental Protection Agency. Field applications of *in-situ* remediation technologies: Chemical oxidation. EPA 542-R-98-008, 1998.

113. J. Leethem, *Contaminated Soil, Sediment and Water Journal, AEHS* , 2001, 54–58.

114. J. Salanitro et al., *Environ. Sci. Technol.* **34** (2000) no. 19, 4152–4162.

115. R. Wilson, D. Mackay, K. Scow, *Environ. Sci. Technol.* **36** (2002) no. 2, 190–199.

116. Union Carbide Corp., EP-A 0 229 994, 1987 (M. Nagji).

6.2
Other Octane Enhancers

Kuno Schädlich, Peter Schug, Werner Dabelstein, Arno Reglitzky

6.2.1
Introduction

Historically, the quality of motor gasoline and the technical development of the spark ignition Otto engine were closely connected. Since its invention in 1876, the Otto engine technology has been continuously improved, a process which is still going on.

Of particular importance for this development was the parallel improvement in gasoline knock resistance. Knock resistance is one of the most important gasoline properties and its improvement allowed the engine manufacturers to increase the compression ratio. This in turn resulted in higher engine efficiency and improved exhaust emission performance. The knock resistance is measured in terms of gasoline octane numbers in accordance with worldwide standardized test procedures.

Figure 6.1 shows, for Germany as an example, the close relationship between octane number and engine compression ratio over the course of the 20th century. Over this period a variety of measures were applied to improve refinery technologies to meet the increasing engine octane demand. These measures incorporated the integration of complex processing plants, from which a range of gasoline component streams with different octane levels became available. In 1929, *alkyllead compounds* were introduced in the United States as efficient octane enhancers. Thereafter these were used generally to improve the gasoline octane quality until environmental concerns led to their use being stepwise restricted worldwide. The resulting shortfall of octane was compensated by improvements in refinery processes, as well as by the use of alternative octane enhancers such as fuel oxygenates. Starting in the 1970s a number of *fuel oxygenates* became available and were increasingly used for gasoline blending.

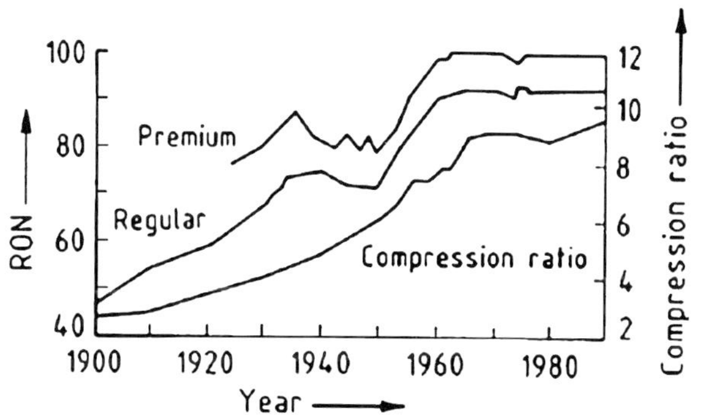

Fig. 6.1 Trend of research octane number and compression ratio in Germany.

More recently, efforts to reduce air pollution led to the enforcement of more restrictive exhaust emission limits for vehicles. In this context, passenger cars with catalytic converters were introduced. Since the addition of lead to gasoline results in the deactivation of the catalytic converters, unleaded gasoline of high octane quality was consequently made available.

6.2.2 Technical Information

6.2.2.1 Combustion in Otto Engines (Chap. 3)

During the compression stroke the air – fuel mixture is compressed to 1.5–2.5 MPa and the temperature rises up to 1000 K, depending on the engine compression ratio. The mixture composition and the engine condition are controlled in such a way that the upper flammability limit of the mixture is not exceeded, i.e., no self-ignition occurs. Shortly before the piston reaches the upper dead point, combustion is initiated by an electrical discharge between the electrodes of the spark plug. The appropriate spark timing is essential for engine efficiency. Best efficiency is achieved when maximum chemical energy released from the combusted charge is transformed into mechanical energy during the expansion stroke. The spark transports the necessary ignition energy to form a self-sustained flame kernel in the air – fuel mixture. Subsequently this flame kernel expands and flame fronts propagate spherically into the combustion chamber. The speed of flame propagation depends on a variety of parameters.

First of all the process is affected by the initial status of the air – fuel mixture with respect to temperature, pressure, and turbulence. Since the combustion process is exothermic, temperature and pressure increase rapidly. This accelerates the flame propagation, which in turn affects the status and turbulence of the burned gas as well as the unburned charge ahead of the flame. The entire combustion consists of a complex interaction of fluid dynamics, reaction kinetics, radical diffusion, and radiation processes.

6.2.2.2 Knock Phenomena

Average flame speeds in the engine combustion process range from 10 to 30 m/s. In contrast, flame speeds of 300 to 500 m/s are measured under knocking conditions. The knock phenomena typically occur at the end of the normal combustion process (Fig. 6.2). They originate from zones of unreacted charge where the critical conditions in terms of temperature and pressure are exceeded. This causes spontaneous self-ignition which in turn results in a sharp pressure increase.

The detailed processes occur very rapidly and involve the complex coupling of fluid dynamic and reaction kinetic processes. Despite a large number of experimental and theoretical studies the phenomenon itself is not fully understood [1].

The phenomenological process, however, is readily detectable. It is characterized by a typical high-frequency noise emitted from engine parts. Knocking

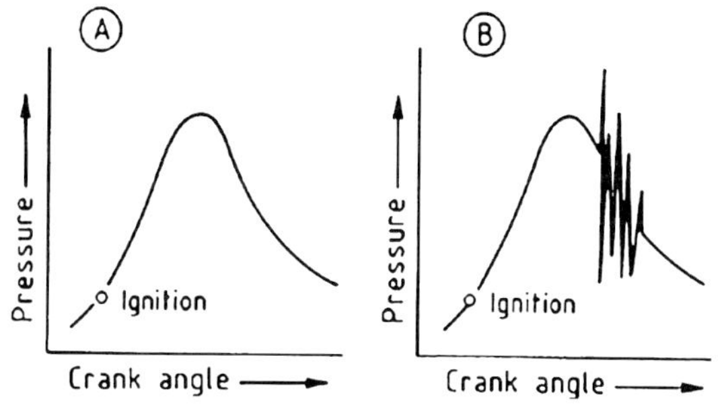

Fig. 6.2 Pressure profile versus crank angle during combustion A) Normal combustion; B) Knocking combustion.

combustion results in high mechanical stress and thermal overload of the engine. Finally, it leads to severe damage to pistons, piston rings, seals, and/or cylinder heads.

6.2.2.3 Octane Number Measurement

The knock resistance of gasolines is measured by their research octane number (RON) and motor octane number (MON). Both are determined in single cylinder engines according to standardized test procedures (Chap. 3.3.1.1). Measurement of MON is performed under higher engine severity than the RON measurement in terms of intake air, temperature, and engine speed.

6.2.3 History

Until the mid-1920s there was only one gasoline grade available in most countries, predominantly produced from straight-run crude oil products which were selected according to their composition of aliphatic and aromatic components. In order to meet the increasing octane demand, e.g., indicated by the introduction of a premium grade in 1924, several refinery processes were developed. Today, a typical refinery produces up to twenty individual product streams used for gasoline blending (Chap. 3.). The octane enhancement of tetraethyllead (TEL) was discovered in 1922 in the United States by T. Midgley, Jr. [2]. It was produced by the Ethyl Gasoline Corporation and first introduced onto the U.S. market in 1929. In Germany it became available shortly before World War II as an octane enhancer. TEL and TML (tetramethyllead) are so-called "lead fluids" and are still used in Yemen, Afganistan, and North Korea. However, leadalkyl compounds have disappeared from gasoline in most countries due to environmental concerns and because lead deactivates catalytic converters.

Other octane enhancers–except fuel oxygenates–never achieved any significant degree of application. The current dominance of unleaded gasolines in the market, however, causes octane shortfalls. This situation has led to renewed attention to additional octane sources, including both new and known oxygenates and alternative octane enhancers. Therefore, octane enhancers may play a more pronounced role in the future provided that environmental aspects are satisfied, such as toxicological properties and full compatibility with modern exhaust emission control systems.

6.2.4 Types of Octane Enhancers

Octane enhancers are substances which result in octane improvements when added to gasolines. This definition also includes all hydrocarbon-based refinery streams of sufficient octane quality. These streams, however, must be regarded as conventional gasoline blendstocks and are not covered in this chapter, in spite of the fact that some are traded as octane enhancers, e.g., aromatic refinery fractions such as toluene.

Alternative ways to improve octane quality involve the use of special chemical compounds. Since these compounds have to be readily soluble in gasolines they must be hydrocarbon-based. They may, however, contain functional groups with heteroatoms such as oxygen, nitrogen, sulfur, phosphorus, halogens, or metal atoms. Depending on the compound concentration in the fuel, components and additives can be distinguished. *Components* are compounds blended in quantities which also affect gasoline properties other than octane number, such as heating value, density, or vapor pressure. In contrast, *additives* are applied in the ppm-concentration range so that other fuel properties are essentially unaffected. This discrimination reflects the fact that additives are much more efficient than components. Additives consist exclusively of organometallic compounds. When combusted they form solid oxidation products (ash) in addition to gaseous combustion products.

6.2.5 Ashless Octane Enhancers

Components are always added in higher quantities, and must, therefore, consist of "ashless" compounds. Only oxygen- or nitrogen-containing functional groups are suitable for octane enhancement. Sulfur-, phosphorus-, or halogen-containing compounds either promote knocking or achieve only insignificant octane improvements.

6.2.5.1 Blend Octane Numbers

When a component is blended to a base gasoline, the resulting octane number of the mixture cannot be calculated from the octane numbers of the neat gasoline and the component. Furthermore, the octane number of the mixture is not proportional to the component concentration. In order to characterize the octane efficiency of a component, therefore, "blend octane numbers" (B-ON) are defined according to the following equation:

$$\mathrm{B-ON} = \frac{\mathrm{ON_M} - \mathrm{ON_{BG}}(1-x)}{x}$$

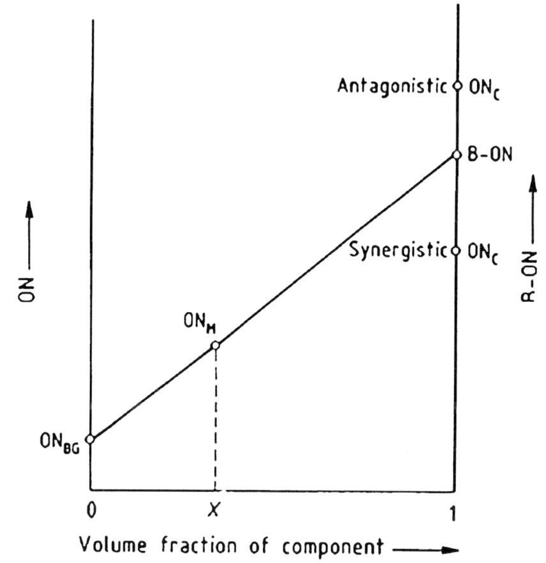

Fig. 6.3 Graphic interpretation of blend formula.

where B-ON is the blend octane number of the component; ON_M the octane number of the mixture; ON_{BG} the octane number of the base gasoline; and x is the volume fraction of the component in the blend. All parameters on the right-hand side of this equation must be determined experimentally. This equation attributes the entire change in octane number to the component. Graphically (Fig. 6.3), it reflects a linear extrapolation from the base gasoline ON_{BG}, via the mixture ON_M, to a fictive B-ON value of the neat component. If the calculated B-ON is compared with the measured octane number of the neat component (ON_C), two alternatives are possible. If the B-ON is larger than the measured ON_C, the mixture has a "synergistic" property, if the B-ON is smaller than the ON_C, the mixture shows an "antagonistic" behavior.

An example of the nonlinear MON blend behavior and the determination of the B-MON value is shown in Fig. 6.4 for straight-run gasoline – MTBE mixtures. This example shows that the extrapolated B-MON value based on $x = 0.2$ is higher than the experimentally determined MON value of neat MTBE, indicating a synergistic effect of MTBE.

In general, B-ON values of octane enhancers and refinery components depend on the component concentration, as well as the octane level and the composition of the base gasoline. These dependencies on the base gasoline are relatively small for refinery components. For some octane enhancers, however, in particular oxygenates such as alcohols, the B-ON values vary significantly with gasoline composition. Therefore, in refinery blending, different B-ON data are used for the same oxygenate when blended to different gasoline grades. Normally, B-ON data derived from paraffinic gasolines are higher than those from gasolines with high olefinic and/or aromatic contents. Most components achieve higher B-RON than

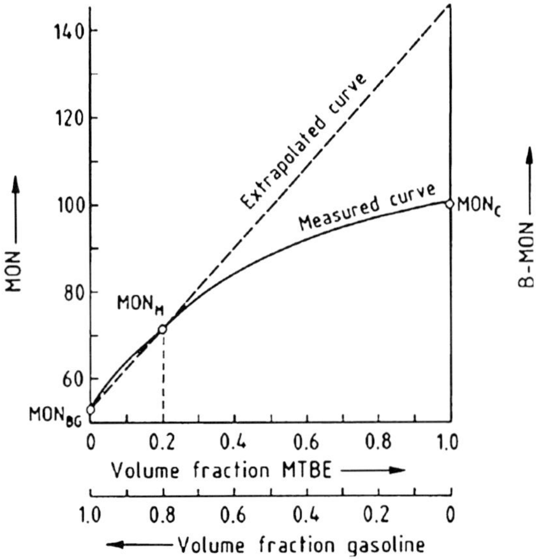

Fig. 6.4 Example of MON blend behavior of straight-run gasoline – MTBE mixtures.

B-MON values. Both depend on the component concentration as can be seen in Fig. 6.4 for B-MON. In practice, this effect can be neglected because the concentration range used is normally small.

Some other properties such as density, volatility, and vapor pressure also exhibit nonlinear blend behavior in mixtures. Blend properties are determined analogously to B-ON and play an important role in the evaluation of the economic potential of a component.

6.2.5.2 Nitrogen-Containing Components

The influence of nitrogen-containing compounds on octane performance has been studied extensively [3]. Depending on the functional group and the molecular structure, they provide either proknock (i.e., they promote knocking) or antiknock properties.

Whereas *organic nitrates* and *aliphatic amines* were generally found to promote knocking, *aromatic amines* achieve relatively high octane enhancements. Examples of B-ON values determined for aniline and some aniline derivatives are shown in Table 6.1. All measurements were performed in the same base gasoline with a component addition of 2 vol %. The octane enhancement efficiency varies with molecular structure. Relative to aniline, differences are small when hydrogen atoms of the aromatic ring are substituted by methyl or ethyl groups. Substitution with longer aliphatic chains results in reduced efficiency. Whereas primary and secondary amines have good octane enhancing properties, tertiary amines, e.g., *N,N*-dimethylaniline are ineffective.

A serious *disadvantage* of nitrogen-containing components results from the fact that nitrogen is partially oxidized in the combustion process. This increases nitrogen

Table 6.1 Blend-octane numbers of nitrogen-containing components (values determined from 2 wt % addition to premium base gasoline)

Component	B-RON	B-MON
3,4-Dimethylaniline	370	320
3,5-Dimethylaniline	340	310
p-Toluidine	340	305
p-Ethylaniline	320	300
Diphenylamine	310	300
Aniline	310	290
p-tert-Butylaniline	300	260
N-Methylaniline	280	250
Indoline	300	150
N,N-Dimethylaniline	95	84

oxide emissions in the exhaust gases. There is also experimental evidence that some of the components may cause deposit formations in fuel intake manifolds and/or combustion chambers, when used at the required concentrations. Of the listed components, only *N*-methylaniline was used as an octane enhancer for a short period after World War II in Germany.

6.2.5.3 Oxygen-Containing Components

Organic oxygen compounds such as alcohols, ketones, ethers, esters, and phenols are commonly characterized as "oxygenates" or "fuel oxygenates". The majority of the oxygenates act as octane enhancers; their efficiency depends, however, on the functional group and on the individual structure of the molecule. Some derivatives of esters, ketones, and phenols achieve relatively high B-ON values, as indicated in Table 2. Because of other inappropriate properties, e.g., high boiling points, high acidity, low material compatibility, and high toxicity they are not used as octane enhancers.

Of the oxygenates, only alcohols and ethers are used to any significant extent. They are preferred over other octane enhancers because of lower costs, availability, handling, and their physical and chemical properties, which are similar to gasoline components. Names, abbreviations, formulae, and molecular mass of the most important alcohols and ethers suitable for fuel blending are listed in Table 3.

Table 6.2 Blend octane numbers of selected oxygenates

Component	B-RON	B-MON
3,4-Dimethylphenol	180	180
Phenol	190	160
m-Cresol	160	150
o-Cresol	160	140
Dimethyl succinate	130	130
Acetylacetone	120	110
Acetone	107	98

Table 6.3 Important fuel oxygenates

Oxygenate	Abbreviation	Formula	M_r	CAS registry no.
Alcohols				
Methanol	MeOH	CH_3OH	32.04	*[67-56-1]*
Ethanol	EtOH	C_2H_5OH	46.07	*[64-17-5]*
1-Propanol	PrOH	C_3H_7OH	60.97	*[71-23-8]*
Isopropanol (2-propanol)	IPA	C_3H_7OH	60.97	*[67-63-0]*
1-Butanol	NBA	C_4H_9OH	74.12	*[71-36-3]*
sec-Butanol (2-butanol)	SBA	$C_2H_5CH(OH)CH_3$	74.12	*[78-92-2]*
Isobutanol (2-methyl-1-propanol)	IBA	$(CH_3)_2CHCH_2OH$	74.12	*[78-83-1]*
tert-Butanol (2-methyl-2-propanol)	TBA	$(CH_3)_3COH$	74.12	*[75-65-0]*
Ethers				
Dimethyl ether (oxybismethane)	DME	CH_3OCH_3	46.07	*[115-10-6]*
Methyl *tert*-butyl ether (2-methoxy-2-methyl-propane)	MTBE	$CH_3OC(CH_3)_3$	88.15	*[1634-04-4]*
Ethyl *tert*-butyl ether (2-ethoxy-2-methylpropane)	ETBE	$C_2H_5OC(CH_3)_3$	102.18	*[637-92-3]*
Diisopropyl ether (2,2′-oxybispropane)	DIPE	$(CH_3)_2CHOCH(CH_3)_2$	102.18	*[108-20-3]*
tert-Amyl methyl ether (2-methoxy-2-methyl-butane)	TAME	$CH_3OC(C_2H_5)\ (CH_3)_2$	102.18	*[994-05-8]*
Isopropyl *tert*-butyl ether [2-methyl-2-(1-methylethoxypropane)]	PTBE	$(CH_3)_2CHOC(CH_3)_3$	116.28	*[17348-59-3]*
Di-*sec*-butyl ether (2,2′-oxybisbutane)	DSBE	$C_2H_5CH(CH_3)OCH(CH_3)C_2H_5$	130.23	*[6863-58-7]*
sec-Butyl-*tert*-butyl ether [2-(1,1-dimethyl-ethoxybutane]	BTBE	$C_2H_5CH(CH_3)OC(CH_3)_3$	130.23	*[32970-45-9]*

6.2.5.3.1 **EC Directive on Fuel Oxygenates**

The use of oxygenates in gasolines was established during the 1970s and has developed rapidly to a large-scale business. Prior to any official regulation, the German automobile and oil industry reached an interindustrial agreement to restrict the methanol content of fuel to 3 vol % with the obligation to use an adequate cosolvent component (e.g., TBA) to ensure appropiate water tolerance of methanol-containing fuels. For the EC the European Council passed a directive in December 1985, regulating the use of oxygenates in gasolines after several years of intensive negotiations of the ministries of commerce supported by experts [4]. Meanwhile, the technical annex of the directive has been implemented in the gasoline standards of all EC member states.

The purpose of the directive was to ensure:

1) The performance of the existing car population
2) The use of gasoline extender and octane-enhancing components in addition to materials derived from crude oil
3) The compensation of octane shortfall due to lowering of alkyllead concentration
4) A harmonization of gasoline quality in Europe

The annex lays down definitions and blend compositions of oxygenates. The most important information is summarized in Table 6.4, providing the maximum volumetric additions of individual oxygenates and the maximum gravimetric oxygen content of the blends.

6.2.5.3.2 **Blend Octane Numbers of Oxygenates**

The octane efficiency of oxygenates cannot be characterized in a simple way because it depends on a variety of factors. Therefore, B-ON ranges are usually defined to account for their possible variation. This makes it difficult to compare octane efficiencies of different oxygenates.

Table 6.4 EC directive on fuel oxygenates

Oxygenate %	**A,* vol %**	**B,** vol**
Methanol (stabilizers to be added)	3	3
Ethanol (stabilizers may be added)	5	5
Isopropanol (IPA)	5	10
tert-Butanol (TBA)	7	7
Isobutanol (IBA)	7	10
Ethers, with five or more C atoms	10	15
Other oxygenates, defined	7	10
Total oxygen content, wt %	2.5	3.7

* A = limits which have to be accepted by all EC countries.

** B = limits which may be adopted by individual member states but above which labeling is mandatory.

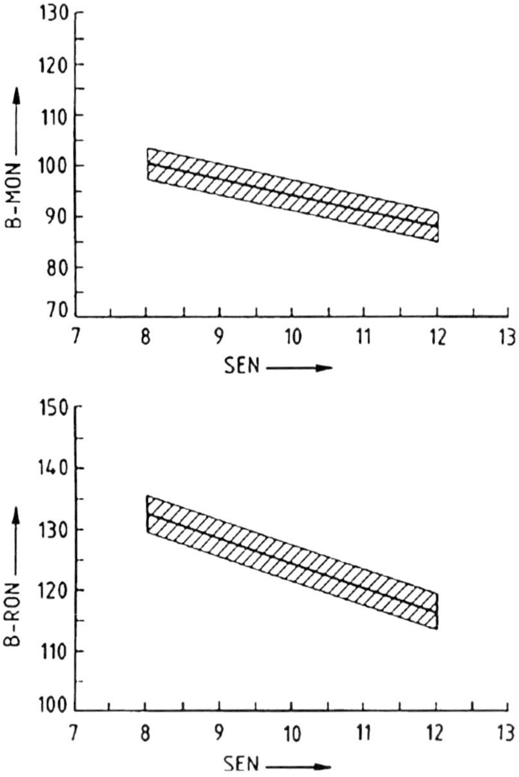

Fig. 6.5 Correlation between B-ON and SEN for methanol.

To overcome this problem, Veba Oil conducted a series of investigations with the objective to characterize the octane blend behavior of some oxygenates. In the progress of this work, the gasoline sensitivity (SEN, i.e., the difference between RON and MON) was found to be a suitable correlation parameter [5]. Empirically it may be considered as an approximate gauge for the gasoline composition; a high paraffinic content corresponds to low SEN, high aromatic and/or olefinic contents result in high SEN.

The measurements were performed in a variety of different unleaded gasolines, the compositions of which were widely varied. Oxygenate concentrations ranged from 10 to 20 vol %. Examples of the graphic correlations based on linear regression of B-ON data versus gasoline SEN are shown in Fig. 6.5 for methanol and in Fig. 6.6 for PTBE. Both B-RON and B-MON decrease with increasing gasoline SEN. Similar blend behavior was found for the majority of the oxygenates investigated. Typically, lower alcohols such as methanol and ethanol show a strong dependence of B-ON on SEN. For higher alcohols and ethers this dependence is less pronounced; in a few cases, e.g., for TBA, no significant dependence was observed.

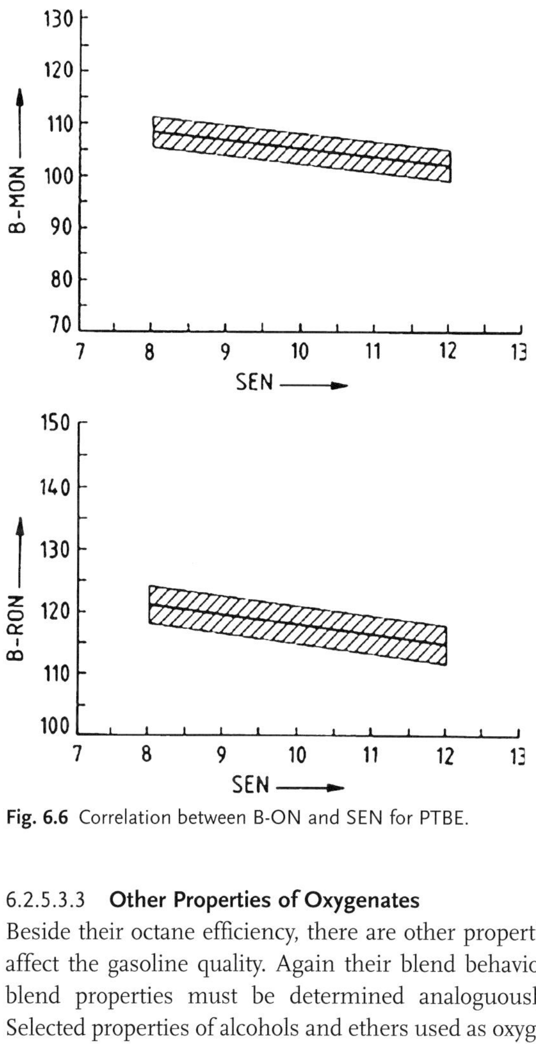

Fig. 6.6 Correlation between B-ON and SEN for PTBE.

6.2.5.3.3 **Other Properties of Oxygenates**

Beside their octane efficiency, there are other properties of oxygenates which also affect the gasoline quality. Again their blend behavior may not be linear so that blend properties must be determined analoguously to the B-ON procedure. Selected properties of alcohols and ethers used as oxygenates are listed in Table 6.5.

Density Densities of alcohols are generally higher than the specified density region for gasolines, whereas those of ethers range within it. Blend densities of alcohols and ethers show only moderate differences when compared with the densities of the neat components.

Vapor Pressure Due to their polarity, lower alcohols such as methanol and ethanol possess blend vapor pressures (B-VP) which exceed the vapor pressure of the neat components, because association of alcohol molecules is hindered in the mixture. Therefore, their B-VP values strongly depend on the component concentration. In contrast, B-VP data of ether components hardly show any dependence on concentration. They are significantly lower than those of lower alcohols and are similar to the vapor pressure of the neat components.

Table 6.5 Properties of oxygenates

Oxygenate	mp, °C	bp, °C	d_4^{20}	n^{20}	Heat of combustion, kJ/mol	RVP*	Composition, wt% C	H	O
MeOH	−97.7	64.70	0.7912	1.3284	−725.7	0.81	37.48	12.58	49.93
EtOH	−114.1	78.29	0.7894	1.3614	−1367.6	0.70	52.13	13.13	34.73
IPA	−87.8	82.26	0.7755	1.3772	−2006.9	0.72	59.95	13.42	26.63
SBA	−114.7	100.00	0.8069	1.3972	−2660.6		64.81	13.60	21.59
IBA	−88.5	107.66	0.8016	1.3958	−2668.5	0.63	84.81	13.60	21.59
TBA	25.6	82.80	0.7866	1.3877	−2644.0	0.64	64.81	13.60	21.59
DME	−141.5	−24.80	0.6689	1.3018	−1460.5		52.14	13.13	34.73
MTBE	−108.6	55.20	0.7406	1.3690	−3362.1	0.65	68.13	13.72	18.15
ETBE	−94.0	73.10	0.7399	1.3756	−4014.9	0.29	70.53	13.81	15.66
DIPE	−85.5	68.30	0.7235	1.3681	−4042.6	0.33	70.53	13.81	15.66
TAME		86.30	0.7700	1.3885	−4010.8	0.77	70.53	13.81	15.66
PTBE		87.50	0.7400	1.3791	−4667.2	0.20	72.35	13.88	13.77

* RVP Reid vapor pressure ASTM D 323.

Volatility The addition of oxygenates affects the boiling characteristics of the gasolines. When ethers are blended, only moderate changes in the distillation curves are observed. Blends of lower alcohols, however, cause significant deformations of the distillation curves due to the formation of azeotropes. This rises the front-end volatility, i.e., an increased distillate fraction in the temperature range of the alcohol boiling points is obtained.

6.2.5.3.4 **Production**

The important processing data of alcohols and ethers are summarized in [6]. Further details of production and economics are described in [7].

Alcohols *Methanol* is produced catalytically from synthesis gas. Apart from its use as a gasoline blending component, large amounts of methanol are also employed as feedstock for the etherification with isobutene to form MTBE (Chap. 6.1.).

Synthetic ethanol is produced by the hydration of ethylene with an acid catalyst such as sulfuric acid and phosphoric acid in reactor columns or in ion-exchanger fixed-bed reactors. *Bioethanol* (see Chap. 3.8.4.) is produced by fermentation of monosaccharides such as glucose, fructose, mannose, or xylose derived from molasses, sugarcane, sugar, corn, or potatoes. Costs of bioethanol are higher than those of synthetic alcohol because the process economics depend essentially on the feedstock price.

Higher alcohols, i.e., 2-propanol, 1-butanol, 2-butanol, and 2-methyl-2-propanol can also be produced by hydration with acid ion-exchanger resins [8]. *Tert*-butanol (TBA) is formed as a coproduct in the production of propylene oxide by epoxidation of propene with *tert*-butyl hydroperoxide (TBHP). In the case of high MTBE demand in the market, TBA is sometimes dehydrated to isobutene which is then fed to the MTBE process [9].

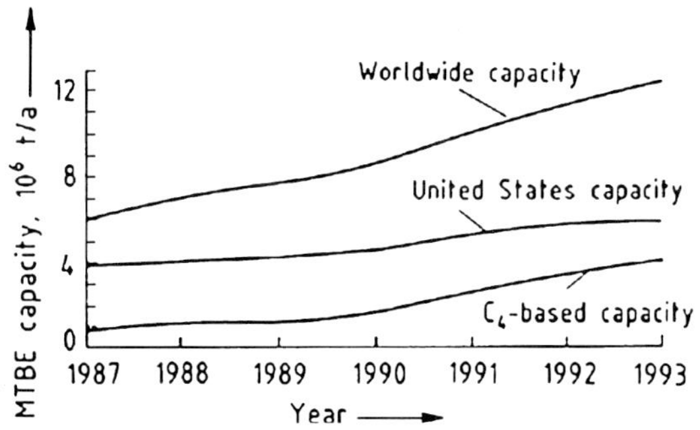

Fig. 6.7 Worldwide MTBE capacity trends (1989 survey).

Ethers The ethers are chiefly manufactured by exothermic reaction of branched olefins bearing the double bond at the branching point with alcohols such as methanol, ethanol, 2-propanol, or butanol according to:

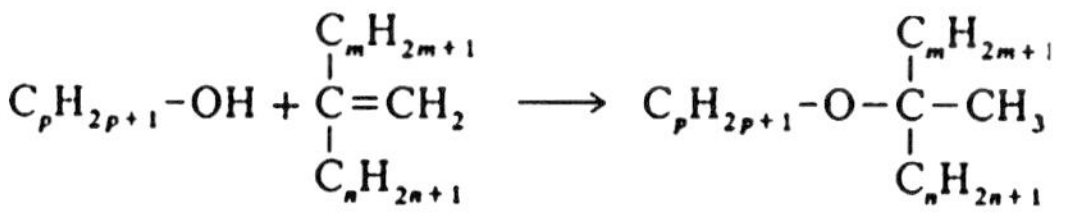

[for MTBE $m = n = p = 1$; for PTBE $m = n = 1$, $p = 3$; for *tert*-amyl methyl ether (TAME) $m = 2$, $n = p = 1$].

MTBE is the most important ether used as an oxygenate. MTBE capacity (see Fig. 6.7) and plant construction have increased steadily because MTBE blending offers simple and economic flexibility to fill any octane gap in the refiner's gasoline pool [10]. More than 40% of the total onstream capacity of MTBE is used as merchant MTBE, which is available to supply refiners in need of additional octane quality [11]. For production of MTBE, see Chap. 6.1.

Reaction of ethanol with isobutene produces ethyl *tert*-butyl ether (ETBE), which has similar properties to MTBE. This process offers an outlet of bioethanol [12, 13]. Reaction of 2-propanol with isobutene yields *PTBE* [14, 15], an oxygenate with excellent blend properties such as low vapor pressure and low oxygen content in combination with high octane quality.

Based on a similar technology methylbutenes (amylenes) of C_4 to C_6 streams can be etherified with methanol to *TAME* [16, 17], a blending ether with a high octane number. The BP Etherol process [18] uses unsaturated C_5 compounds of cracker fractions. Branched olefins react with methanol to form higher ethers. However, most of the branched olefins have terminal double bonds with low reactivity. To increase the yield of 3-methyl-2-butene (isoamylene) the double bond is shifted in the presence of a palladium-impregnated cation-exchanger resin. The isomerization proceeds according to:

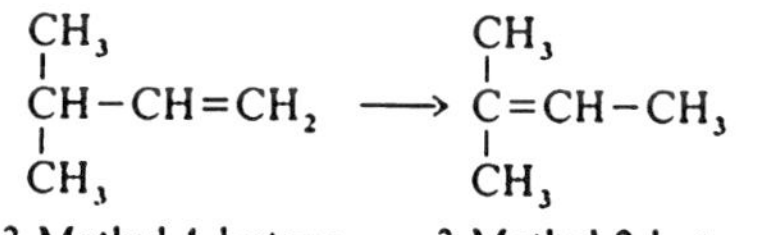

3-Methyl-1-butene **3-Methyl-2-butene (isoamylene)**

Isoamylene can then be reacted with methanol to form TAME.

6.2.5.3.5 **Environmental Protection**

The compulsory use of oxygenates in gasolines is becoming widespread in the *United States*, because several investigations have shown that oxygenates reduce carbon monoxide emissions. This effect offers a means of bringing CO emissions down to required standards. Depending on the area and season in the United States, the minimum oxygen content in the fuel is in the range of 2.0–3.0 wt% oxygen (Clean Air Act Amendment) [19]. Addition of TBA and methanol may reduce the CO emissions up to 40% and the hydrocarbon emissions up to 20% in comparison with conventional gasoline [20]. Insufficient supply of oxygen in air – fuel mixtures causes incomplete combustion of the fuel, accompanied by CO formation. When oxygen is supplied by oxygenates that are blended to the gasoline, this chemically available oxygen reduces the formation of carbon monoxide. Governments of several states made efforts to establish the relationship between fuel oxygen content and CO exhaust concentrations. In 1988, the state of Colorado mandated a minimum of 1.5 wt% oxygen in gasoline from January to March; this program was continued from November 1988 until March 1989, this time a minimum of 2.0 wt% oxygen was required. The 1988 program was estimated to lower CO air concentrations by 8–10% along the Front range of the Rocky Mountains [10]. In the United States emission measurements with different vehicles show CO reduction up to 32%, during the Federal Test Procedure (FTP) cycle (a standardized driving cycle for emission tests) [21]. The tests included vehicles with different exhaust and closed-loop systems and demonstrated a statistically significant CO reduction due to the oxygenated fuels, depending on the system and oxygen content (Fig. 6.8). The gasolines were blended with MTBE and ethanol. Emissions of unburned hydrocarbons and nitrogen oxides were virtually unaffected.

Under *European conditions*, studies were made with ethers and alcohols on the influence of oxygen content on the limited pollutants such as carbon monoxide, hydrocarbons, and nitrogen oxide [5]. Blend components were methanol with IPA – TBA as cosolvents and the ethers MTBE, PTBE, and BTBE. Another investigation was conducted by the European Fuel Oxygenates Association (EFOA) using mixtures

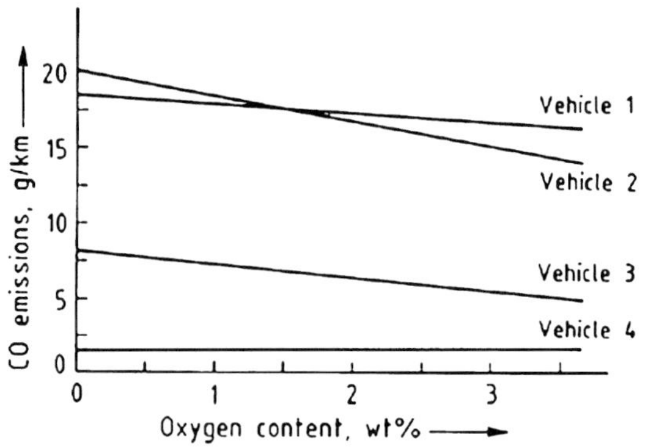

Fig. 6.8 CO emissions FTP 75.

of ethers and alcohols to study the dependence on oxygenates systematically [22]. Oxygen content was varied up to 3.6 wt %. The results showed that oxygenates reduce CO emissions to a large degree. The reduction increases with oxygen content but is not a function of the type of oxygenate.

6.2.5.3.6 **Toxicology and Occupational Health**

PTBE. The Ames test does not show any mutagenity. According to an acute toxicity study in rats, PTBE is considered to be nontoxic up to a dose of 2000 mg per kilogram body weight.

6.2.5.3.7 **Storage and Transportation**

The higher ethers MTBE, PTBE, and TAME are stable in the presence of air under atmospheric conditions. Formation of peroxides is not observed during contact with oxygen. Containers, sealing materials, and also elastomers in automobiles have to meet the same requirements as for contact with gasoline.

6.2.6
Metal-Containing Additives

A large number of organometallic compounds are known to act as octane enhancers. Generally, compounds containing Pb, Cu, Ni, Th, Fe, and Mn achieve high octane improvements, whereas those containing Co, Zn, Bi, V, Se, and Ce are slightly less effective. For practical use, most of them must be excluded because of high costs, instability, toxicity, impact on catalytic converters, or other undesired properties.

Their use in gasolines also requires an official approval in most European countries and in the United States. In Germany, the application of metal-containing additives–other than lead–is restricted by the gasoline lead act (Benzin – Blei – Gesetz).

The octane enhancement of organometallic additives is explained by a catalytic effect, because during combustion the metal compound is oxidized to form very fine solid particles of metal oxide. These particles have large surface areas with active centers which can react with free radicals formed by partial oxidation of the gasoline. Therefore the spontaneous combustion is delayed and knocking does not occur. The consumed centers are reactivated by oxygen and can react again with free radicals. Due to this cycle only small concentrations of additive are necessary to cause high octane enhancement.

A metal-free component is unable to delay the spontaneous combustion catalytically. The organic compounds are consumed irreversibly by the intermediate reactants which are responsible for knocking. Therefore the effectivity of antiknocking components is limited and high concentrations are needed to get a higher octane quality.

A disadvantage of metal-containing additives results from the formation of solid combustion products that lead to deposit accumulation in the combustion chambers. For the lead fluids this problem was solved by addition of lead scavengers.

Table 6.6 Comparison of TEL and TML

Chemical and physical properties	Tetraethyllead $(C_2H_5)_4Pb$	Tetramethyllead $(CH_3)_4Pb$
Molecular mass, g/mol	323.5	267.3
Lead content, %	323.5	267.3
Density, g/cm^3	1.66	1.995
bp, °C	200 (decomp.)	110
Vapor pressure (20 °C), Pa	34	31.5×10^2

However, this is not feasible for most other organometallic additives, thus their use must be restricted to very low concentrations. Additives that achieve sufficient octane improvements in the ppm concentration range and that meet other technical and economical requirements are methylcyclopentadienylmanganese tricarbonyl (MMT) and dicyclopentadienyliron (ferrocene).

6.2.6.1 **Alkyllead Compounds**

The strong octane-boosting effect of tetraethyllead (TEL) was discovered in 1921 in the United States [23]. Its wide use began in the United States in 1929 and in Germany shortly before World War II. Curiously, investigations at that time regarded lead as relatively environmentally friendly.

The effect of lead is based on an intervention in the mechanism of the combustion process. In precombustion reactions, organic peroxides are formed, which decompose to form free radicals. The latter initiate knocking reactions. These highly reactive radicals can be deactivated by very finely divided lead compounds formed from the alkyllead compounds in the combustion chamber [24,25].

In addition to tetraethyllead, which was used exclusively at first, tetramethyllead (TML) came into use from 1960 onward. Mixed alkyllead compounds, such as diethyldimethyllead or ethyltrimethyllead, and mixtures of TEL and TML were also employed, because a broad-spectrum effect can be achieved through a combination of different characteristics (Table 6.6). Thus, the comparatively low boiling point of TML is good for increasing the front octane number. (Fig. 6.9).

The increase in octane number that can be achieved with alkyllead compounds depends strongly on fuel composition. The *lead susceptibility*, a measure of the response behavior to the addition of lead, decreases in the following order: paraffins $\gg$ cycloparaffins $>$ aromatics $>$ olefins $>$ diolefins ($\approx$ 0). Accordingly, fuel components rich in paraffins, such as straight-run gasoline, alkylate, and isomerate (Chap. 3), react very well to lead addition, whereas olefin-rich catalytically cracked gasoline has a poor response (Fig. 6.10).

In connection with the addition of alkyllead compounds to gasoline, the term "lead bonus" was sometimes used. It meant that highly leaded fuels, in comparison with unleaded or low-lead ones, have a disproportionately high road octane number, although the RON and MON determined in test engines are equally high [26–30]. Investigations carried out with the introduction of unleaded fuels showed,

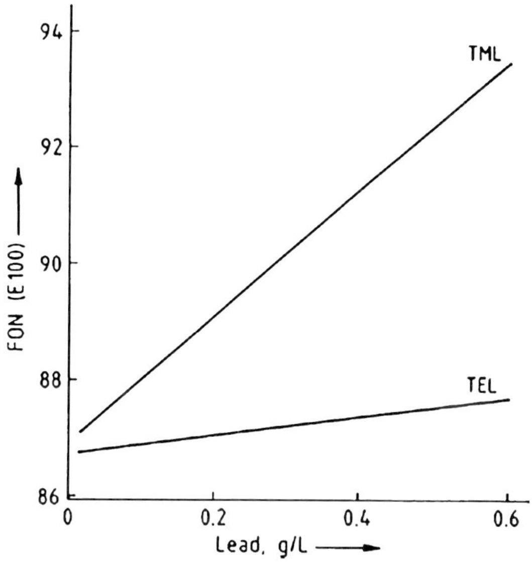

Fig. 6.9 Effect of tetraethyllead and tetramethyllead on the gasoline front octane number (FON).

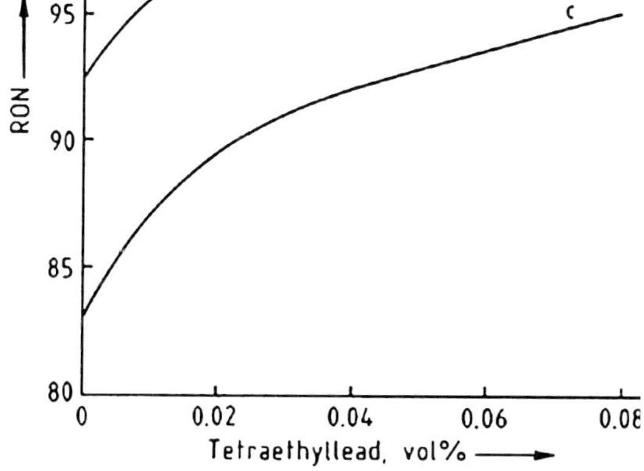

Fig. 6.10 Effect of tetraethyllead on some base fuels
(octane quality of the base fuel types are widely varying
in practice). a) Olefin-rich fuel;
b) Aromatic-rich fuel;
c) Paraffin-rich fuel.

however, that low-lead and unleaded fuels behave almost identically with regard to the road octane number [31] in most cases.

Usually, alkyllead compounds have always been used in combination with scavengers. These are organochlorine and -bromine compounds, namely, 1,2-dichloroethane and 1,2-dibromoethane. In the combustion chamber, scavengers cause the formation of lead chloride and lead bromide whose good volatility at high temperature ensures that ca. 70–80% of the lead is not deposited there. However, investigations in 1989 have shown that scavengers can be removed without adverse affects when the lead concentration in the fuel is low (e.g., 0.15 g/L).

On the other hand, treatment of leaded fuels with an antioxidant is necessary. Tetraethyllead, in particular, is prone to oxidation over long storage periods, leading to the separation of solid residues that can lead to blockages in the fuel system.

Originally, the addition of lead to gasoline was considered to be safe to the environment. This opinion has steadily changed, however [32–34], so that a decrease in the maximum permitted lead content has been legislated virtually worldwide. In the United States, Japan, Canada, Australia, Europe, and other areas leaded fuels have practically disappeared from the market. From a toxicological point of view, however, not only the lead but also the scavengers connected with it are suspected to be health risks, because these can form traces of highly toxic halogenated dibenzodioxins and -furans in the combustion chamber [35]. However, not only the poor compatibility of leaded fuels with the environment but also their incompatibility with catalytic converter technology (catalyst poisoning, see Chap. 7) has driven the move to unleaded gasoline in many parts of the world.

Alkyllead compounds also protected unhardened valve seats, present in older engines. Here, lead compounds formed in the combustion chamber inhibited erosion and wear ("recession") of the valve seats in two ways: (1) by avoiding high-temperature corrosion and (2) by their effect as solid lubricants. However, nontoxic alkali-metal additives are now available for this purpose, e.g., Powershield (Lubrizo), Formula Shell (Shell) and Valve Master (Du Pont). Practically the only suppliers of alkyllead compounds are Ethyl and Octel.

6.2.6.2 Methylcyclopentadienylmanganese tricarbonyl

Methylcyclopentadienylmanganese tricarbonyl (MMT) *[12108-13-3]*, is an organometallic additive used as an octane enhancer at low concentration. MMT is a metallocene compound in which the delocalized electrons of the cyclopentadienyl group are partially donated to the manganese atom and form covalent bonds. Carbon monoxide has a lone electron pair on the carbon atom which is donated to the manganese atom to form the carbonyl bond. In Canada it has been blended to unleaded gasoline for many years at a concentration of 0.0165 g Mn per liter of gasoline. In the United States, MMT is used in combination with alkylleads in leaded gasoline. Ethyl Corporation (Orangeburg) is the U.S. producer of MMT.

History Ethyl Corporation discovered MMT in 1954 and introduced it into the U.S. market as a supplement to TEL in 1958. Evaluation in unleaded gasoline was made in 1969. From 1974 to 1979 MMT was used in unleaded gasoline.

The effect of MMT was tested on a series of in-use three-way catalysts. Catalyst activity was measured and compared with the activity of the catalyst without MMT operation. Steady-state and light-off temperature measurements indicated a reduction in catalyst efficiency as exposure to MMT was increased [37].

The EPA predicted that over the useful life of vehicles, a greater number of MMT-fueled vehicles would not meet their designed hydrocarbon emission standard compared with corresponding vehicles fueled without MMT. Based on these results and in accordance with Clean Air Act Amendments, the EPA banned MMT in 1979 for unleaded gasoline [38].

In 1983 production of MMT was ca. 1800 t/a.

Trade names: AK-33 X, Antiknock – 33, Combustion improver – 2.

Physical Properties [39] MMT is sparingly soluble in water (70 ppm at 25 °C) and soluble in hydrocarbon solvents. MMT is a dark orange (straw color) liquid with a herbaceous odor; *mp* – 2.2 °C, *bp* 232.8 °C.

Chemical Properties When heated to decomposition MMT emits toxic fumes of CO.

Production The methylcyclopentadiene dimer [*26472-00-4*] is added to molten sodium suspended in diethylene glycol dimethyl ether at 185 to 190 °C. At this temperature, the dimer splits into the monomer and reacts with sodium to form sodium methylcyclopentadienide:

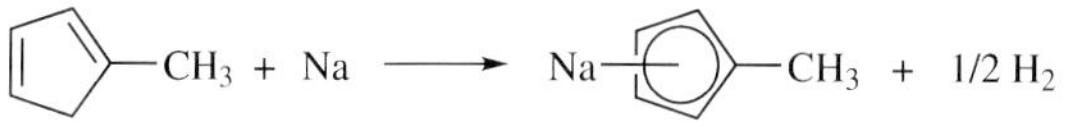

Anhydrous, flaked manganese chloride is added to the reaction mixture. The reaction forms bis(methylcyclopentadienyl)manganese with sodium chloride as byproduct. In the next step carbonylation under a pressure of 4.3 to 4.5 MPa at 190 °C forms MMT. Pure MMT is subsequently separated by fractional distillation.

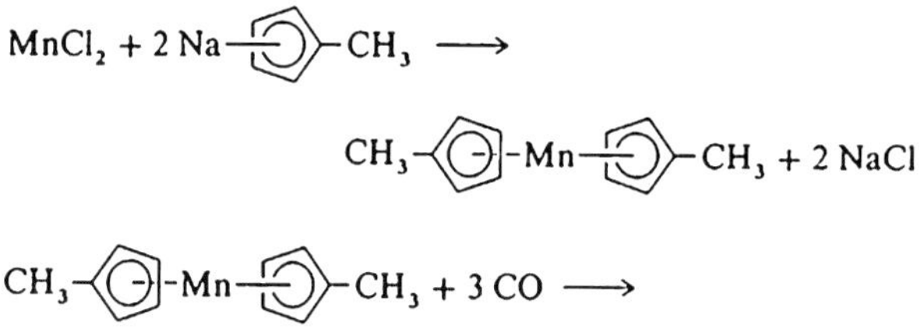

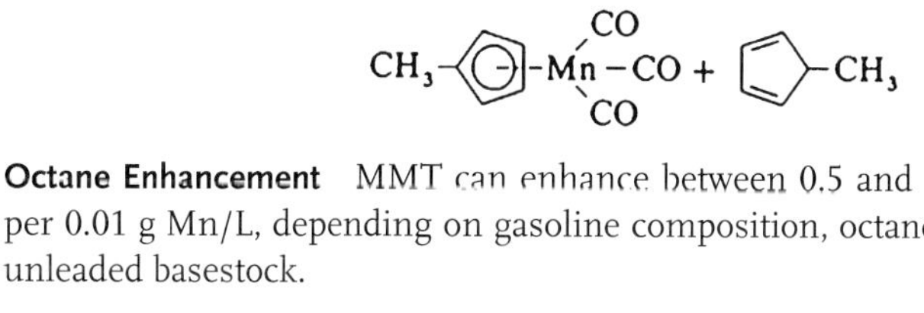

Octane Enhancement MMT can enhance between 0.5 and 1.0 (RON + MON)/2 per 0.01 g Mn/L, depending on gasoline composition, octane level, and leaded or unleaded basestock.

Toxicology and Occupational Health MMT is a poison by ingestion, inhalation, skin contact, intravenous, and intraperitoneal routes (Table 6.7). MMT is included

Table 6.7 Toxicity data of MMT* [40]

Test	Species					
	Rat	Mouse	Dog	Rabbit	Guinea pig	Cat
Oral						
LD_{50}, mg/kg	50	230				
LDLo, mg/kg			620	95	900	
Skin						
LD_{50}, mg/kg				140		
Inhalation						
LD_{50}, mg/m^3 (4 h)	76	58.6				
LCLo, mg/m^3			489(2 h)	1237(1 h)	1237(1 h)	223(7 h)
Intraperitoneal						
LD_{50}, mg/kg	23	152				
Intravenous						
LDLo, mg/kg			10			
LD_{50}, mg/kg				6600		

* American Conference of Governmental Industrial Hygienists–Threshold Limit Values: 0.1 mg Mn/m^3 (skin) time weighted average.

in the EPA's Extremely Harzardous Substances List. Manganese and some compounds are on the Community Right To Know List.

6.2.6.3 Ferrocene

In the early 1930s iron pentacarbonyl was used commercially as an octane-enhancing additive in Europe in concentrations similar to those of lead at that time. During combustion in engines iron pentacarbonyl formed excessive amounts of iron oxides which caused engine wear and influenced spark plug operation.

Ferrocene is another iron compound with similar octane-enhancing properties. Ferrocene [*102-54-5*] or dicyclopentadienyliron, Fe $(C_5H_5)_2$, belongs to the metallocenes; the iron atom is centered between two parallel cyclopentadiene rings. Ferrocene was first synthesized by Miller, Teboth and Tremaine in 1952. The sandwich or doublecone structure was first elucidated by G. Wilkinson and E. O. Fischer.

Physical Properties Important physical data of ferrocene are listed below [41]:

M_r	186.04
Density	1.49 g/cm^3
mp	173 °C
bp (at 102.3 kPa)	249 °C
Triple point	183 °C
Vapor pressure	
40 °C	0.0066 kPa
100 °C	0.34 kPa
Heat of fusion (at 175 °C)	17.8 kJ/mol

Heat of vaporization (at 175 °C)	47.2 kJ/mol
Heat of sublimation (at 25 °C)	70.2 kJ/mol
Start of decomposition	465 °C

Ferrocene is readily soluble in all hydrocarbon and hydrocarbon – oxygenate fuel blends and does not alter specified gasoline properties except for octane numbers.

Chemical Properties Ferrocene is resistant to water, air, light, and heat.

Production The modern electrochemical synthesis of ferrocene consists of two stages:

1) Electrochemical step (NaBr added as conductivity enhancer)

$$Fe + 2C_2H_5OH \rightarrow Fe(OC_2H_5)_2 + H_2$$

2) Chemical conversion

$$Fe(OC_2H_5)_2 + 2C_5H_6 \rightarrow Fe(C_5H_5)_2 + 2C_5H_5OH$$

Total:

$$Fe + 2C_5H_6 \rightarrow Fe(C_5H_5)_2 + H_2$$

In the *first stage*, metallic iron is oxidized anodically in ethanol to iron(II)ethoxide. At the cathode, the acidic protons of ethanol are reduced and form gaseous hydrogen which evolves out of the liquid. The iron(II)ethoxide is sparingly soluble and precipitates on the rotating cathode. A fixed scraper separates the iron(II)ethoxide from the cathode surface and suspends it in the liquid. In the *second stage* the ethoxide reacts with cyclopentadiene to form ferrocene and ethanol, which is recycled to the first stage. Ferrocene is filtered from the solution and purified by sublimation.

Octane Enhancement The octane improvements depend–as in the case of other octane enhancers–on the ferrocene concentration and on the composition of the base gasoline. Figure 6.11 represents a compilation of octane measurements performed in about 50 different base gasolines with ferrocene concentrations up to 30 ppm [42]. The octane response of RON is higher than that of MON.

The composition of the different unleaded fuel grades tested was widely varied. The fuel blends included all major refinery components as well as available oxygenates within their maximum volumetric addition presently specified in Europe. Generally it was found that the lower the octane number of the base gasoline, the higher the octane gain. Several measurements of octane response versus concentration have shown that response is more pronounced in the low ferrocene concentration region and that levelling-off effects are obtained at higher concentrations.

Toxicology The toxicological data of ferrocene are listed in Table 6.8.

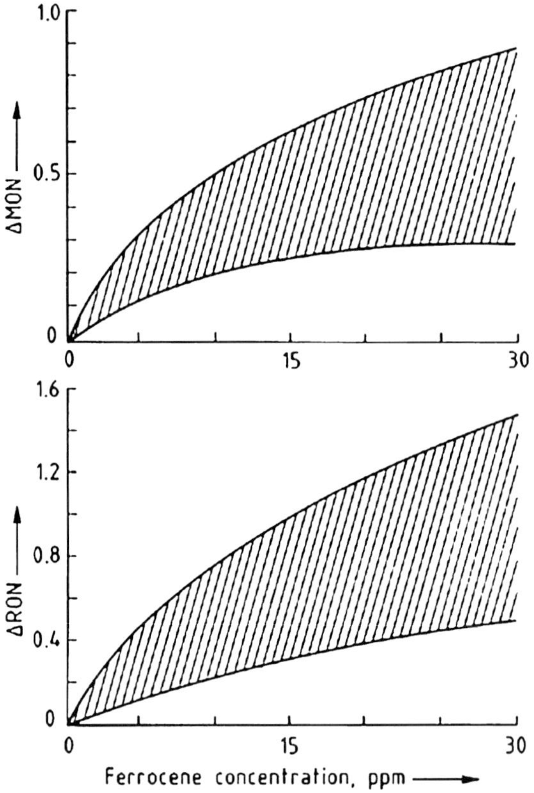

Fig. 6.11 Ferrocene octane enhancement in unleaded gasolines.

Table 8 Toxicity data of ferrocene [40]

Test	Species Rat	Mouse
Oral		
LD_{50}, mg/kg	1320	832
Intraperitoneal		
LD_{50}, mg/kg	500	335
Intravenous		
LD_{50}, mg/kg		178

6.3 References

1. *CEC Symp. Knocking of Combustion Engine*, Wolfsburg, Germany Nov. 1981.
2. T. MIDGLEY JR., Ind. Eng. Chem. **31** (1939) no. 5, 504–506.
3. M. H. GRAHAM, L. HESS: Organic Nitrogen Compounds as Antiknock Additives Report IR 52–12 Ethyl-Corporation, June 1952.

4. M. Allion, *EFOA 4th Conf.*, Brussels 1990.
5. K. Gottlieb, H.-K. Schädlich, P. Schug: "Erzeugung von oktanzahlsteigernden höheren Ethern und ihre anwendungstechnische Erprobung unter praxisnahen Bedingungen", *BMFT-Forschungsbericht 85 034* (1989).
6. L. F. Hatch, S. Matar: *From Hydrocarbons to Petrochemicals*, Gulf Publ. Co., Houston 1981.
7. SRI reports: P. L. Morse, Rep. no. 53, "Synthetic Ethanol and Isopropanol", (1969); H. W. Scheeline, S. Terada, S. L. Soder, Rep. no. 131, "Methyl-Tertiary-Butyl-Ether and Tertiary-Butyl-Alcohol", (1974); P. L. Morse, Rep. no. 111, "Methanol as a Chemical Raw Material", (1979), Stanford Research Institute, Menlo Park, California.
8. J. Schulze, M. Homann: *C 4-hydrocarbons and Derivatives*, Springer Verlag, Berlin, 1989.
9. H. E. Eilerts de Haan: "MTBE and Methanol Blends in Europe", *EFOA 3rd Conf.*, ARCO Chemical Europe, Frankfurt 1988.
10. G. H. Unzelmann, Oil Gas J., **87** (1989) Apr. 17, 44.
11. Oxygenated Fuels Association, Inc., Feb. 10, 1989, Communication to EFOA.
12. H. Bruderreck, BML-Report FE 83 NR 003 (1985).
13. D. Schliephake (ed.): *Nachwachsende Rohstoffe*, Verlag J. Kordt, Bochum 1986.
14. VEBA OEL, EP 006 315, 1982 (K. Gottlieb, H. Bruderreck, F. H. Wehmeier).
15. T. Kaitale, E. Merikallio, E. Rautianen, *EFOA 2nd Conf.*, Rome 1987.
16. W. J. Piel: *National Conference on Octane & Oxygenates*, San Francisco (1989).
17. *Oil Gas* J. **86** (1988) Nov. 21, 41.
18. H. Wittcoff, CHEMTECH 1990, Jan., 48.
19. EFOA, Newsletter no. 4, July 1989.
20. ARCO, *Report on Methanol*, 1983.
21. J. A. Gething, J. S. Welstand, J. C. Horn, SAE-paper 890 216 (1989).
22. H. Guttmann, K. P. Schug: "Results of Exhaust Emission Measurements with Oxygenates Motor Gasolines", SAE-paper no. 900 274, 1990.
23. D. Blackmore, A. Thomas (eds.): *Fuel Economy of the Gasoline Engine*, Macmillan, New York 1977, pp. 79.
24. R. Sharma, J. Bardwell, *Combust. Flame* 9 (1965) 106–108.
25. K. Salooja, *Combust. Flame* **9** (1965) 211–217.
26. A. Bell, J. Keene, K. Reders, SAE-Paper no. 710 624, 1971.
27. D. Gray, A. Azhari, SAE-Paper no. 720 084, 1972.
28. H. Bigley, J. Benson, SAE-Paper no. 730 013, 1973.
29. N. Mariotti, A. Ramsden, Int. Automob. Tech. Congr., paper B 2-1, 1974.
30. D. Hornbeck, R. Labruyere, L. Stinson, G. Unzelmann, SAE-paper no. 750 936, 1975.
31. K. Reders, W. Dabelstein, A. Reglitzky, W. Lange, *Mineralöltechnik* (1989) no. 13, 38.
32. H. Schellhaus, W. Obländer, K. Neuschwander, *Schriften. Ver. Wasser- Boden- Lufthyg.* **42** (1974) 245–258.
33. Conference Report on International Symposium, Airborne Lead in Perspective, National Academy of Sciences, Conference Report, Washington, D.C. 1972.
34. EEC Commission: Health Aspects of the Environmental Pollution with Lead, Amsterdam 1972/1973.
35. E. Bessey, Report on the BMFT (Bundesministerium für Forschung und Technologie)— FVV (Forschungsvereinigung Verbrennungskraftmaschinen)–DGMK (Deutsche Gesellschaft für Mineralöl und Kohle) Gemeinschaftsprojekt "Emission halogenierter KW," DGMK-Conference Report 8901, 1989, pp. 85–119.
36. G. Schoonveld, R. Riley, S. Thomas, S. Schiff, Autom. Eng. 95 (1987) no. 1, 72–76.
37. R. G. Williams, W. L. H. Watkins, R. G. Griffins, SAE-paper no. 890 582 (1989).

38. Environmental Protection Agency; Regulation of Fuel and Fuel Additives, *MMT Suspension of Enforcement,* Ann Arbor 1979.

39. R. K. Hinderer, *Am. Ind. Hyg. Assoc. J.* **40** (1979) 164.

40. N. I. Sax, R. J. Lewis: *Dangerous Properties of Industrial Material,* 7th ed., van Nostrand– Reinhold, New York 1988.

41. Chemische Betriebe Pluto, *Ferrocene-Product-Information,* Herne, Germany 1989.

42. K. P. Schug, H.-J. Guttmann, A.-W. Preuss, K. Schädlich, SAE-paper no. 900 154 (1990).

7
Automobile Exhaust Control

Martin Votsmeier, Thomas Kreuzer, Gerhard Lepperhoff

7.1
Introduction

The widespread use of automobiles offers a degree of mobility to the individual that was unthinkable just a few centuries ago. However, with increasing car and truck population today it is obvious that the massive use of combustion-engine vehicles is not free of problems, one of the major problems being air pollution. In Germany in 1994 more than 50% of nitrogen oxide and CO emissions and more than 40% of hydrocarbon emissions originated from road traffic. A total of 4.1×10^6 t of carbon monoxide, 1.3×10^6 t of nitrogen oxides, and 0.74×10^6 t of hydrocarbons were emitted on German streets in 1994 [1]. The formation of the major pollutants in combustion engines and their impact on environmental and human health are discussed in Section 7.2.

Legal regulations limiting emissions of the major pollutants carbon monoxide, hydrocarbons, nitrous oxides, and diesel soot were passed in the 1970s in the USA and Japan and later also in Europe and most other parts of the world. As a consequence substantial improvement in air quality has been observed. For example, in California the number of smog-alert days decreased from 121 in 1977 before the introduction of catalytic converters to only one smog alert day in 1997, although car use has doubled since 1977. The 1996 regulations already reduced emissions per kilometer by 96% compared to 1970 levels, with further improvements in 2000.

Section 7.3 gives a brief introduction to engine measures that allow reduction of emissions. While improved engine technology has led to some reduction in raw emissions, the dramatic reduction in tailpipe emissions, as reflected by the increasingly stringent legal limits, would not have been possible without the introduction of catalytic exhaust gas treatment systems [2–5]. Section 7.4 gives a general introduction to the working principle of catalytic exhaust gas treatment and especially focuses on the physical design of the catalyst system. The most

Handbook of Fuels: Energy Sources for Transportation. Edited by Barbara Elvers

ISBN: 978-3-527-30740-1

effective type of catalyst system today is the three-way catalytic converter that simultaneously reduces concentrations of CO, NO, and hydrocarbons. Three-way catalysts are used with spark-ignition gasoline engines and allow the reduction of pollutant emissions by 96% compared to the pollutant level before the introduction of catalytic converters. Three-way catalysts are described in Section 7.5.

Carbon dioxide is an unavoidable combustion product of fossils fuels and the only way to reduce CO_2 emissions from internal combustion engines is to construct more energy efficient engines. Due to its role as a major greenhouse gas and due to concerns about global warming, energy efficiency of automobiles has become a major concern for engine design. Today the most fuel efficient engines run with excess air and therefore do not allow application of the proven three-way catalyst. This poses the challenge of developing catalyst concepts that allow fuel-efficient lean-burn operation of the engine while retaining the low pollutant levels reached by three-way catalysts. Aftertreatment systems for lean exhaust gas are treated in Section 7.6.

7.2
Pollutant Formation and Limitation

In a combustion engine a hydrocarbon fuel is oxidized by atmospheric oxygen. Under ideal conditions only CO_2, water, and heat would be generated (Eq. 1).

$$C_mH_n + (m + 0.25n)O_2 \rightarrow mCO_2 + 0.5nH_2O \qquad (1)$$

Under engine conditions, however, ideal combustion is impossible because of the lack of chemical equilibrium and the inhomogeneous gas phase. Typical combustion products resulting from incomplete combustion are:

- Carbon monoxide (CO), a combustion intermediate
- Unburned and partially oxidized hydrocarbons (HC)
- Oxides of nitrogen (NO_x)
- Particulate matter

Carbon monoxide (CO) This formed as an intermediate reaction product during combustion of hydrocarbons. Incomplete combustion due to low combustion temperature, lack of oxygen, or too short a reaction time leads to the emission of CO, a poisonous gas that displaces oxygen from the blood.

Hydrocarbons (HC) Incomplete combustion of hydrocarbon fuels leads to the emissions of unburned and partially oxidized hydrocarbons [6].

In spark-ignition engines hydrocarbons in the exhaust originate mainly from regions in which the flame extinguishes or regions not reached by the flame (quench zones). Typical examples are layers near the combustion chamber walls and crevices, e.g., in the piston ring groves (wall quenching). With very lean mixtures, the flame can extinguish prematurely (flame quenching) at the end of the expansion because of low gas temperatures in the combustion chamber. In highly turbulent mixtures, exhaust gas can surround combustible gas zones before they are reached by the flame front. The HC emissions rate can be largely caused by the combustible gas losses in two-stroke gasoline engines and, to a lesser extent, in supercharged four-stroke engines.

In compression-ignition engines, HC emissions originate in the outer zones of the injection spray, where the fuel has thoroughly mixed with air to produce an nonignitable mixture. Emissions also result from the spray center, which has a very rich mixture, as well as from the chamber wall areas. Finally, near the end of the expansion, fuel vaporizes out of the injection nozzle.

The hydrocarbons emitted from internal combustion engines are always a mixture of various compounds: paraffins, olefins, acetylenes, and aromatics. The composition depends on the combustion system, the operating conditions of the engine, and the fuel.

Oxides of nitrogen (NO_x) These are generated at combustion temperatures by oxidation either of atmospheric nitrogen or nitrogen contained in the fuel. During the combustion process mainly NO is formed; NO_2, which is more toxic than NO, is formed by the relatively slow oxidization of NO in the atmosphere. In combustion processes with an excess of air significant NO_2 can be expected. NO_x reacts with hydrocarbons in sunlight to form ozone and photochemical smog. NO_x can increase respiratory illnesses and is a contributor to acid rain. Tropospheric ozone causes breathing difficulties and damages plants. In the upper atmosphere NO_x is involved in a catalytic cycle that leads to the destruction of the ozone layer.

Three mechanisms of NO_x formation must be considered [7]:

1) Thermal NO_x
2) Prompt NO_x
3) Fuel NO_x

Fuel NO_x. Fossil fuels contain organic nitrogen compounds (e.g., amides, amines, heterocyclic compounds) in concentrations of <1 % (coals have higher concentrations). Equation (2) is an example of such a reaction pathway.

$$\begin{aligned} &C_nH_mNH_2 + \left(n + \frac{m}{4} + 1\right)O_2 \rightarrow \\ &\quad nCO_2 + \left(\frac{m}{2} + 1\right)H_2O + NO \end{aligned} \tag{2}$$

Thermal NO_x. The direct oxidization of N_2 by oxygen at high temperatures, the so-called thermal NO_x mechanism, was first described by ZELDOVIC. At temperatures above 1300 °C NO is formed by the following reaction sequence:

$$O + N_2 \rightarrow NO + N \tag{3}$$

$$N + O_2 \rightarrow NO + O \tag{4}$$

In rich zones of the flames (oxygen deficiency) the following reaction takes place above 1300 °C:

$$N + OH \rightarrow NO + H \quad (5)$$

The amount of NO formed largely depends on:
- The air-to-fuel ratio in the combustion zone
- The temperature in the combustion zones
- The time for which the reaction partners remain at combustion temperature

Prompt NO_x. FENIMORE described a mechanism in which reactive radicals formed during hydrocarbon combustion react with molecular nitrogen to form NO:

$$CH + N_2 \rightarrow HCN + N \quad (6)$$

$$C + N_2 \rightarrow CN + N \quad (7)$$

$$CN + H_2 \rightarrow HCN + H \quad (8)$$

$$CN + H_2O \rightarrow HCN + OH \quad (9)$$

$$HCN, CN + O \rightarrow NO + \text{organic residue} \quad (10)$$

This mechanism requires rich combustion conditions and therefore does not play a significant role in a well-controlled engine combustion processes.

Particulate Emissions Particulates are tiny solid and liquid particles. They occur in the exhaust of diesel engines in sizes from 0.01 to 1 μm and consist of carbon-rich particles (soot), hydrocarbons, and sulfates. In a diesel engine, due to incomplete mixing, zones of air deficiency arise and lead to soot formation. In parallel, in zones with excess air, soot is partially burned. Soot emissions are strongly increased if the air–fuel mixture is enriched to near stochiometry.

Recent research has shown that especially ultrafine particles (<100 nm in mean diameter) and nanosized particles (<50 nm in mean diameter) are potentially dangerous due to their capability to enter deep into the respiratory tract. Furthermore, these small particles supposedly can penetrate cell membranes.

Carbon Dioxide Carbon dioxide (CO_2), although the most commonly emitted combustion gas, is normally not considered as a pollutant since adverse health effects are not known. However, CO_2 should be regarded as a serious pollutant in terms of its global-warming potential. CO_2 has been identified as a greenhouse gas, i.e., a gas-phase species with strong infrared adsorption bands that lets short-wavelength light from the sun pass through but adsorbs the light reflected from the earth in the troposphere. Due to the large CO_2 emission rate and its long lifetime its contribution to the greenhouse effects is estimated to be 50 %. Since CO_2 is an intrinsic reaction product of fuel combustion, the only way to reduce CO_2 emissions is to improve fuel economy.

Sulfur Compounds Any sulfur contained in the fuel is emitted through the tailpipe in the form of sulfur compounds, mostly as SO_2. Under atmospheric conditions the SO_2 is oxidized to SO_3, which leads to the formation of sulfuric acid. Sulfuric acid originating from the combustion of sulfur-containing fuels is a major contributor to acid rain. Currently the only feasible way to reduce emissions of sulfur compounds from internal combustion engines is to reduce the sulfur content of the fuel.

If the engine is operated under rich conditions SO_2 can be reduced to H_2S in the catalytic converter. Since H_2S has a characteristic unpleasant smell, it is the task of the engine management system to avoid driving conditions that lead to the formation of significant concentrations of H_2S.

7.3 Developments in Engine Technology

Engine technology is a wide field and is not the main focus of this article. For this reason here only a few developments in engine technology that have direct impact on catalytic emission control are outlined. A more complete treatment of the development of engine technology can be found elsewhere [8].

Traditionally diesel engines show some advantage in fuel economy compared to spark-ignition gasoline engines. The fuel economy of diesel engines was recently improved even further by the introduction of high-pressure direct-injection technologies, so that today the diesel engine exhibits an advantage in fuel economy over the conventional gasoline engine of about 25–30 %.

Several effects contribute to the fact that conventional gasoline engines show poorer fuel economy than diesel engines, the major reason being that these engines must be run with a stochiometric air – fuel mixture and that at low load the air intake of the engine has to be adjusted to the fuel by placing a throttle in the air intake. Especially at low load the throttle creates a vacuum in the intake manifold that imposes extra work on the cylinders. Other effects that contribute to the limited efficiency of current gasoline engines are temperature loss to the walls and friction. Several possibilities exist to improve the fuel economy of gasoline fuel cars, all of which try to minimize the energy inefficiency caused by the operation of the throttle at low engine load:

- Downsizing
- Electromechanical valve control
- Direct injection

The operating points met in urban driving can be shifted towards the more fuel efficient high-load region by downsizing the engine. To retain the same overall maximum power the specific power of the downsized engine must be increased by supercharging the engine. Best fuel economy at low load is obtained if a turbocharger with variable compression ratio is used.

An alternative technology is electromechanical valve control. Here the valves can be controlled individually and independent of the cylinder movement so that the air intake can be controlled by the timing of valve opening instead of the energy-inefficient throttle. Both downsizing and electromechanical valve control operate the engine with a stochiometric air – fuel mixture and therefore can be used in combination with the well-established three-way catalyst. The improvement in fuel economy expected from these technologies is limited to about 15 % relative to conventional gasoline engines.

Ultimate improvements in fuel economy can probably only be achieved if direct injection or combinations of the above-mentioned technologies with direct injection are employed. In a gasoline direct-injection engine the fuel is directly injected into the cylinder, as is in diesel engines. This allows operation of the engine with an oxygen-rich air – fuel mixture under partial load. In this way the disadvantages related to throttling the air intake can be completely avoided. Additional fuel economy is gained because modern gasoline direct-injection engines minimize wall heat losses and cylinder friction. Due to the lean operation at partial load, exhaust from a gasoline direct-injection engine can not be cleaned with a conventional three-way catalyst. New catalyst technologies must be developed for the treatment of the lean-burn direct-injection gasoline exhaust. Catalysts for the treatment of lean exhaust from gasoline lean-burn and diesel engines are described in Section 7.6.

7.4 Catalytic Exhaust Aftertreatment, General Concepts

Even though advanced engine technology offers substantial potential for the reduction of emissions of hazardous compounds, engine improvements alone do not allow a reduction of emissions to a level necessary to fulfill current and future emission legislation. For this reason aftertreatment of the raw exhaust gasses emitted by the engine is unavoidable. Under typical operation conditions the thermodynamic equilibria for the main reactions transforming exhaust gas pollutants into harmless compounds are on the product side, so that automotive exhaust purification essentially is a kinetic problem. For this reason catalysts that enhance the reaction rates are the method of choice for exhaust aftertreatment [2–6].

7.4.1 The Physical Design of the Catalytic Converter

Exhaust gas catalysis is a heterogeneous surface process, and to reach acceptable conversion within a reasonable catalyst volume a carrier structure must be used that exposes the exhaust gas flow to a high geometric surface area. In the early days of exhaust gas catalysis pellets were used for this purpose. In recent years pellets have been completely replaced by monolithic honeycomb carrier structures. Monolithic

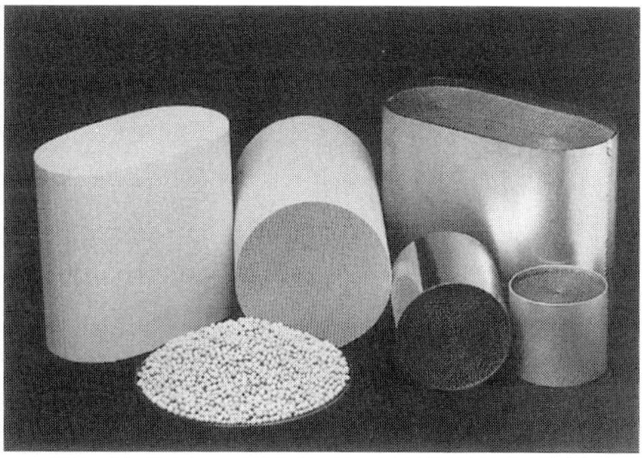

Fig. 7.1 Various designs of emission control catalysts (beads, ceramic monoliths, metal monliths).

converters consist of a single body that possesses a multitude of parallel channels, with the catalytically active material deposited along the walls of the channels. Figure 7.1 shows various designs of emission control catalysts. Honeycomb converters exhibit several advantages over pellet converters, the most important being the lower pressure drop and the reduced complexity of the converter housing design. A typical converter design with a ceramic monolith is shown in Fig. 7.2.

Monolithic supports are produced either from ceramic or metal, each of which has specific advantages.

Ceramic monoliths are most commonly produced by an extrusion process from synthetic corderite 2 $MgO \cdot 2\, Al_2O_3 \cdot SiO_2$. This material has a number of advantageous properties for this application, such as high melting point (>1300 °C),

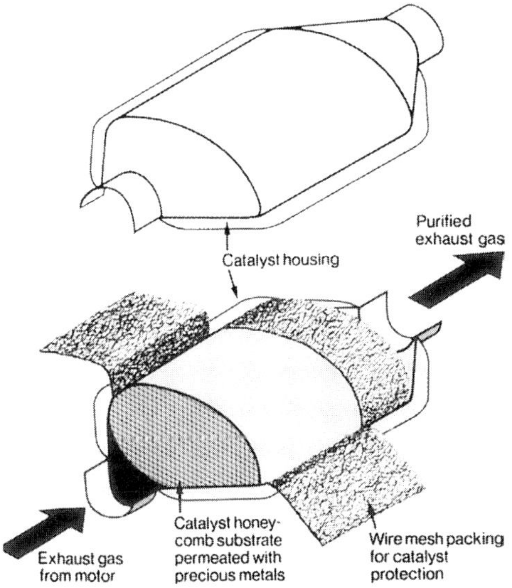

Fig. 7.2 Design principle of a ceramic monolithic based converter for aftertreatment of exhaust gases.

high mechanical strength, and most importantly a very low thermal expansion coefficient that results in excellent thermal shock resistance. Ceramic supports can be produced in a variety of different forms that differ in their outer dimensions, cell shape, cell density, and wall thickness. Today the most commonly used supports have square channels, a cell density of about 62 channels per square centimeter, and a wall thickness of about 0.15 mm. Thinner walls increase the open frontal area and accelerate heating up during vehicle start. Higher cell densities improve the geometrical contact surface within the channels. For these reasons there are ongoing efforts to develop supports with thinner walls and higher cell densities.

In the last few years monoliths made of metal were introduced to the market. Their main advantage is that they can be prepared with thinner walls. This allows higher cell densities with increased frontal area and thus reduced pressure drop. The low heat capacity of the thin metal walls and the higher thermal conductivity result in faster heat up compared to ceramic catalysts. However, metals are prone to corrosion under exhaust conditions, and their thermal expansion behavior is less compatible with the catalytic material.

7.4.2 The Washcoat

Ceramic and metallic monolith structures have a geometric surface area in the range of 2.0–4.0 m^2 per liter of support volume. This is much too low to adequately perform the catalytic conversion of the exhaust gas components. For this reason the monolith channel walls are coated with a thin layer of inorganic oxides with a very high internal surface area. This layer of carrier oxides is called the washcoat. The washcoat enhances the surface of the catalyst by a factor of about 10 000. Figure 7.3 shows individual channels of a ceramic monolith coated with washcoat. The main components of the washcoat are inorganic oxides and mixed oxides, and it also contains the catalytically active metals, mainly the platinum group metals (PGMs) Pt, Pd, and Rh, for which the oxides act as a support.

Fig. 7.3 Close-up view of a washcoated ceramic monolith.

Fig. 7.4 Transmission electron micrograph of PGMs on a washcoat particle.

The inorganic carrier oxides serve several functions vital for operation of the catalytic converter. First, they must maintain a high surface area even under severe aging conditions at temperatures above 1000 °C. Second, the carrier oxides have to prevent the catalytically active precious metal particles from sintering to larger particles at high temperatures. This is accomplished by tailoring the oxide composition such that the precious metal particles are tightly bonded on the carrier oxide surface. Figure 7.4 shows a high-resolution transmission electron micrograph of washcoat particles with finely dispersed PGM particles. Third, the carrier oxide improves the catalytic function of the precious metal particles by favorable electronic interaction between the precious metal clusters and the carrier oxide surface. To fulfill all these requirements modern washcoat systems are very complex chemical systems. The washcoat matrix consists of specifically designed mixed crystal lattices, and various chemically different phases are selectively combined to optimize performance and selectivity. The manufacturing process controls the optimum placement of the precious metals with respect to primary particle size and location on the different carrier phases.

Alumina is the most common carrier oxide used in automotive exhaust gas catalysis. Typically γ-, δ-, and θ-Al_2O_3 are used with internal surface areas of 50–250 $m^2\ g^{-1}$. Frequently the Al_2O_3 is doped with rare earth or alkaline earth elements. The role of the dopant is to increase the aging stability of the carrier and to optimize the carrier–precious metal interaction to prevent precious metal sintering and to enhance the catalytic activity of the precious metal particles. Besides alumina other common washcoat constituents are cerium and zirconium oxides. Cerium oxide mostly serves an oxygen storage component (see Section 7.5.2). Recently, zirconium cerium mixed oxides with optional additional dopants have been used.

The washcoat is usually prepared as an aqueous slurry and applied to the monolithic carrier by a dipping or spraying process followed by blowing out of excess washcoat slurry, drying, and calcination. Tight process and materials control is essential due to the high cost of the ingredients, especially the PGMs.

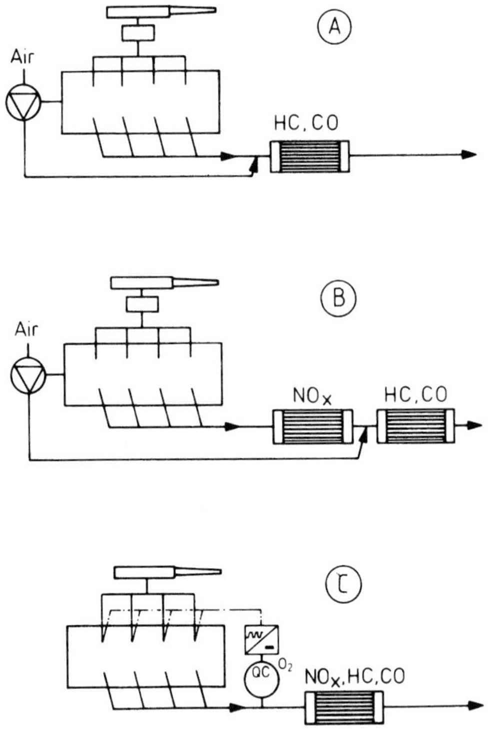

Fig. 7.5 Processes for catalytic automotive-exhaust purification A) Single-bed oxidation catalyst; B) Dual-bed catalysis; C) Three-way catalyst with closed-loop control.

7.5 Catalytic Aftertreatment of Stoichiometric Exhaust Gas

Up to now the majority of passenger cars and light-duty vehicles, especially in the USA, are propelled by spark-ignition engines operated under stoichiometric conditions. Due to their large population and, compared to diesel engines, their high raw emissions these vehicles accounted for a large part of the air pollution in urban areas. Therefore, spark-ignition engines were the first to require catalytic exhaust aftertreatment to fulfill emissions regulations. Most of the commercial experience in automotive catalysis till now has been gained in the field of stochiometric spark-ignition engines.

Different concepts of catalytic pollution control for stoichiometric spark ignition engines have been used over time (Fig. 7.5). During an initial period following the introduction of precious metal exhaust gas catalysts in 1975, NO limits could be reached by engine measures such as exhaust gas recirculation, and legislation only required the treatment of CO and HC. At this time engines were run slightly rich to reduce NO emissions. Secondary air was injected in front of the catalyst (Fig. 7.5 A), and the catalyst was subjected to oxygen-rich exhaust gas so that CO and hydrocarbons could be oxidized by the following reactions:

$$2CO + O_2 \rightarrow 2CO_2 \tag{11}$$

$$C_mH_n + (m + 0.25n)O2 \rightarrow mCO_2 + 0.5nH_2O \tag{12}$$

The next step in legislation in 1977 called for a reduction in nitrogen oxides. For this purpose a reduction catalyst was introduced that operated directly on the CO- and hydrocarbon-rich exhaust gas and removed nitrogen oxide by the reactions:

$$CO + NO \rightarrow 0.5N_2 + CO_2 \tag{13}$$

$$C_mH_n + (2m + 0.5n)NO \rightarrow mCO_2 + 0.5nH_2O + (m + 0.25n)N_2 \tag{14}$$

Secondary air was injected downstream of the reduction catalyst (Fig. 7.5 B), and CO and hydrocarbon emissions were reduced by an additional oxidation catalyst.

A further improvement in exhaust gas control technology for stoichiometric spark-ignition engines was achieved by the development of three-way catalysts that simultaneously convert all three regulated pollutants.

7.5.1 Three-Way Catalysis

In Fig. 7.6 the conversion efficiencies of a typical closed-loop three-way catalyst (Fig. 7.5 C) for the major pollutants are plotted as a function of the normalized air

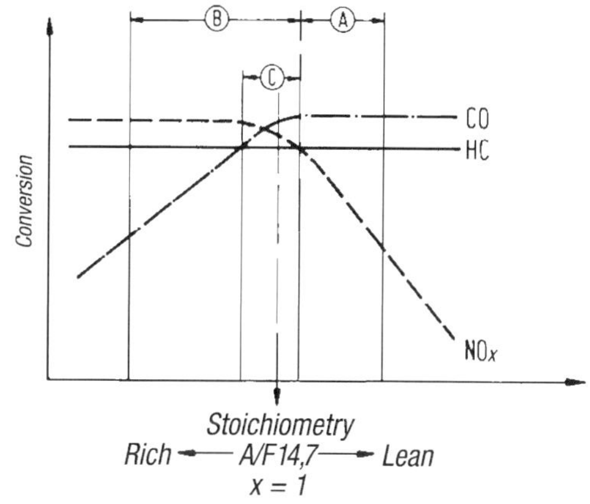

Fig. 7.6 Range of λ values in which various catalytic aftertreatment concepts are operated A) Single-bed oxidation catalyst; B) Dual-bed catalysis; C) Three-way catalyst with closed-loop control.

to fuel (A/F) ratio lambda. Under oxygen-rich conditions ($\lambda>1$) CO and hydrocarbon conversions are high, while NO conversion is prevented by insufficient supply of reductants. Under fuel-rich conditions ($\lambda<1$) the catalyst shows high NO conversion but low conversion of CO and hydrocarbons. The conversion ranges for NO and CO/hydrocarbons overlap at $\lambda = 1$. In a small λ range around stoichiometry the catalyst exhibits acceptable conversion for all three pollutants. Efficient operation of the three-way catalyst requires the engine to be run exactly within this λ window at all times and was therefore only made possible by the introduction of λ sensors and active-feedback λ control.

The λ sensors applied for feedback control of the A/F ratio are electrochemical solid-state sensors that measure the concentration of free oxygen after complete conversion of the exhaust gas mixture on a catalytically active layer coated on the sensor element. The active material of the sensor itself is the anion conductor ZrO_2, and the potential between a reference gas and the exhaust gas across the ZrO_2 is determined by the equilibrium concentration of free O_2 after complete conversion on the catalytic surface of the λ sensor [4]. The λ sensor gives a signal of about 50 mV at $\lambda > 1$, and about 800 mV for $\lambda < 1$ with a sharp jump at $\lambda = 1$. In a basic design of a three-way catalytic system, a λ sensor is situated in the exhaust gas before the catalytic converter and returns a feedback signal which is used to control the electronic fuel injection such that derivations from $\lambda = 1$ are compensated.

7.5.2
Oxygen Storage in Three-Way Catalysts

Due to the finite time response of the feedback control the A/F ratio in the feedback-controlled mode performs regular λ variations around $\lambda = 1$. These perturbations occur with a frequency around 1 Hz and an amplitude of about 1–10 % of the A/F set point. Comparison with the conversion plot in Fig. 7.6 reveals that A/F variations of 1–10 % are large compared to the width of the three-way conversion window, and that during such perturbations the A/F value is outside the optimal range for most of the time.

For this reason three-way catalysts contain oxygen components that buffer the A/F variations. The compound most commonly used for this purpose is CeO_2. The redox transition between Ce^{IV} and Ce^{III} acts as an oxygen buffer [9]. During lean excursions Ce_2O_3 is oxidized by excess O_2, and this results in the desired correction of the gas phase A/F ratio towards stoichiometry.

$$Ce_2O_3 + 0.5O_2 \rightarrow 2CeO_2 \quad (15)$$

In a susequent rich excursion the stored oxygen in turn is used to oxidize the excess CO and hydrocarbons, again shifting the gas phase composition towards stoichiometry.

$$2CeO_2 + CO \rightarrow Ce_2O_3 + CO_2 \quad (16)$$

Later it was discovered that ZrO_2/CeO_2 mixed oxides show superior O^{2-} ion mobility than pure CeO_2 and hence have improved oxygen-storage behavior, and the oxygen-buffer performance can be further improved by doping with additional rare earth elements. For this reason in modern three-way catalysts pure CeO_2 has been widely replaced as an oxygen storage material by specifically designed mixed crystal lattices.

A further benefit of including CeO_2 or CeO_2/ZrO_2 mixed oxides in the washcoat is their ability to catalyze the water gas shift and steam reforming reactions:

$$CO + H_2O \rightarrow CO_2 + H_2 \tag{17}$$

$$C_mH_n + (2m)H_2O \rightarrow mCO_2 + (2m + 0.5n)H_2 \tag{18}$$

These reactions contribute to the removal of CO and hydrocarbons, and the H_2 acts as a very active reductant for NO:

$$NO + H_2 \rightarrow 0.5N_2 + H_2O \tag{19}$$

Due to the presence of oxygen storage components λ fluctuations at the exit of the three-way catalyst are by far less pronounced than upstream of the catalyst. Consequently, a λ sensor downstream of the catalyst is perturbed to a much lesser extent by λ fluctuations and allows more precise determination of the long-term average A/F ratio. For this reason modern engine management systems use a dual λ sensor concept. A front sensor located upstream of the catalyst is used for short-term λ feedback control, and a second, rear sensor located downstream of the catalyst in a secondary control loop controls the long-term average A/F ratio.

Current regulations require the performance of the exhaust gas converter system to be continuously monitored during vehicle operation. Ideally this on board diagnostics (OBD) requirement would be fulfilled by directly monitoring the output of one or several of the regulated pollutants. However, currently detectors for the major exhaust gas pollutants are expensive and voluminous devices, and there are no sensors available that from the point of view of cost, space requirements, and sensitivity are suitable for use in serial-production three-way catalytic converter systems. For this reason legislation currently accepts the monitoring of oxygen-storage capacity as an indicator of catalyst performance. The idea behind this is that oxygen-storage capacity is a key factor determining the performance of the three-way catalyst system and that major catalyst aging mechanisms impact oxygen-storage capacity in approximately the same way as the catalytic performance of the catalyst. Generally, on board diagnostics for the oxygen-storage capacity use a dual λ sensor configuration. A common scheme makes use of the natural λ fluctuations imposed by the A/F feedback control and determines the oxygen-storage performance from the comparison of up- and downstream modulation amplitude, phase shift, or related quantities.

7.5.3 Precious Metals in Three-Way Catalysis

Much effort has been spent on replacing platinum group metals by less expensive and more abundant materials. Numerous publications describe the results obtained with catalysts that contain, for example, the oxides of Cu, Cr, Fe, Co, and Ni. However, until today no breakthrough has been achieved, and automotive exhaust treatment still relies on precious metals, mostly Pt, Pd, and Rh, as catalytically active components. Each of these three major precious metals has specific advantages and disadvantages.

Platinum and palladium are generally used for oxidation catalysts and to promote the oxidation reactions (11) and (12) in three-way catalysts. The characteristic advantage of Pt over Pd is its higher resistance to poisoning by sulfur and lead. This promoted the preferred use of Pt in the early days of exhaust gas catalysis.

The specific advantage of Pd is its higher resistance to high-temperature aging, especially under oxygen-rich conditions, and its better performance at lean cold start. Both make Pd the precious metal of choice for close-coupled start catalysts. One drawback associated with the use of Pd in three-way catalysts is that its three-way conversion window is generally narrower. This difficulty could be overcome by improved λ feedback control and specifically designed washcoat compositions.

Rhodium is employed in three-way and reduction catalysts due to its superior performance in promoting the NO reduction reactions (13) and (14) and its high selectivity for the formation of N_2 in these reactions, whereas Pt and Pd tend to form NH_3 and N_2O.

A combination of technical and price considerations led to the preferred use of different precious metals for different generations of three-way catalysts. In the early days of three-way catalysis there was no alternative to the application of Pt or Pt/Rh catalysts due to the contamination of the fuel with sulfur and lead. Starting from about 1980 preferences shifted towards Pd/Rh or even Pd-only systems. This can be explained by the increased fuel quality and by the relatively low price of Pd at that time. More recently, due to the high demand from the automotive catalyst industry, the cost advantage of Pd-based systems has disappeared, and interest has focused again on Pt-based systems. In general, the supply situation is considered more critical for Pd than for Pt, since Pd is a byproduct of Ni production, so that Pd production does not flexibly adjust to changes in Pd demand, while Pt production capacity, at least in the long term, is expected to increase according to the growing demand.

7.5.4 Recent Developments in Three-Way Catalysis

Today three-way catalysts, after having reached their operating temperature, show very high conversion of the major polutants CO, HC, and NO. More than 80% of the overall emissions, especially of the hydrocarbon emissions, occur during the warm-up phase directly following the cold start of the vehicle. Consequently, the key to further reductions of overall emissions is to improve the cold-start performance of the catalytic converter.

One approach to accelerate the warm-up of the catalytic converter is to employ a start catalyst that is mounted very close to the outlet of the engine and is therefore heated very rapidly after starting the engine. This close-coupled catalyst can either be used in addition to or can even completely replace the traditional underfloor catalyst. Due to the high temperatures encountered in the close-coupled position, very temperature resistant washcoat formulations had to be developed for this application.

An alternative approach is to actively heat the catalyst, either electrically or by a burner system mounted in front of the catalyst. A general drawback of actively heated catalytic converters is the high cost associated with these systems.

A third approach to reduce hydrocarbon emissions during cold start is the application of hydrocarbon adsorption traps. Here during coldstart hydrocarbons are adsorbed by an adsorber material, generally a zeolite. When the catalytic converter has reached a higher temperature, the hydrocarbons are desorbed and converted on the catalyst. Hydrocarbon adsorber systems are frequently employed in combination with actively heated catalysts.

7.6 Catalytic Aftertreatment of Lean Exhaust Gas

Diesel engines and lean-burn gasoline engines become more and more important due to current efforts to increase fuel economy and reduce emissions of the greenhouse gas CO_2. The oxidation of the pollutants CO and hydrocarbons is in principle facilitated by the excess of oxygen in lean exhaust gas. However, NO reduction on Rh is inhibited by high oxygen concentrations. This makes NO_x removal the technical challenge in the treatment of lean exhaust gas. Several strategies have been developed to remove NO_x from lean exhaust gas. In principle these strategies can be applied for diesel and gasoline lean-burn engines. Consequently, first the different techniques for NO_x removal under lean conditions are discussed, followed by sections treating the specific problems and solutions associated with gasoline lean-burn and diesel engine operation, including the additional measures necessary to control soot emissions from diesel engines.

7.6.1 NO_x Removal from Lean Exhaust Gas

DeNO_x Catalysts Certain catalyst materials are known to selectively promote the reduction of NO by hydrocarbons even in the presence of excess oxygen. Two classes of DeNO_x catalysts have been intensively studied: zeolites exchanged with transition metals (typically Cu/ZSM-5) and platinum group metals.

One major difficulty encountered with DeNO_x catalysts is their relatively narrow temperature window of operation: At low temperatures conversion is too slow, while at higher temperature the selectivity is lost, and hydrocarbons are consumed by reaction with excess oxygen.

Further technical problems are caused by the limited temperature stability of current catalysts. This requires the catalysts to be positioned in an underfloor position and limits the light-off performance of the converter system.

Efficient conversion of NO_x requires the presence of HC in high enough concentrations. One option is to increase the hydrocarbon concentration in the raw exhaust by appropriate adjustment of the injection timing. This configuration exhibits severe problems in meeting HC limits, especially during start up of the vehicle, because it is not possible to use a start-up oxidation catalyst close to the manifold. Alternatively, injection of secondary fuel into the exhaust line upstream of the $DeNO_x$ catalyst can be used.

Due to the technical problems associated with the application of $DeNO_x$ catalysts the system is currently not widely used in serial applications. One exception are the first-generation gasoline direct injection vehicles that are sold in Japan, which use specially developed $DeNO_x$ catalysts based on iridium as the catalytically active material.

Selective Catalytic Reduction (SCR) with Ammonia Selective reduction of NO_x by NH_3 is a well-established process for the exhaust treatment of stationary sources. In this process NH_3 or an appropriate precursor is added to the exhaust gas, and a catalyst promotes the selective reduction of NO by NH_3 in the presence of excess O_2 by the following reaction [10]:

$$NO + NH_3 + 0.25O_2 \rightarrow N_2 + 1.5H_2O \tag{20}$$

For automotive applications NH_3 is generated on board by catalytic or thermal decomposition of urea, a nontoxic and relatively easy to handle chemical. The amount of NH_3 added to the exhaust stream must be matched to the engine NO emission at each operating point. Vehicle tests have shown that this technology possesses promising potential, at least for applications with heavy-duty truck diesel engines.

Initially, the introduction of NH_3 SCR has met some resistance, mainly because of the need to handle and store a chemical on board, and because of the expenses associated with building a new infrastructure to distribute it. However, it is now becoming apparent that the advantages inherent to the system, mainly the possibility to reduce NO_x and simultaneously optimize fuel economy and particulate emissions, may well outweigh the disadvantages. NH_3 SCR has been introduced into heavy duty truck serial production in 2005. A typical system for onboard NH_3 selective catalytic reduction is shown in Fig. 7.7.

The NO_x Adsorber Catalyst The operating principle of NO_x absorbers is based on the fact that highly basic alkaline earth and alkali metal oxides form stable nitrates at temperatures below about 600 °C. In an NO_x adsorber/converter system NO is first oxidized to NO_2, which then reacts with the basic oxides in the adsorber to form a

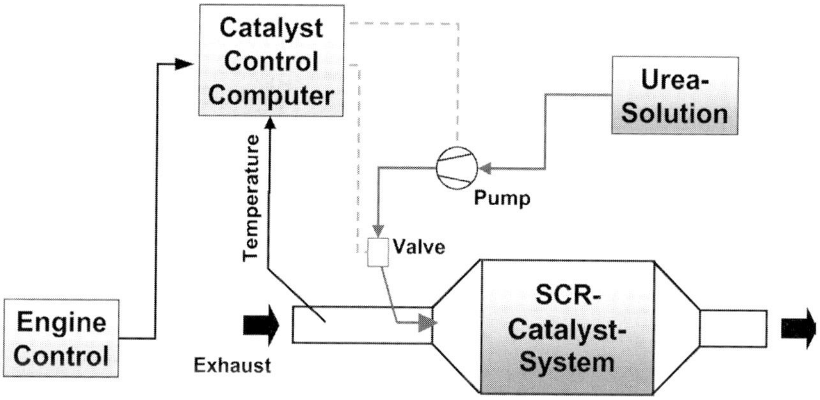

$$m_{Harnstoff} = f\,(T, Md, n, DMd, Dn, NO_x)$$

Fig. 7.7 A typical system for NO_x control by area onboard SCR.

nitrate. Oxides frequently employed for this purpose are BaO and K_2O. For BaO the reaction reads:

$$BaO + 2NO_2 + \frac{1}{2}O_2 \rightarrow Ba(NO_3)_2 \tag{21}$$

Of course, the storage capacity of the oxide is soon exhausted and the storage catalyst must be regenerated. For this purpose the engine is switched to rich operation for a few seconds. Finely divided precious metal on the adsorber catalyst promotes the reduction of nitrate by CO, hydrocarbon, or H_2 to form N_2:

$$Ba(NO_3) + 5CO \rightarrow BaO + 5CO_2 + N_2 \tag{22}$$

$$Ba(NO_3)_2 + 5H_2 \rightarrow BaO + 5H_2O + N_2 \tag{23}$$

In this way the oxide is reformed and can serve as an NO_x adsorber during the following cycle of lean operation (Fig. 7.8). The regeneration cycle is triggered by the engine management computer system that monitors the NO_x level of the adsorber catalyst either by model-based computation of the total NO_x fed into the adsorber during each cycle or by applying a sensor downstream of the adsorber. For this reason the adsorber catalyst becomes an integral part of the engine- and fuel-management system.

Currently the major technical problem associated with the application of NO_x storage catalysts is the adsorption of sulfur oxides by the same mechanism as NO_2. Since the resulting sulfates, owing to their higher stability, are not destroyed in an average rich regeneration cycle, sulfur accumulates on the catalyst and lowers the NO_x adsorption capacity. For this reason the catalyst must be regenerated from sulfur from

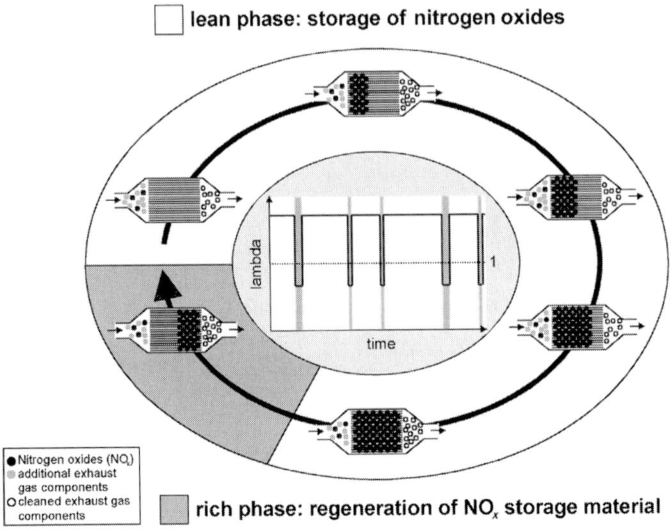

Fig. 7.8 NO_x storage cycle.

time to time, e.g., by running the engine rich at elevated temperatures for several minutes. The sulfur regeneration cycle is fuel-expensive, and since the execution frequency of the cycle is determined by the fuel sulfur content a high sulfur concentration directly translates into a corresponding fuel penalty during operation of the NO_x adsorber catalyst.

7.6.2
Catalytic Aftertreatment of Lean-Burn Gasoline Exhaust

Catalytic converters for direct-injection lean-burn engines are derived from three-way catalyst systems [11]. However, under the operating conditions when the engine is operated at lean burn the three-way converter does not convert NO. For this reason special measures must be taken to allow NO conversion under oxygen-rich conditions. The different strategies of NO treatment in lean exhaust gas are described in Section 7.6.1.

The first direct-injection lean-burn vehicles offered on the Japanese marked were equipped with special $DeNO_x$ catalysts (see Section 7.6.1) based on iridium as the catalytically active material. However, such systems do not exhibit high enough temperature stability to withstand the severe aging conditions during U.S. and especially European driving.

Currently, the preferred technology for NO_x treatment of gasoline direct-injection lean-burn engines is the NO_x absorber. A typical converter system for gasoline lean burn exhaust gas treatment is shown in Fig. 7.9. The exhaust gas first passes through a start catalyst that removes CO and hydrocarbons during light off and oxidizes NO to NO_2. The start catalyst is followed by the NO_x adsorber catalyst and a conventional three-way catalyst.

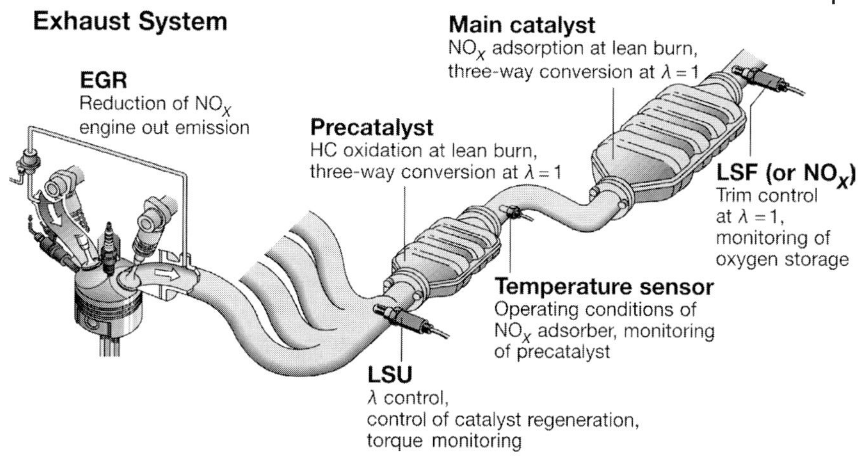

Fig. 7.9 Aftertreatment system for a gasoline direct-injection engine.

7.6.3
Catalytic Aftertreatment of Diesel Exhaust

Compared to gasoline engines, diesel catalytic exhaust treatment is complicated by two additional problems: the relatively high sulfur concentration in diesel fuel and the emission of soot and other particulate matter by diesel engines. For these reasons, and because of the intrinsically lower gaseous pollutant concentrations compared to spark ignition engines, catalytic aftertreatment for diesel engines was implemented much later than for spark-ignition vehicles. The only diesel exhaust technology widely used in serial applications today is the diesel oxidation catalyst. This technology has been generally applied to passenger cars in Europe and to some medium- and heavy-duty trucks in the USA since about 1991.

The function of a diesel oxidation catalyst is the reation of CO and hydrocarbons with the excess oxygen in the diesel exhaust (reactions 11 and 12). Additionally the mass of particulate matter is reduced by the oxidation of liquid and polynuclear aromatic hydrocarbons adsorbed on the particulates. Though diesel oxidation catalysts use washcoats that are similar to those in three-way catalysts, technical requirements are very different for diesel oxidation catalysts, and washcoat compositions optimized to meet these requirements have been developed.

The key is to design a catalyst that selectively catalyzes the oxidation of CO and hydrocarbons and simultaneously suppresses the oxidation of NO and SO_2. Especially SO_3 formation must be avoided because SO_3 deactivates alumina-based washcoats by the formation of sulfates and because SO_3 condenses as sulfuric acid on the particulates and thus increases their mass.

One additional problem faced in diesel exhaust catalysis is clogging of the monolith channels by accumulation of particulate matter. This problem is generally solved by applying monoliths with higher cell diameter.

Thermal aging is a less severe problem with diesel catalysts because of the intrinsically lower exhaust temperature of diesel engines. However, due to the low temperature of diesel exhaust, diesel oxidation catalysts must be optimized to become catalytically active at very low temperatures, frequently below 150 °C.

7.6.4
Removal of Particulate Emissions

Besides NO, CO, and hydrocarbons diesel engines emit particulate matter as a fourth major regulated pollutant. In contrast to the catalytic removal of the gaseous pollutants removal of particulate matter is a physical separation process. The exhaust gas is forced to flow through a filter material with pores smaller than or similar in size to the particles, which are trapped within the pores or on the surface of the filter. The currently most common design for particulate filters is the wall-flow monolith design (Fig. 7.10). This design is derived from the flow-through monoliths used for catalytic exhaust treatment by plugging the channels alternatively at the inlet and at the outlet of the monolith. In this way the exhaust gas is forced to flow through the porous walls of the monolith and the particulate matter is deposited along the walls.

Since accumulated particulate matter blocks the pores of the trap and causes excessive pressure drop, the filter must be regenerated by burning the trapped particulate matter with the oxygen present in diesel exhaust. Ignition of the trapped particulate matter requires temperatures above about 550 °C. Such temperatures are rarely reached in the exhaust of diesel vehicles. One solution to this problem is to apply filter traps that catalytically lower the ignition temperature of the soot. Regeneration of the particulate trap at temperatures below about 300 °C is achieved in this way. Alternatively fuel additives have been used that catalytically lower the ignition temperature of the deposited particulate matter. The extra effort involved in handling, storing, and dosing the extra fuel additive onboard the vehicle are obvious. As a further alternative so-called continuously regenerating trap

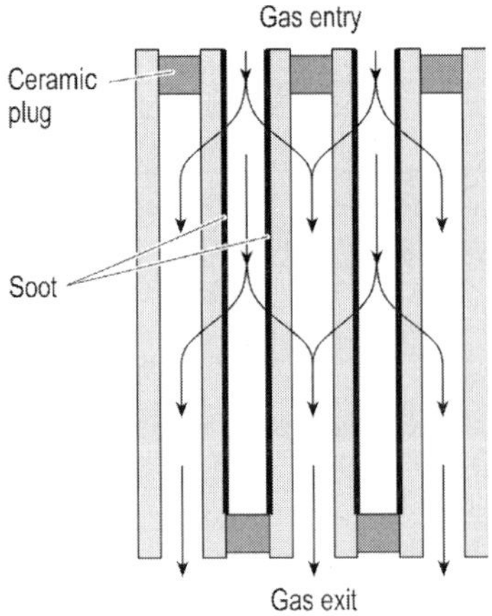

Fig. 7.10 Wall-flow monolith particulate filter.

(CRT) systems have been designed. Here the naturally occurring NO is oxidized by a strong oxidation catalyst to NO_2, which is used to continuously oxidize the deposited particulate matter on the filter trap.

7.7 Conclusion and Outlook

Catalytic exhaust gas treatment is a well-established technology that has allowed a significant reduction of air pollution. Since the first introduction of exhaust catalysts in the 1970s, aftertreatment technology has continuously improved, so that today automotive exhaust catalysis has become one of the most innovative fields of cataytic science and one of the most important applications of heterogeneous catalysis. Today, pollution from vehicles has been reduced by more than 90% compared to the level before the introduction of catalytic exhaust control. The near future will see even further reductions.

A major remaining challenge for the automotive industry is the lowering of fuel consumption and CO_2 emissions. This is also a challenge for catalyst development, since the most fuel-efficient engines operate with excess air, and the conventional three-way catalyst does not allow NO_x reduction under these conditions. New technologies are being developed that allow fuel-efficient lean operation of the engine and at the same time retain the high emission standards of conventional three-way catalysts.

Early automotive catalysts were more or less add-ons to vehicles. Future emission control technologies will be significantly more complex. One driving force for this increased complexity is the need to apply storage-type catalysts. While conventional catalysts can be operated continuously, storage-type catalysts require cyclic operation with alternating storage and regeneration phases. This unsteady operation requires the catalytic system to be designed as an integral part of the overall fuel- and engine-management system. Consequently, automotive emission control has evolved into an interdisciplinary field in which technological and commercial success can only be achieved in an integrated development process with close cooperation of engine-design engineers, fuel-management experts, and catalysis specialists.

7.8 References

1. http://www.umweltbundesamt.de
2. E. S. Lox, B. Engler: "*Environmental Catalysis*", G. Ertl, H. Knözinger, J. Weitkamp (eds.): *Handbook of Heterogeneous Catalysis*, VCH, Weinheim, 1997.
3. K. C. Taylor in A. Crucq, A. Frennet (eds.): *Catalysis and Automotive Pollution Control*, VNR, New York; 1987.
4. C. Hagelüken (ed.): *Autoabgaskatalysatoren* Expert Verlag, Renningen, 2001.

5. R. M. Heck, R. J. Farrauto: *Catalytic Air Pollution Control*, VNR, New York 1995.
6. N. A. Henein, D. J. Patterson: *Emissions from Combustion Engines and Their Control*, Ann Arbor Science, Ann Arbor, Michigan, 1972.
7. J. A. Miller, C. T. Bowman: "*Mechanism and Modeling of Nitrogen Chemistry in Combustion*", *Prog. Energy Combust. Sci.* 15 (1989) 287 – 338.
8. Verbrennungsmotoren Band I und II, Vorlesungsdruck, Prof. Dr.-Ing. S. Pischinger, Lehrstuhl für Verbrennungskraftmaschinen, RWTH Aachen, 2002.
9. J. Kaspar, M. Graziani, P. Fornaserio: "*Ceria-Containing Three-Way Catalysts*", in K. A. Gscheidner, Jr., L. Eyring (eds.): *Handbook on the Physics and Chemistry of Rare Earths* Vol. 29, Elsevier Science, Amsterdam, 2000.
10. J. Gieshoff, A. Schäfer-Sindlinger, P. C. Spurk, J. A. A. van den Tillaart, G. Carr: *Improved SCR for Heavy Duty Applications* SAE 2000-01-0189, Detroit, 2000.
11. W. Strehlau, J. Leyrer, E. S. Lox, T. Kreuzer, M. Mori, M. Hoffmann: *New Developments in Lean NOx Catalysis for Gasoline Fueled Passenger Cars in Europe*, SAE 962047, Detroit, 1996.

8
Aviation Turbine Fuels

Geoff J. Bishop

8.1
History

Turbine-powered aircraft were used by both *British* and *German* air forces for military operations during the latter part of World War II. The British engines were operated on illuminating kerosine containing 1% light lubricating oil (jet engine oil) [1]. The German engines burned a wider boiling mixture of gasoline and heavier distillate fractions. Fuel development in Germany stopped at war's end. Continuing British research has tightened some requirements over the years, yet today's primary British specifications still describe kerosine-type fuels for both commercial and military aircraft, the main difference being in the additive content.

The initial United States military fuel was a low freezing point kerosine called JP-1, but its limited availability forced the development of a wider boiling fuel that included gasoline as well as kerosine fractions [2]. This "wide-cut" fuel called JP-4 was, until recently, the major military grade. However, safety concerns associated with its flammability led to its phase-out in favor of the kerosine based grade JP-8. The switch from JP-4 to JP-8 was largely completed by the end of 1996.

Commercial aircraft have been burning mostly kerosine-type fuels since 1953 when the early turboprop aircraft came into service, and this has remained the major civil fuel in the turbojets and later the turbofans which power most of the world's civil aircraft fleet. The principal grades are Jet A in the United States, TS-1 in Russia and Jet A-1 in the rest of the world. The wide-cut fuel Jet B, analogous to the military grade JP-4, is now rarely used in civil aviation.

8.2
Fuel Types and Specifications

The most widely used civil aviation turbine fuel grades are defined by a small number of key specifications as listed in Table 8.1. These specifications are similar in most respects although a small number of significant differences remain.

Handbook of Fuels: Energy Sources for Transportation. Edited by Barbara Elvers

ISBN: 978-3-527-30740-1

Table 8.1 Major civil jet fuel grades and their specifications

Grade	Area of usage	Major specification
Jet A-1	world-wide	ASTM D1655, DEF STAN 91-91
Jet A	USA only	ASTM D 1655
TS-1	Russia and former Eastern Bloc countries	GOST 10227
Jet Fuel no.3	China	GB 6537
Jet A-1	Canada	CAN/CGSB-3.23

The US Jet A fuel, defined by specification ASTM D1655 [3], differs from Jet A-1 only in freezing point. The British DEF STAN 91-91 specification (formerly well known as DERD 2494 whilst under the jurisdiction of the UK Ministry of Defence) [4] is recognized as a civil jet fuel specification and is used as such in many countries around the world. The Russian GOST 10227 specification also defines a grade designated RT which is very similar to western Jet A-1, as is the Chinese no.3 Jet Fuel.

Because of the worldwide nature of commercial aviation, the close correspondence between specifications is a necessity to discourage the proliferation of many different national or local specifications. The major jet fuel suppliers subscribe to a document entitled "Aviation Fuel Requirements for Jointly Operated Systems" (AFQRJOS) which defines a "Check List" which comprises the most restrictive requirements of the ASTM D 1655 and DEF STAN 91-91 fuel specifications. Thus fuel meeting the Check List automatically meets all the requirements of the parent specifications. This document was developed to cater for the situation where many different fuel suppliers deliver fuel into the common storage facilities which predominate at most international airports.

Further efforts are underway to minimize the differences between the major fuel specifications with the eventual aim of developing a single, international specification to define aviation turbine fuel. An International Civil Jet Fuel Specification Liaison Group was formed in 1989 to promote this activity and a growing number of national specification bodies are contributing to its progress.

This liaison group issued an "International Specification Guide for Aviation Turbine Fuels", but this has now been incorporated into the "IATA Guidance Material for Aviation Turbine Fuel Specifications" [5].

8.2.1 Specification Requirements

Details of the major fuel specifications for Jet A-1 type fuel are shown in Table 8.2. Although test methods in most western countries are either identical or very similar, no international standardization exists and direct comparisons between limits may be misleading (refer to the notes applying to Table 8.2). Test method designations published by the principal standards organizations are given in Table 8.3.

8.2.2 Fuel Properties

Unlike most chemicals but like other petroleum fuels, aviation turbine fuels are not defined by composition, molecular structure, and purity. Instead their characteristics are limited by boundary conditions imposed by inspection tests. Some of these tests are fundamental, others empirical; many reflect a solution to past engine or aircraft problems. However, any fuel meeting all requirements is acceptable for aircraft operation. The specifications are dynamic, but changes are conservative because of the potential risk to aviation operations. Long equipment life ($\geqq$ 20 years) and certification restrictions (see Section 8.5) also mitigate against drastic specification changes.

The individual specification properties or requirements can be divided rather arbitrarily into bulk and trace properties. *Bulk properties* are those requiring significant composition changes, involving more than 5 vol% of the total fuel, to effect a change in the property. *Trace properties* tend to respond to changes $<$ 1 vol % or even 1 ppm or less. Changes in bulk properties affect product yield, whereas changes in trace properties do not. The grouping has no direct bearing on fuel performance. Instead it helps assess effects of crude composition and manufacturing steps on product yield [8].

8.2.2.1 Bulk Properties

Density The mass–volume relationship is important because the quantity of fuel loaded onboard an aircraft is metered volumetrically, but fuel consumption is measured by mass. This relationship is also used extensively in commercial sales of fuel. Although in the past there have been several different ways of defining this relationship, such as specific gravity, relative density, or API gravity, all the major jet fuel specifications now define density at 15 °C in units of kg/m^3. Changes of density with temperature are given in [9]. Density is commonly measured using hydrometers or digital density meters. In-line density measurement is also possible and offers the opportunity for fuel delivery by mass instead of volume.

Specific Energy Specific energy or heat of combustion is directly related to density. The fuel's specific energy, its density, the aircraft's performance characteristics, and the predicted flight conditions are employed to calculate the amount of fuel required for a given flight or conversely to estimate the aircraft's range with maximum fuel aboard. Specific energy is determined by measuring the total heat released by a unit mass of fuel and subtracting from this value the heat of condensation of the water vapor formed during the combustion. However, this measurement is time consuming, so several empirical relationships have been developed to calculate specific energy from combinations of other inspection tests. The most common is based on the product of aniline point and density, ASTM D 4529 [3]; others use the density, boiling range, and aromatic content, ASTM D 3338 [3], or the hydrogen content and density, IP 355 [6]. All three methods include corrections for fuel sulfur content. Although these calculation methods are used routinely, a method of measurement, ASTM D 4809 [3], is still specified as the referee method for use in cases of dispute.

Table 8.2 Major fuel specifications for civil Jet A-1 type fuel

	United States	United Kingdom	International	Russia [a]	China [b]
Specification	ASTM D1655	DEF STAN 91-91	Checklist	GOST 10227-86	GB 6537-94
Issue	06c	5	21	Amendment 4	effective June 95
Date	Oct. 2006	Mar. 2006	Jun. 2006	Apr. 96	Aug. 94
Grade	Jet A-1	Jet A-1	Jet A-1	TS-1	No.3 Jet Fuel
Composition					
Acidity, total, mg KOH/g (max)	0.10	0.015	0.015	0.009 [c]	0.015
Aromatics, vol % (max)	25	2250 [d]	2250 [d]	22 mass %	20.0
Olefins, vol % (max)	NR [d]	NR [d]	NR [d]	iodine no.2.5 max	5.0
Sulfur, total, mass % (max)	0.30	0.30	0.30	0.25	0.20
Sulfur, mercaptan (thiol), mass % (max)	0.003	0.003	0.003	0.005	0.002
or					
Doctor test	negative	negative	negative	NR [d]	negative
Volatility					
Distillation:					
Initial boiling point (IBP), °C (max)	NR [d]	report	report	150	report
10 % vol recovered at, °C (max)	205	205	205	165	205
20 % vol recovered at, °C (max)	NR [d]	NR [d]	report	NR [d]	report
50 % vol recovered at, °C (max)	report	report	report	195	232
90 % vol recovered at, °C (max)	report	report	report	230	report
Final boiling point (FBP), °C (max)	300	300	300	98% (250 °C)	300
Residue, vol % (max)	1.5	1.5	1.5	NR [d]	1.5
Loss, vol % (max)	1.5	1.5	1.5	NR [d]	1.5
Flash point [e], °C (min)	38	40 (38)	40 (38)	33 (28)	38
Density at 15 °C , kg/m^3	775–840	775–840	775–840	775 (min) (20 °C)	775–830 (20 °C)
Fluidity					
Freezing point, °C (max)	− 47	− 47	− 47	− 50 [f]	− 47
Viscosity at 20 °C , mm^2/s (min)	NR [d]	NR [d]	NR [d]	1.25	1.25
Viscosity at − 20 °C, mm^2/s (max)	8.0	8.0	8.0	8 (− 40 °C)	8.0
Combustion					
Specific energy (net), MJ/kg (min)	42.8	42.8	42.8	42.9	42.8
Smoke point, mm (min)	25	25	25	25	25
or					
Smoke point, mm (min)	18	19	19	NR [d]	20
and naphthalenes, vol% (max)	3.0	3.0	3.0	NR [d]	3.0
or					
Luminometer no. (min)	NR [d]	NR [d]	NR [d]	NR [d]	45

Corrosion					
Copper corrosion (max)	1	1	1	passes	1
Silver corrosion (max)	NR [d]	NR [d]	NR [d]	NR [d]	1
Stability					
Thermal stability:				not comparable	
JFTOT test temp., °C	260 [g]	260	260		260
Filter pressure drop, mm Hg (max)	25	25	25		3.3 kPa
Tube deposit (visual) rating	LT 3	LT 3	LT 3		LT 3
Peacock/abnormal deposits	fail	fail	fail		fail
Contaminants					
Existent gum, mg/100mL (max)	7	7	7	5	7
Water reaction					
Interface rating (max)	NR [e]	1b	NR [d]	1	1b
Separation rating	NR [d]	NR [d]	NR [d]	NR [d]	report
Microseparometer (MSEP) (min)	85 (+ SDA = 70)	85 (+ SDA = 70)	85 (+ SDA = 70)	NR [d]	NR [d]
Particulates, mg/L (max)	NR [d]	NR [d]	NR [d]	none	report
Ash content, mass % (max)	NR [d]	NR [d]	NR [d]	0.003	NR [d]
Electrical conductivity [h], pS/m	50–600	50–600	50–600	50–600	50–450 (20 °C)
Additives					
Antioxidant (hydroprocessed fuels)	optional	mandatory	mandatory	mandatory	optional
Antioxidant (other fuels)	optional	optional	optional	optional	optional
Metal deactivator	optional	optional	optional	NR [d]	NR [d]
Static dissipator	optional	mandatory	mandatory	optional	optional
Fuel system icing inhibitor	permitted	permitted	permitted	permitted	permitted
Lubricity improver	permitted	permitted	permitted	mandatory in HP	permitted

[a] Specification limits are in terms of Russian GOST test methods and may not be directly comparable with limits obtained using ASTM test methods.
[b] Specification limits are in terms of Chinese GB/T or SH/T test methods and may not be directly comparable with limits obtained using ASTM test methods.
[c] Value in terms of mg KOH/g calculated from GOST limit which is in terms of mg KOH/100 mL.
[d] NR = no requirement under this heading.
[e] Flash point limits are in terms of ASTM D56 (TAG) test method; limits in parantheses are in terms of other, local, test methods.
[f] The GOST test method records wax appearence point on cooling which is 3 °C below the wax disappearance point on warming, as recorded in the ASTM method. Thus the specification limits for TS-1 and Jet A-1 are equivalent.
[g] If the test at 260 °C produces a failure, a pass at a test temperature of 245 °C is acceptable.
[h] Conductivity limits apply only if fuel contains static dissipator additive.

Table 8.3 National standard aviation fuel test methods

Test name	ASTM [3]	IP [6]	ISO [7]	GOST [a]
Flash point (TAG closed cup)	D56			6356
Distillation	D86	123	3405:1988	2177
Corrosion (copper strip)	D130	154	2160:1990	6321
Color (Saybolt)	D156			
Vapor pressure (Reid)	D323	69	3007:1986	1756
Existent gum	D381	131	6246:1981	1567
Kinematic viscosity	D445	71	3104:1994	33
Water reaction	D1094	289	6250:1982	27154
Sulfur (lamp method)	D1266	107		19121
Density (hyrometer)	D1298	160	3675:1993	3900
Aromatics	D1319	156	3837:1993	6994
Smoke point	D1322	57	3014:1993	4338
Sulfur (high temperature method)	D1552			
Luminometer no.	D1740			17750
Naphthalene hydrocarbons	D1840			17749
Particulate contamination (field method)	D2276	216		
Freezing point	D2386	16	3013:1974	5066
Sulfur (X-ray method)	D2622			
Electrical conductivity	D2624	274	6297:1984	25950
Sulfur, mercaptan	D3227	342	3012:1991	17323
Undissolved water	D3240			
Thermal oxidation stability (JFTOT)	D3241	323	6249:1984	17751
Total acidity	D3242	354		5985
Estimated heat of combustion	D3338			
Estimated hydrogen content	D3343			
Hydrogen content by NMR	D3701	338		
Peroxide no.	D3703			
Flash point (SetaFlash closed tester)	D3828	303	3679/3680:1983	
Water separation (MSEP)	D3948			
Density (Digital meter)	D4052	365		
Sulfur (ND-XRF method)	D4294			
Filter flow at low temperature	D4305			
Estimated heat of combustion	D4529	381	3648:1994	11065
Heat of combustion (bomb)	D4809			21261
Doctor test	D4952	30	5275:1979	
Vapor pressure (automatic)	D5190			
Vapor pressure (mini method)	D5191			
Particulate contamination (lab method)	D5452			
Sulfur (XRF method)	D5453	336	8754:1992	
Freezing point (auto/optical)	D5901			
Freezing point (auto/phase transition)	D5972			
Corrosion (silver strip)		227		
Copper content		225		
Sulfur (Wickbold combustion)		243	4260:1987	

[a] Although ASTM, IP and ISO methods are essentially identical, GOST methods may be quite different.

Volatility A fuel's volatility or its tendency to vaporize affects evaporation rate from tanks, including aircraft fuel tanks. Volatility can cause vapor lock in fuel systems when the combined fuel and vapor volume exceeds pump capacity; cavitation can result when the fuel vapor pressure exceeds the liquid system pressure between gear teeth or in plain bearings. Volatility also directly affects flammability and causes safety problems both inside and outside of tankage [10]. Because of the complex composition of turbine fuel, its volatility cannot be described by a single fundamental property. Therefore, volatility is normally expressed using a combination of boiling range, flash point, or vapor pressure.

The volatility of kerosine type fuels is controlled by flash point and distillation range (Table 8.2). Wide-cut fuel is regulated by a combination of distillation range and vapor pressure.

Combustion Combustion in a modern jet engine is a highly turbulent, high-pressure, and high-intensity mixing process. Engine designers carefully control air flow, fuel flow, and fuel atomization to maximize combustion efficiency and to minimize smoke, carbon deposits, and the formation of partially oxidized gases, such as carbon monoxide. Fuel affects all these as well as the intensity of radiation emitted by the flame [11-13]. No single fuel property requires a greater degree of empiricism than the estimation of fuel combustion characteristics. *Smoke point,* and to a lesser extent the *luminometer,* have been favored because as actual combustion tests, they respond to a combination of other fuel properties including volatility, composition, and viscosity. However, a direct comparison of these laminar, wick-fed flames with the high pressure, turbulent combustor flame is not possible and thus other alternative controls also appear in the specifications. *Naphthalenes content* was added because multi-ring aromatics have shown adverse combustion effects, in terms of flame radiation and smoke formation, out of proportion to their concentration [14]. During the 1980s, the use of hydrogen content, which can be measured very accurately by NMR (ASTM D 3701), gained favor as a possible replacement for the less precise measurements of aromatics and naphthalenes contents. However, it was found that the total hydrogen content of the fuel could not predict reliably the combustion behavior of all fuels[15] and it is now unlikely that it will be adopted for civil fuels, although it continues to find limited use in certain military fuel specifications.

Total aromatic content (ASTM D1319, [3]), covered under "Elastomer Compatibility", does not correlate particularly well with engine combustion, possibly because of the particular separation and classification of hydrocarbons by the method [16].

In general, correlations between engine combustion performance and fuel inspection tests will always be difficult to establish because of the differing performance of the same fuel in different engines, the extreme expense of controlled engine tests, and the long time required to obtain engine experience in the commercial world, where only a limited quantity of minimum quality fuel is burned [16].

Elastomer Compatibility A major requirement of jet fuels is their compatibility with materials of construction, particularly with the elastomers used in diaphragms, fuel tanks, fuel tank coatings and sealants, linings, and a wide variety of static and dynamic seals in all parts of the fuel system. The fuel property most indicative of elastomer compatibility has been *total aromatic content.*

Low-Temperature Operability The extremely low temperatures at high altitudes can cause several different fuel problems. During fuel cooling, hydrocarbon molecules precipitate as wax crystals. *n*-Hydrocarbons are the most insoluble, and in fact measurement of their type and concentration can indicate the insoluble fraction expected at low temperature [17].

The *freezing point* of the fuel is defined as the temperature at which the last wax crystal melts on warming the fuel following a specified cooling process. Traditionally, this has been determined visually but now several automated methods exist which employ variously, light scattering, phase transition measurement or flow through a fine filter (ASTM D 4305) [3].

With the increasing use of twin-engined aircraft on long over-water routes, the so-called extended operations (ETOPS) flights, and with the introduction of very long range aircraft such as the Boeing 747–400, there has been increasing focus on fuel low-temperature properties. Studies underway may question the usefulness of "freezing point" as currently measured and could lead to its replacement by a more meaningful, flow-based test.

Low fuel temperature can also affect engine starting, which depends heavily on fuel atomization. Adequate atomization is determined largely by fuel viscosity. Typical viscosities versus temperature are plotted in Figure 8.1. Because fuels within the same grade have the same viscosity-temperature slope, a single maximum viscosity at a standard temperature adequately controls this property.

8.2.2.2 **Trace Properties**

Instead of describing trace properties individually, a better overview is obtained by grouping these properties into the following performance categories.

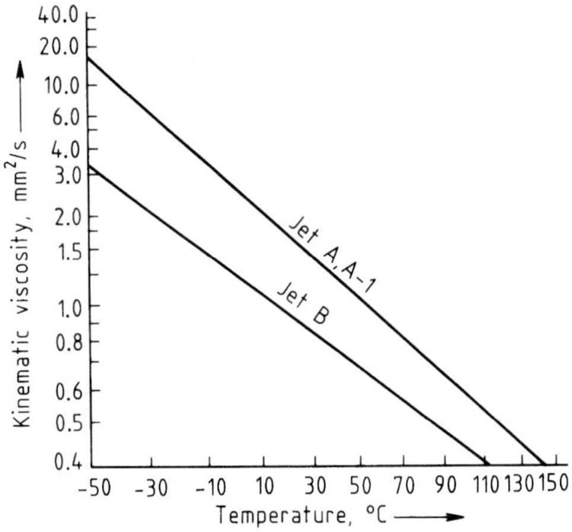

Fig. 8.1 Typical viscosities vs. temperature [18].

Oxidative Stability Jet fuel's exposure to oxidative conditions over a wide temperature and time range can result in a wide variety of problems [19].

Ambient temperature oxidation can form peroxides that degrade certain elastomers, e.g. Buna N, which are used in aircraft fuel systems. Long-term storage can cause soluble and insoluble oxidation products labeled "gums". Peroxide formation is usually seen in fuels which have been produced in the refinery by catalytic hydrogen processes (see Section 8.3) which remove certain trace compounds containing sulfur or nitrogen which are natural anti-oxidants. Anti-oxidancy is restored by the addition of oxidation inhibitor additives (Section 8.3.2). Gum formation is rare but is monitored and controlled by the Existent Gum test.

High-temperature oxidation is of major concern. Oxidation products–gums, lacquers and insoluble deposits–can form in the hot areas of an aircraft's fuel system causing a number of problems [20]. The heat exchanger of the fuel-cooled oil-cooler can lose efficiency; flow control valves in the fuel control unit can stick or jam, coke deposits can form in the burner feed arms which can, in the worst case, cause plugging of the fuel injector nozzles. These deposits represent such a small fraction of the fuel that the oxygen dissolved in the fuel is sufficient to cause their formation. Although deoxygenation is effective on a laboratory scale, in large scale practice it is impossible to reduce the oxygen content of the fuel to a level low enough to prevent oxidation. The problem is attacked by designing aircraft fuel systems to minimize hot spots; indeed, thermal management is one of the most important criteria in modern engine and aircraft design. In addition, the thermal stability of fuel supplies is controlled by a test in the fuel specification (Table 8.2) using an instrument called the Jet Fuel Thermal Oxidation Tester (JFTOT), which is a small-scale simulation of a heat exchanger in which formation of lacquer and deposit are measured (ASTM D 3241) [3].

Certain metals, notably copper, are oxidation catalysts and when dissolved in the fuel at extremely low levels, e.g. 20 μg/kg (ppb), can promote thermal instability and deposit formation. Traces of copper can be picked up by the fuel from copper-containing alloys sometimes encountered in the refinery or during transportation. The deleterious effects of this dissolved copper can be alleviated by dosing the fuel with a metal deactivator additive (Section 8.3.2).

Corrosivity Corrosion of fuel system metals can be a problem. Copper or bronze in early fuel pump bearings and other components needed protection [1]. Silver surfaces in highly loaded portions of variable displacement pumps were attacked and ultimately removed by shear forces [21]. *Dihydrogen sulfide* and elemental *sulfur* were identified as culprits. As little as 0.5 mg/kg (ppm) of either material caused the difficulties. As a result, all the major fuel specifications contain a test for copper corrosion. Some specifications have, in addition, a test for silver corrosion.

Solutions to other corrosion difficulties have addressed the offending fuel constituent rather than the metal involved. *Thiols* cause cadmium corrosion [1] and are limited in fuel specifications. *Sulfur,* which includes all organosulfur compounds, causes high-temperature turbine blade corrosion, particularly in the presence of sodium [22]. However, such low sulfur concentrations cause corrosion that virtually complete sulfur removal from the fuel would be needed to solve the problem, and blade metallurgy was changed instead. Therefore the current sulfur specification maxima are not based on any corrosion criteria.

Organic acidity, primarily as naphthenic acids, has been blamed for some diesel engine corrosion, but the low-acidity level permitted by the jet fuel specifications appears to have prevented similar aviation difficulties. The primary role of organic acids is their involvement in the formation of surface-active agents as discussed in the next paragraph.

Water Separation Free or immiscible *water* is a potential major hazard in aircraft. When cooled below 0 °C the resulting ice can block orifices, screens, or filters. In large quantities water will "put out the fire" by interrupting fuel flow. Water can lead to microbial growth in stagnant parts of the fuel system. Therefore, every effort is made to remove all immiscible water by special filter-separators before the aircraft is fueled. Satisfactory operation of these filters depends on the maintenance of internal hydrophobic surfaces. Trace levels of surfactants render these surfaces ineffective [23]. Although the organic acids normally in jet fuel are not strong surfactants, they react with sodium hydroxide or other basic materials. The resultant salts or soaps can have detrimental effects [1], [24], [25]. The lower organic acidity maximum in the British specification offers greater protection in this regard than does the higher level in ASTM D 1655, but surfactants are not limited to those formed by organic acids, and additional tests are used for surfactant detection and control.

The Water Reaction test gives a semiquantitative indication of insoluble interface material formed when fuel is shaken with buffered water. In a much more sensitive test a fuel–water emulsion is pumped through two layers of coated fiberglass from a filter-separator. The ability of the filter element to strip all water from the emulsion is measured.

The earliest test, the Separometer (ASTM D 2550 [3]), was developed to detect low levels of sodium sulfonate [33]; but this test as well as the later versions, the Miniseparometer (ASTM D 3602) and the Microseparometer (ASTM D 3948), are widely used in the United States and elsewhere to test for the presence of various surfactants. Because of the difficulties in correlating results from these tests with field performance, there is only limited industry agreement on the definition of acceptable fuel by these tests. ASTM D 2550 and D 3602 have now been discontinued.

Static Electricity Whenever a hydrocarbon liquid flows with respect to another surface, an electrical charge is generated in the liquid and an equal but opposite charge is imposed on that surface. One set of charges is swept along with the fuel while the opposite charges, which accumulate along the wall surface, usually leak to ground (earth). This charge separation results in a rise in voltage in the moving fuel [26]. High speed pumping of aviation fuel through pipes and filters can generate very high charges and when the charged fuel enters a tank, a substantial voltage difference may be produced between the surface of the liquid and the tank walls and this may result in a static discharge [27]. If the vapor space above the liquid is in the flammable range, such a spark discharge can cause a fire or explosion.

Charge build-up can be reduced by reducing the flow-rates of the fuel or designing and constructing facilities–pipework, filters, valves, etc.–to minimize charge generation and/or provide time for the charge to relax to ground before entering a tank [28]. Another approach is to add a static dissipater additive to the fuel (Section 8.3.2) to increase the electrical conductivity of the fuel to facilitate charge relaxation.

8.2.3 Nonspecification Properties

Some aviation fuel properties are not limited by specifications but are useful in equipment design or performance.

Absolute Vapor Pressure Reid vapor pressure, specified for Jet B fuel, is not an absolute vapor pressure because it is measured with a vapor–liquid ratio of 10:1; it is determined only at 38 °C, and also it includes the partial pressures of water vapor and air. With low-vapor-pressure products, such as Jet A or A-1, absolute or true vapor pressure is used instead. Figure 8.2 shows typical true vapor pressure-temperature curves for different jet fuel grades [18].

Water Miscibility Even the very low concentration of dissolved water in aviation turbine fuels is hazardous if this water separates and freezes in a critical aircraft location. Because dissolved water cannot be avoided in vented systems and is not removed by airport fuel handling equipment, aircraft fuel systems either heat the

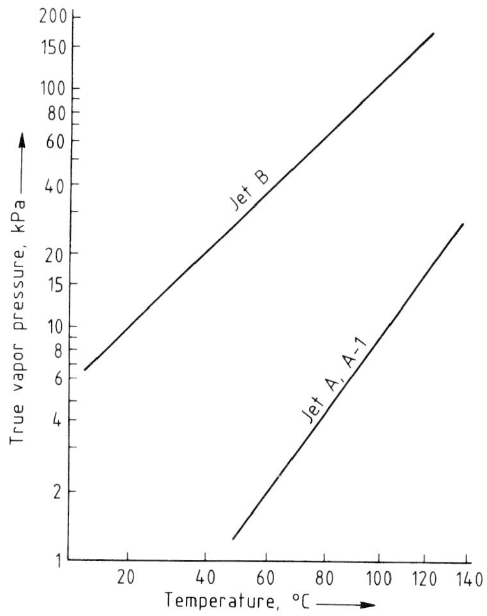

Fig. 8.2 True vapor pressures [18].

Table 8.4 Typical nonspecification properties [18]

	20 °C	40 °C	60 °C
Water solubility, vol%			
Jet A, A-1	0.0049	0.0088	0.0163
Jet B	0.0076	0.0138	0.0255
Specific heat, kJ $kg^{-1}K^{-1}$			
Jet A, A-1	1.95	2.04	2.13
Jet B	2.06	2.14	2.23
Thermal conductivity, $Wm^{-1}K^{-1}$			
Jet A, A-1	0.1155	0.1124	0.1085
Jet B	0.1155	0.1124	0.1085
Dielectric constant at 400 Hz			
Jet A, A-1	2.13	2.10	2.07
Jet B	2.06	2.03	2.00
Surface tension, mN/m			
Jet A, A-1	23.3	21.7	20.2
Jet B	21.8	20.5	18.4

fuel above 0 °C or depend on anti-icing additives (see Section 8.3.2). Typical water saturation levels for commercial jet fuels are listed in Table 8.4 [18].

Lubricity Various aircraft fuel system components–fuel pumps, servo valves, fuel control units–rely on the fuel to lubricate their moving parts. The effectiveness of a jet fuel as a lubricant in such equipment is referred to as its "lubricity". Differences in component design and materials result in varying degrees of equipment sensitivity to fuel lubricity. Similarly, jet fuels vary in their level of lubricity. The chemical and physical properties of jet fuel cause it to be a relatively poor lubricating material under high temperature and load conditions where lubrication may depend on adsorbed films of trace polar compounds present in the fuel, so-called "boundary lubrication". Severe hydroprocessing (see Section 8.3) removes trace components, resulting in fuels which tend to have lower lubricity than straight-run or wet-treated fuels. This lack of lubricity can be remediated by the addition of corrosion inhibitor/lubricity improver additives (Section 8.3.2). Such additives are mandatory in most military jet fuel specifications but they have not found widespread use in civil aviation. It is generally acknowledged that the ultimate solution to the fuel lubricity "problem" is to design fuel system equipment such that it is insensitive to fuel lubricity. Until recently, civil jet fuel specifications did not contain a requirement for a minimum lubricity level, not least because there was no universally acceptable method with which to test the fuel. However, one method, the Ball-on-Cylinder Lubricity Evaluator (BOCLE), ASTM D5001 [3] has now been accepted in some specifications for defining a minimum acceptable level of lubricity specifically for those fuels containing 95% hydroprocessed material and where at least 20% is severely hydroprocessed.

Other Properties Detailed data on other properties, including gas solubility, flammability, surface tension, heat of vaporization, enthalpy, specific heat, thermal conductivity, and dielectric constant, are contained in the *CRC Handbook of Aviation Fuel Properties* [18]. Some typical values for several properties are given in Table 8.4.

8.3 Production

8.3.1 Fuel

Since its earliest days virtually all aviation turbine fuel has been refined from selected petroleum crudes by atmospheric distillation followed by limited processing to reduce thiol or total sulfur. Crude selection is vital for such simple processing because the only other variable available for controlling bulk properties is adjusting the product's *boiling range*. Desirable crudes are low in sulfur (sweet crudes), low in *n*-hydrocarbons (nonwaxy crudes), low in aromatic content, and high in product boiling between 160 and 290 °C.

Adjusting either end of the boiling range has predictable results. Lowering the initial boiling point improves combustion and low-temperature properties but decreases handling safety (see Table 8.5). Raising the final boiling point degrades combustion and low-temperature performance and has no effect on handling safety. Because the crude also yields both lighter and heavier fractions, the demand for these products affects distillation cut points and therefore the volume of jet fuel available [29].

Total sulfur can be reduced by relatively low-pressure hydrotreating. Thiol sulfur is reduced (sweetened) by low-pressure hydrotreating or wet chemical treating, such as Merox, Linde, Doctor, or Bender treating. Flash point may be adjusted so that the same product can be sold as either aviation turbine fuel or commercial illuminating or heating kerosine.

Some heavier fractions are hydrocracked to make jet fuel, but the 160–290 °C fraction from catalytic cracking is not suitable for jet fuel unless cracking is followed by high-pressure hydrotreating. A limited amount of kerosine is hydrotreated to reduce aromatic content and increase smoke point by aromatic saturation. In the United States the finished jet fuel is often filtered through towers of activated Attapulgus clay to adsorb surfactants and to remove particulate fines.

8.3.2 Additives

All aviation fuel additives must be tested and approved. The extent of testing is at the discretion of the aircraft and engine manufacturers who issue the approval. Additive composition plays a major part in approval procedures, with ash-forming materials receiving the greatest scrutiny. Recommended Practice D 4054 (ASTM) [3] covers the

Table 8.5 Alteration of boiling range

Lowering initial temperature	**Raising final temperature**
Increases	
smoke point	density
hydrogen content	aromatic content
J/mass	naphthalenes
vapor pressure	viscosity
Decreases	
density	smoke point
flash point	hydrogen content
aromatic content	J/mass
naphthalenes	
freezing point	
J/vol	
viscosity	
Leads to	
better combustion quality	poorer combustion quality
better low-temperature properties	poorer low-temperature properties
poorer handling safety	unchanged handling safety
higher gravimetric specific energy	lower gravimetric specific energy
lower volumetric specific energy	higher volumetric specific energy

evaluation of compatibility between additives and aircraft fuel system materials. Approved additives and their permitted levels are normally listed in the fuel specifications. Unapproved additives cannot legally be used in aircraft operation.

Oxidation Inhibitors Oxidation inhibitors prevent peroxide formation and improve storage stability in certain fuels. They are generally secondary and tertiary hindered phenols and are listed in the fuel specification by chemical composition. They are effective primarily during storage, not during high-temperature use. Antioxidants are mandatory for hydroprocessed fuels in the major military fuel specifications and in civil Jet A-1 covered by the DEF STAN 91-91 specification and Joint System Checklist documents, but only optional in the ASTM D 1655 specification. Dosage rate is 17–24 mg/L. Annual consumption of oxidation inhibitors in aviation turbine fuel is difficult to estimate because the same additives are widely used in other petroleum products.

Metal Deactivators Metal deactivators are intended to passivate trace metals by chelation. The only approved deactivator is *N,N'*-disalicylidene-1,2-diaminopropane [94-91-7]. It is permitted up to 5.7 mg/L, but is rarely used.

Corrosion Inhibitors and Lubricity Improvers Unlike the preceding additives, corrosion inhibitor formulations are proprietary. They usually contain dibasic acids that function as liquid phase rust inhibitors. The polar ends of the molecule adsorb onto metal surfaces and in ground handling and distribution systems, e.g., steel pipelines, they act to protect surfaces from corrosion. In aircraft fuel systems, the tendency to adsorb on to metal surfaces is utilized for lubricity improvement rather than corrosion inhibition. Because these additives are surface-active (commonly referred to as "surfactant"), they can display the adverse side-effect of impairing the water separation properties of the fuel. Furthermore, their surfactancy means that they can be lost from the fuel during distribution (as, of course, they should be if they are to protect metal surfaces from corrosion); consequently, if added for lubricity enhancement, these additives are often injected at the airport, just prior to aircraft fueling. Lists of approved additives are published in the Qualified Products Lists (QPLs) associated with the U.S. MIL-I-25017 and UK DEF STAN 68-251 specifications for this type of additive.

Static Dissipater Additives (SDAs), also referred to as Conductivity Improving Additives, are added to jet fuel to increase electrical conductivity, and thereby reduce the tendency of a fuel to accumulate dangerous electrostatic charge during handling [28]. There are only two additives approved for use in aviation fuel, Shell ASA-3 and Octel (formerly Du Pont) Stadis 450. Production of the Shell additive ceased in 1991 leaving Stadis 450 as the only additive currently available. The concentration is adjusted to obtain an electrical conductivity within the range 50–600 pS/m but with a maximum permitted dosage of 3 mg/L. A further addition up to 2 mg/L is permitted if conductivity has been lost during fuel distribution. The minimum conductivity limit is set to ensure charge relaxation sufficient to alleviate static discharges, while a maximum limit is applied to avoid causing anomalies in aircraft fuel tank contents gauges which operate on an electrical capacitance principle. The use of SDA is mandatory in military JP-4 and JP-8 fuels and in civil Jet A-1 outside the United States; its use is optional in Jet A fuel within the United States.

Anti-icing additive, also referred to as Fuel System Icing Inhibitor (FSII), prevents the freezing of trace levels of free water in fuel at the low temperatures encountered in aircraft fuel tanks at high altitudes and avoids blockage of fuel lines and filters by ice crystals). Traditionally, ethylene glycol monomethyl ether (2-methoxyethanol) [*109-86-4*], has been used for this purpose but, owing to its toxicity, this has largely been superseded by diethylene glycol monomethyl ether (2-(2-methoxy)ethanol) [*111-77-3*]. Normal dosage rate is between 0.1–0.15 vol %. Such use is mandated in military fuels but in civil aviation its use is restricted to small aircraft which do not have heaters in their fuel systems. The additive, which is only sparingly soluble in the fuel, preferentially dissolves in any free water present (partition coefficient about 200:1) where it acts as an antifreeze. This affinity for water can be problematic in fuel distribution systems and additive loss must be considered and possibly corrected. FSII has the additional property of acting as a biostat, preventing the growth of bacteria and fungi in fuel tanks (see below).

Biocides Microorganisms that metabolize hydrocarbons can be present in both ground and aircraft fuel systems [1]. Certain anaerobic bacteria that reduce sulfate to hydrogen sulfide have caused copper and silver corrosion. The formation of slimes, fungal mats, and corrosive byproducts has caused wing tank corrosion and capacitance gage failures. The only generally approved fuel-soluble biocide is Biobor JF (Hammonds), a blend of 95% dioxaborinanes and 5% naphtha. Recommended dosage to kill microorganisms is 226 mg/L; maintenance level is 108 mg/L. An alternative material, which is approved for most applications, is Kathon FP1.5 (Rohm & Haas) which is based on a mixture of a methyl-isothiazolinone [*2682-20-4*] and a chloromethyl isothiazolinone [*26172-55-4*]. Treatment levels are between 100 and 300 mg/L. Although treatment with biocides kills the microbial growths in fuel tanks, subsequent tank cleaning is still essential to remove all the dead material which might otherwise lead to filter blockage.

Thermal Stability Additives In the early 1990s, the USAF perceived that future aircraft designs would demand a fuel with enhanced thermal oxidative stability. They decided that the most economic way of achieving this would be to use fuel additives to increase the thermal stability of the existing fuel type, JP-8, rather than developing a special fuel. It was calculated that an increase of 100 °F (56 °C) in the thermal stability level of the fuel would be required and this led to the new fuel being referred to as "JP-8 + 100". The extensive research program to develop and test such additives was referred to as "HiTTS", standing for High Temperature Thermal Stability [30]. The main problems at high temperature are fouling of the injector feed arm and coking in the fuel supply rings to the augmenter (afterburner or reheat) section. An additive package comprising a detergent/dispersant, an antioxidant, and metal deactivator proved to be very effective in preventing deposit formation in the hot areas of existing aircraft fuel systems. Already the benefits in terms of reduced maintenance costs are being appreciated. However, because the dispersant/detergent element of the additive package is very surface active, it cannot be added to the fuel in storage because it would encourage water and dirt pick-up and also disarm the filter/water separators in the airport fueling system. Instead, the additive package is injected into the fuel as the aircraft is being fueled.

8.4
Handling, Storage, and Transportation

System Descriptions It is essential to prevent aviation fuel from being contaminated with other petroleum products between the time it is manufactured in the refinery until the moment it is put on-board an aircraft. Consequently fixed storage facilities are designed to keep aviation fuel physically isolated from other products. However, large pipelines, marine transports and sometimes even road tankers are multiproduct systems. Each transportation mode therefore has an elaborate protocol to protect the quality of all the products it carries. Pipelines carry compatible products both preceding and following jet fuel; the interface between products is cut into the non-aviation grade. Marine and road transports depend on cleaning procedures, with the amount and type of cleaning dependent on the preceding product. When aviation fuel has been transported through a non-dedicated (i.e., a multiproduct) system, samples are taken for recertification testing in which those properties most sensitive to contamination, e.g., flash point, freezing point, gum, etc., are retested and the results compared with the original values reported on the refinery Certificate of Quality to confirm that no inter-product contamination has occurred. At the airport, the fuel is delivered from storage to the aircraft either via underground pressurized piping (hydrant system) through mobile hydrant dispensers or by specialized tank trucks called fuelers. Both types of fueling vehicle are equipped with filter/water separators or filter monitors to ensure only clean, dry fuel is pumped into the aircraft.

Contamination Removal Equipment Multiproduct transportation systems and long distances between refinery and airport furnish opportunities for water, rust, dust, surfactants, and other contaminants to enter the fuel. Specialized equipment and procedures remove these unwanted materials [31]. Water and solids settle in storage, assisted by suction lines that float just beneath the liquid level. Filter/water separators [32] and filter monitors [33] remove free water and fine solids, while activated Attapulgus clay filters can be used to adsorb surfactants.

The operation of this equipment is checked regularly with a variety of test equipment. Solids are detected and measured per ASTM D 2276 and ASTM D 3830 [3]. Surfactant removal by clay filters can be observed with the Microseparometer (ASTM D 3948 [3]). A variety of tests exist for the detection of low levels of undissolved or free water, with most of the tests containing a water-sensitive dye; ASTM D 3240 [3] measures water levels $\geq$ 1 ppm.

8.5 Legal Aspects

All aircraft, including all flight components, are certified for flight by a government agency which promulgates and enforces the certification rules. These rules have the force of law. Aircraft and engines use fuel meeting certain specifications during certification. These specifications become part of the aircraft's operating certificate. The fuel must therefore completely meet the certification specification even though a deviation may not affect operation adversely. Normally aircraft and engine manufacturers designate industry fuel specifications (e.g., ASTM D 1655 [3]), but manufacturers can certify to their own proprietary specification. In such a case any points of difference from the industry norms must be specified. Once certified, aircraft must be operated by the rules of the country in which they are registered.

8.6 Environmental Aspects

Aircraft exhaust emissions are controlled by regulations issued by the International Civil Aviation Organisation (ICAO). These regulations define the permitted levels of emissions of nitrogen oxides (NO_x), carbon monoxide, unburned hydrocarbons, and smoke for each engine type. At present, emissions are controlled only for the landing/take-off (LTO) cycle, which is defined as engine idle, taxi out, take-off, and climb to 900 m, descent from 900 m, taxi in, and idle. Exhaust emissions at altitude are not restricted although there are numerous studies underway to determine the effects of high altitude emissions on global warming and ozone depletion [34–36]. Future regulation is expected but this is likely to be met by further advances in combustion chamber design rather than from significant changes in the composition or properties of aviation fuel.

8.7 Economic Aspects

Compared with gasoline and diesel fuel production, aviation turbine fuel represents a small proportion of the crude barrel. In 1995, worldwide production of kerosine was 285×10^6 t of which ca.190×10^6 t was for jet fuel with the remainder for non-aviation uses (heating, lighting, cooking, etc.). This total kerosine volume represents ca. 9% of world oil consumption [37]. World air travel is predicted to grow, on average, by 5.5% per year over the next decade and 4.9% over the coming twenty years [38–40]. Highest growth rates are in the "Pacific Rim" countries. By 2006, air travel will rise by nearly 75%; by 2016, it will reach two-and-a-half times current levels. Fuel volumes will also grow significantly but at a lower rate due to improvements in aircraft fuel efficiency. The United States is the largest single consumer of jet fuel with about 50% of the total world volume; this share is expected to decrease in the future as the Far Eastern economies continue to grow.

A forecast of world aviation fuel demand in 10^6 t is given below [37]:

1990	1993	1995	2000	2005	2010	2014
161	175	189	255	330	420	500

8.8 Future Trends

8.8.1 Petroleum-Derived Fuels

Despite the predicted increase in demand for aviation fuel, there should be no overall supply problem. Both refining capacity and processing technology are available to supply these volumes. Whether the quality of jet fuel will change significantly in the future is open to debate. In the early 1980s, predictions tended to be gloomy with the expectation that fuel quality would inevitably deteriorate due to decreasing availability of sweet crudes and the increasing use of heavy and high sulfur crudes. This led to design studies aimed at developing engines which could cope with the resultant poorer combustion quality [41]. However, these predictions were not borne out during the 1980s and 1990s because new oilfields continued to be discovered. Advances in process technology will be able to maintain, or even improve on, fuel quality for the foreseeable future.

Conventional petroleum crude will continue to form the major source for jet fuel production well into the 21st century. Experimental production of jet fuel from oil shale in the United States during the 1980s, although successful, was terminated because it was economically non-viable in a world of continuing low crude oil prices. However, this technology remains on the shelf and could be resurrected should the economic climate change significantly. Production of jet fuel from tar sand derived crudes is practiced in Canada, but gives only a very small proportion of the total fuel production.

Synthesis of liquid fuels from natural gas can be achieved by the Shell Middle Distillate Synthesis process (SMDS) [42], [43]. Whilst producing kerosine of a very high quality in terms of combustion performance, the SMDS product itself is unlikely to be suitable as a jet fuel because its complete lack of aromatics causes concern in the areas of compatibility with elastomers and lubricity. However, the product remains an excellent blending component for upgrading lower quality conventional kerosines which would otherwise be unsuitable as jet fuel

8.8.2 Alternative Fuels

Kerosines derived from *coal* using the Fischer–Tropsch process have been synthesized in South Africa by SASOL for a number of years. The process produces kerosine which could be considered suitable as jet fuel although it has never been supplied as such. Although energy intensive, increasing jet fuel demand in southern Africa has prompted the producers of this synthetic product to seek approval from the aircraft and engine manufacturers for its use as jet fuel. However, like SMDS kerosine, many of the coal-derived fractions contain no aromatics and will need to be blended with other streams before being acceptable. Although this production route has been confined to South Africa, at some time in the future it may be applied in other countries with large coal reserves, e.g., China. *Oxygenated liquids* like alcohols or esters have been examined as potential aviation fuels or fuel constituents. The pure materials penalize aircraft operation by their much lower specific energy content, whereas blends with petroleum fuels may suffer from water extractability and system material compatibility difficulties [29]. Thus the widespread adoption of oxygenated compounds, particularly methanol, as commercial aviation fuel appears unlikely.

It is often said that the long-term future may lie with *cryogenic fuels* such as liquid methane or liquid hydrogen. Successful flights have been made using methane in Russia, while the "Cryoplane" project [44] is aimed at proving the feasibility of a liquid-hydrogen powered aircraft. However, a major barrier to these cryogenic fuels is that they cannot be used in conventional aircraft, and a new generation of aircraft specifically designed for them would need to be developed. Furthermore, completely new production, storage, distribution, and handling systems would have to be constructed all around the world to ensure supply at all major airports. Proponents of cryogenic fuels tend to underestimate the capital investment involved in establishing a supply infrastructure and it is questionable whether these fuels will ever become commercially viable [45]. *Methane* may be more advantageously used as the starting material in the manufacture of conventional liquid hydrocarbon fuels rather than as a liquid fuel itself [46]. *Liquid hydrogen* has excellent combustion qualities and produces no CO_2 and, at first sight, appears to be more environmentally friendly than conventional kerosine-based jet fuel. However, unless it is produced via nuclear or hydroelectric power, its production from petroleum resources would continue to consume energy and produce CO_2. Moreover, combustion of hydrogen in aircraft engines would still not avoid the production of NO_x and, furthermore, produces 2.5 times as much water vapor as kerosine, which will also contribute to the greenhouse effect.

The massive investments required for new fuels together with the long lead time in design and certification will encourage the use of fuels similar to those currently used for the longest possible time.

8.9 References

1. M. SMITH: *Aviation Fuels*, G. T. Foulis Ltd., Henley-on-the-Thames 1970, 495 pp.
2. V. B. GUTHRIE (ED.): *Petroleum Products Handbook*, McGraw-Hill, New York 1960, pp. 5-1 to 5–28.
3. *1997 Annual Book of ASTM Standards*, **vols 05.01–05.03**, ASTM, West Conshohocken, PA 1997.
4. DEFENCE STANDARD 91-91(DERD 2494), ISSUE 5, MINISTRY OF DEFENCE, LONDON, 2006.
5. *IATA Guidance Material for Aviation Turbine Fuels.*, 5th ed., IATA Montreal, 2003.
6. *Standard Methods for Analysis and Testing of Petroleum and Related Products*, Institute of Petroleum, London, 2006.
7. International Organisation for Standardisation, Case Postale 56, CH-1211 Geneva 20, Switzerland.
8. *Manual on the Significance of Tests for Petroleum Products MNL 1*, 7th ed., ASTM, 2003.
9. *Petroleum Measurement Tables*, **vol. I–XII** ASTM, West Conshohocken, PA.
10. *Aviation Fuel Safety 1975," CRC Report 482*, Co-ordinating Research Council, Atlanta 1975.
11. C. A. MOSES, D. W. NAEGELI: *American Society for Mechanical Engineers Preprint 79-GT-178*, March 1979.
12. J. M. CASPER, E. E. EKSTEDT, W. J. DODD, M. W. SHAYESON, *NASA Report CR-159855*, Cleveland Sept. 1980.
13. R. E. VOGEL, D. L. TROTH, A. J. VERDOUW, *U.S. Air Force Report AFAPL-TR-79-2072*, Wright-Patterson, Ohio, April 1980.
14. E. A. DROEGEMUELLER, R. K. NELSON: *SAE Preprint 47 A*, SAE, Warrendale, Pa., April 1958.
15. T. T. BOWDEN, J. H. PEARSON, *The Effect of Fuel Composition Upon Combustion Performance in a Rolls-Royce Tyne Combustor, ASME Paper 85-GT-39*, ASME, March 1985.
16. NATIONAL RESEARCH COUNCIL, *Aviation Turbine Fuels-An Assessment of Alternatives*, National Academy Press, Washington, D.C., 1982.
17. W. A. AFFENS, R. N. HAZLETT, D. E. SMITH, *NRL Report 8692*, Naval Research Lab., Washington, D.C., July 1983.
18. *"Handbook of Aviation Fuel Properties", CRC Report* 635, Coordinating Research Council, Atlanta, 3rd ed. 2004
19. *"CRC Literature Survey on the Thermal Oxidation Stability of Jet Fuel," CRC*

Report 509, Coordinating Research Council, Atlanta, April 1979.
20. *"Thermal Oxidation Stability of Aviation Turbine Fuels," ASTM Monograph 1*, ASTM, West Conshohocken PA 1991.
21. S. A. W. Thompson, S. R. Hills, N. P. H. Brown, A. Lewis, *J. Inst. Petr. (London)* **50** (1964) 72.
22. V. Patarini, N. S. Bornstein, M. A. DeCrescente: *ASME Preprint 78-GT-82*, American Society of Mechanical Engineers, New York April 1978.
23. J. A. Bert, H. R. Porter: *Paper 11 a–63*, American Petroleum Institute, Washington, D.C., May 13 1963.
24. R. N. Hazlett, H. W. Carhart, *Filtr. Sep.* 9 (1972) 456.
25. *"Water/Fuel Separation Characteristics: A Bibliography," CRC Report 460*, Coordinating Research Council, Atlanta July 1973.
26. W. M. Bustin, W. G. Dukek: *Electrostatic Hazards in the Petroleum Industry*, Research Studies Press, London 1983.
27. *"Generation and Dissipation of Electrostatic Charge During Aircraft Fueling-A Selected Literature Survey," CRC Report 466*, Coordinating Research Council, Atlanta 1964.
28. ASTM D 4865 -96: 1996.
29. K. H. Strauss: *AIAA Preprint AIAA-81-0770*, May, 1981.
30. S. P. Heneghan, S. Zabarnick, D. R . Ballal, W. E. Harrison, *" JP-8 + 100: The Development of High-Thermal-Stability Jet Fuel," ASME Journal of Energy Resources Technology* **118** (1996) September, 170–179.
31. *Manual of Aviation Fuel Quality Control Procedures, MNL 5*, 3rd ed., ASTM, West Conshohocken, PA, 2004.
32. *Specifications and Qualification Procedures - Aviation Jet Fuel Filter/Separators, API Bulletin 1581*, 5th ed., American Petroleum Institute, Jan. 2002.
33. *Specifications and Qualification Procedures - Aviation Fuel Filter Monitors with Absorbent Type Elements*, Institute of Petroleum, January 1995.
34. L. J. Archer, *Aircraft Emissions and the Environment: CO_X, SO_X, HO_X and NO_X*, Oxford Institute for Energy Studies, 1993.
35. Air Transport and the Environment, IATA Seminar, Washington, March 1993.
36. Air Transport and Environmental Protection, Airbus Industrie, March 1993.
37. *Aviation Fuels Business Worldwide 1996*, Petroleum Economist, London April 1996.
38. "2006 Current Market Outlook," Boeing Commercial Airplane Group, 2006.
39. Outlook for Commercial Aircraft 1995–2014, McDonnell-Douglas, August 1996.
40. Global Market Forecast 2006–2025, Airbus Industrie, Nov. 2006.
41. J. P. Longwell, J. Grobman, *J. Eng. Power* **101** (1979) 161.
42. The Shell Middle Distillate Synthesis Process, Shell Selected Paper, November 1989.
43. P. J. A. Tijm, J. M. Marriott, The Markets for Shell Middle Distillate Synthesis Products, *Alternative Energy '95*, Vancouver, May 1995.
44. *"The Light Stuff," Chemistry in Britain* 1997 29–31.
45. A. M. Momenthy, Aviation Fuel in a Changing World, Boeing, January 1994.
46. O. Hadaller, A. M. Momenthy, Fuel Reformulation and Jet Fuel, Boeing, November 1994.

9
Marine Fuels

Klaus Reders, Torsten Mundt

9.1
History

Marine fuels for use in diesel engines cover a wide range of different grades, from gas oil fractions often referred to as "distillates" – and including a grade with properties similar to automotive diesel fuel – to heavy residual fuel oils containing refinery residues that remain after crude oil distillation. While the distillate fuels are mainly used in the smaller high-speed engines (speeds higher than 1000 rpm), the residual fuel oils are used in large-bore, low-speed cross-head engines (below 350 rpm) and also in medium-speed engines.

For a long period of time marine diesel engines were solely fueled with the distillate type of fuels. Only after World War II was the potential use of heavy residual fuels investigated. These fuels, with densities up to 1000 kg/m^3 and above and viscosities up to 700 mm^2/s at 50 °C, were originally used only as boiler fuels for steam generation. The low cost of these residual fuels was an incentive for investigations into their use in the large marine diesel engines also. However, initially the cost savings obtained were more than offset by the increased difficulty in fuel handling and treatment (heating and separation of unsuitable material) and even more significantly by the higher rates of engine wear and additional maintenance cost. This situation changed dramatically after the so called "oil crises" in the 1970s. The fuel cost subsequently increased and also the gap in the differential cost between the higher quality distillate fuels and the residual fuels widened, followed by a general move to the cheaper heavier fuels, especially in the low- and medium-speed marine engines.

The higher cost of fuel also required an improvement in the overall efficiency of the marine propulsion systems. As the diesel engine offered more scope (by approaching 50 % total efficiency) than the steam turbine, an increasing number of new ships were equipped with diesel engines fitted with the treatment systems (mainly heating systems and centrifuges) required for the use of heavy residual fuel grades.

Handbook of Fuels: Energy Sources for Transportation. Edited by Barbara Elvers

ISBN: 978-3-527-30740-1

9.2 Specifications

While initially only fuel specifications from the suppliers existed, the need for international specifications was recognized with the increased use of heavy fuel oils in diesel engines. It was for this reason that the British Standard BS MA 100 was introduced in 1982, followed by the ISO Standard 8217 in 1987. Furthermore, there is also an ASTM Standard D 975 and a diesel engine manufacturers and users specification issued by the International Council on Combustion Engines (CIMAC), which was first published in 1986.

The ISO Standard 8217 specifies four distillate fuels (see Table 9.1), the three lighter ones DMX, DMA and DMB, to be used without specific fuel treatment on board, and 15 residual fuels, which require heating to reduce their viscosity and also usually centrifuging to separate water and foreign matter before their use in the engine (Table 9.2).

9.3 Composition

The hydrocarbon composition of marine diesel fuels depends mainly on that of the crude oil used in the refinery (e.g., distillates and distillation residues have compositions controlled by the crude oil) and also on the secondary refinery process units, e.g., aromatics content when the highly aromatic catalytic cracker cycle oil is used as a component in the residual fuels. Whereas the distillate fuels consist mainly of alkanes and cycloalkanes (i.e., paraffins and naphthenes), alkenes (olefinic hydrocarbons), and aromatics, the residual fuels also contain asphaltenes. These are extremely large hydrocarbon molecules with a very high carbon/hydrogen ratio and have complex, mainly aromatic structures and molecular masses between about 5000 and 40 000. In a stable residual fuel, which meets the standard, the asphaltenes are colloidally dispersed (see Section 9.4.2). Depending on the crude oil, the oil molecules also contain oxygen, sulfur, nitrogen, metals such as vanadium, chromium, nickel, aluminum and other nonhydrocarbon atoms in low concentrations.

When a paraffinic crude oil is used to produce a residual fuel oil, an aromatic diluent (such as cat cracker cycle oil) must frequently be added in the refinery to prevent agglomeration of the asphaltene molecules and avoid sludge formation.

Long-chain alkanes in residual fuels may be present to some extent, in the form of wax at ambient temperature. Care must be taken to avoid precipitation and formation of waxy sludges during storage.

The residual fuel oils may also contain impurities such as water, sand, rust from tanks and pipelines, and catalyst fines from refinery processes, which must be separated prior to use in the engine to avoid unacceptable wear in fuel pumps, injectors, and cylinder liners. The chemically bound sulfur present in marine diesel fuels (in the heavier grades up to 5 %) requires the use of specific highly alkaline cylinder lubricants to avoid corrosive wear.

Table 9.1 Requirements for marine distillate fuels

Characteristic	Limit	Category ISO-F- DMX	DMA	DMB	DMC	Test method reference
Appearance		clear between +10 and +25°C	clear between +10 and +25°C	–	–	visual rating
Density at 15 °C, kg/m³	max.	[a]	890.0	900.0	920.0	ISO 3675 or ISO 12185 (see also 6.3[e])
Viscosity at 40 °C, mm²/s [b]	min.	1.40	1.50	–	–	ISO 3104
	max.	5.50	6.00	11.00	14.0	ISO 3104
Flash point, °C	min.	43	60	60	60	ISO 2719 (see also 6.4[e])
Pour point (upper), °C [c]						
Winter quality	max.	—	−6	0	0	ISO 3016
summer quality	max.	—	0	6	6	ISO 3016
Cloud point, °C	max.	−16[d]	—	—	—	ISO 3015 (see also 6.5[e])
Sulfur, % (m/m)	max.	1.0	1.5	2.0	2.0	ISO 8754 (see also 6.6[e])
Cetane number	min.	45	40	35	—	ISO 5165 (see also 6.7[e])
Carbon residue [micro method. 10% (V/V) distillation bottoms], %(m/m)	max.	0.30	0.30	—	—	ISO 10370
Carbon residue (micro method), % (m/m)	max.	—	—	0.30	2.50	ISO 10370
Ash, % (m/m)	max.	0.01	0.01	0.01	0.05	ISO 6245
Sediment, % (m/m)	max.	—	—	0.07	—	ISO 3735
Total existent sediment, % (m/m)	max.	—	—	—	0.10	ISO 10307-1
Water, % (V/V)	max.	—	—	0.3	0.3	ISO 3733
Vanadium, mg/kg	max.	—	—	—	100	ISO 14597
Aluminum plus silicon, mg/kg	max.	—	—	—	25	ISO 10478 (see also 6.8[e])

[a] In some geographical areas. there may be a maximum limit.
[b] 1 mm²/s = 1 cSt.
[c] Purchasers should ensure that this pour point is suitable for the equipment on board. especially if the vessel operates in both the northern and southern hemispheres.
[d] This fuel is suitable for use without heating at ambient temperatures down to −15 °C.
[e] Referring to ISO Standard 8217:1996(E).

Table 9.2 Requirements for marine residual fuels

Characteristic	Limit	Category ISO-F-															Test method reference
		RMA 10	RMB 10	RMC 10	RMD 15	RME 25	RMF 25	RMG 35	RMH 35	RMK 35	RMH 45	RMK 45	RML 45	RMH 55	RMK 55	RML 55	
Density at 15 °C, kg/m³	max.	975.0	9.981.0	981.0	985.0	991.0	991.0	991.0	991.0	1010.0	991.0	1010.0	—	991.0	1010.0	—	ISO 3675 or ISO 12185 (see also 6.3[c])
Kinematic viscosity at 100 °C, mm²/s [a]	max.	10.0	10.0	10.0	15.0	25.0	25.0	35.0	35.0	35.0	45.0	45.0	45.0	55.0	55.0	55.0	ISO 3104
Flash point, °C	min.	60	60	60	60	60	60	60	60	60	60	60	60	60	60	60	ISO 2719 (see also 6.4[c])
Pour point (upper), °C [b]																	
winter quality`	max.	0	24	24	30	30	30	30	30	30	30	30	30	30	30	30	ISO 3016
summer quality	max.	6	24	24	30	30	30	30	30	30	30	30	30	30	30	30	ISO 3016
Carbon residue, % (m/m)	max.	10	10	10	14	14	15	20	18	22	22	22	—	22	22	—	ISO 10370
Ash, % (m/m)	max.	0.10	0.1	0.1	0.10	0.10	0.15	0.15	0.20	0.2	0.20	0.2	0.2	0.20	0.2	0.2	ISO 6245
Water, % (V/V)	max.	0.5	0.5	0.5	0.8	1.0	1.0	1.0	1.0	1.0	1.0	1.0	1.0	1.0	1.0	1.0	ISO 3733
Sulfur, % (m/m)	max.	3.5	3.5	3.5	4.0	5.0	5.0	5.0	5.0	5.0	5.0	5.0	5.0	5.0	5.0	5.0	ISO 8754 (see also 6.6[c])
Vanadium, mg/kg	max.	150	15	300	350	200	500	300	600	600	600	600	600	600	600	600	ISO 14597
Aluminum plus silicon, mg/kg	max.	80	80	80	80	80	80	80	80	80	80	80	80	80	80	80	ISO 10478 (see also 6.8c)
Total sediment. potential, % (m/m)	max.	0.10	0.10	0.10	0.10	0.10	0.10	0.10	0.10	0.10	0.10	0.10	0.10	0.10	0.10	0.10	ISO 10307-2 (see also 6.9[c])

[a] Annex C gives viscosity/temperature table for information purposes only. 1 mm²/s = 1 cSt.
[b] Purchasers should ensure that this pour point is suitable for the equipment on board, especially if the vessel operates in both the northern and southern hemispheres.
[c] Referring to ISO Standard 8217:1996(E).

9.4 Properties

9.4.1 Distillate Fuels

The specification of the lightest distillate fuel DMX is basically similar to that of automotive diesel fuel, however, DMX has no density and no distillation limits. These parameters are indirectly controlled by viscosity and cloud point. The minimum ignition quality (cetane number) of DMX is lower than that of automotive diesel fuel and the maximum sulfur content is higher. This distillate fuel is also referred to as an emergency fuel, because it can be used without any treatment on board down to low temperatures. The remaining heavier distillate fuels DMA (also referred to as gas oil), DMB and DMC, have higher maximum density, viscosity and sulfur levels, whereas low-temperature fluidity and cetane number are reduced. The more viscous fuels are geared to the requirement of medium-speed engines with longer piston strokes, offering more time for the air – fuel charge to ignite and burn during the combustion cycle, and hence the ignition performance (characterizing ignition delay after injection and ease of combustion) of the heavier fuels is less important. The heaviest distillate fuel, DMC, may contain traces of residual fuel oils and therefore specification criteria used in this fuel category to control impurities such as water, vanadium and aluminum/silicon have been included.

9.4.2 Residual Fuels

The 15 residual fuels RMA 10 to RML 55 are grouped according to viscosity, their numbering being identical to the viscosity at 100 °C, rising in steps of 10 mm^2/s, with one intermediate fuel of 15 mm^2/s. Each viscosity group contains either two or three fuels, allowing different concentrations of asphaltenes described by variations in, e.g., densities, carbon residues, or vanadium content. Fuel manufacturers' specifications often use the viscosity at 50 °C and their fuel denotations thus differ from that used in ISO 8217. In the traditional marine business the viscosities at 50 °C are still the most widely used viscosity parameters.

While initially the maximum density of the fuels was restricted to about 0.990 kg/m^3 to guarantee the separation of water by centrifuging, modern centrifuges can actually safely handle fuels with densities greater than 1000 kg/m^3 and for this reason the maximum density of some of the more viscous fuels was raised to 1010 kg/m^3.

The flash point (safety aspect in storage tanks) of all residual fuels is identically specified to be $\geq$ 60 °C. In theory, the fuels should not produce ignitable air – fuel mixtures in the fuel tank atmosphere below this temperature.

Paraffinic wax formation at low temperature is controlled by specification of the pour point. The pour point is 3 °C above the temperature at which the fuel will not flow under its own weight when tested by standard methods. The pour point is a guide to low temperature filterability/pumping limits and wax precipitation in storage tanks. (In automotive diesel fuel the cold filter plugging point, CFPP, a low-temperature filterability test, has shown better correlation with low-temperature fuel filterability performance.) As residual fuel oils are heated anyway to reduce their viscosity, wax formation, even with the high pour points of 30 °C, does not usually cause any problems. However, care must be taken that the fuel is not cooled and heated in cycles around the pour point, because under these conditions wax formation may increase. It should be noted that some paraffinic wax crystals, once formed in residual fuel oils, have melting temperatures approaching 100 °C.

Ignition performance of residual fuels is not specified, because there is sufficient time in the low-speed – long-stroke engines to burn even highly aromatic/ asphaltenic fuels with low ignition performance. It has also been shown that engine and operating parameters (e.g., fuel viscosity when injected, charge temperature and pressure) can compensate for fuel deficiencies in this respect. When changing from one fuel to another, it may, however, be necessary to rank their ignition performance. It is well known that the ignition performance of a diesel fuel decreases with increasing aromatic content. In the absence of both a suitable and simple test method of ignition performance and of aromaticity of residual fuels, the Calculated Ignition Index (CII) and the Calculated Carbon Aromaticity Index (CCAI) have proven useful. Both methods use fuel viscosity and density as parameters. Increasing viscosity and density simultaneously increases the aromatic content and reduces the ignition performance. At a given density reducing viscosity, and at a given viscosity increasing density, increases the calculated aromatic content and reduces the ignition performance of the fuel. Details are described in an appendix of ISO 8217.

The carbon residue of marine diesel fuels is measured to estimate the coke forming and fouling tendency of a fuel. It is generally related to the asphaltene content, but there are considerable differences in this relationship between different fuels. The correlation of carbon residue with formation of deposits in marine engines is poor, and this is not surprising because modern marine engines (specifically low- and medium-speed engines) are built to operate safely with fuels of high asphaltene contents and high carbon residue. However, the carbon residue has been retained in the specifications to characterize the fuel.

The ash content of fuels is specified, because inorganic ash-forming matter, either of natural origin or originating from contamination, can form deposits in the turbocharger and in the engine. Specifically, deposits on the exhaust valve seats can lead to valve burning and finally failures. Whereas insoluble and water-soluble material (e.g., rust and sodium salts from sea water) are generally removed by fuel centrifuging, "natural" oil-soluble ash-forming impurities (mainly various metals and silicon), which tend to concentrate in the distillation residues, cannot be separated on board.

The most common contaminant of marine fuels is water. The water content is often increased after delivery in the ships fuel tanks, either because of leakage into the tank or by tank breathing and condensation. Water bottoms in fuel tanks can lead to rust formation and to the growth of fungi in the water – oil phase. This may cause additional sludge in storage tanks. Fuel centrifuging separates most of the water before the fuel is supplied to the engine.

Sulfur is a natural constituent of crude oil. It is chemically bound to the hydrocarbon molecules and the sulfur concentration varies depending on the crude oil source, ranging from about 0.1 % (e.g., Nigerian crude) to nearly 3 % (e.g., Venezuelan crude). In distillation residues, used for the blending of residual fuels, the sulfur content is even higher. Heavier grades may contain a maximum of 5 %. During combustion, the sulfur is converted into sulfur oxides. Together with the water formed by combustion, the sulfur oxides form sulfuric acids, which may cause severe corrosion of cylinder liners and piston rings. By the use of highly alkaline cylinder lubricants and higher engine temperatures to avoid condensation, the engine wear problem has largely been overcome, however, there remains the emission of sulfur oxides with the exhaust gases. The use of high-sulfur fuels is therefore restricted, or offenders penalized, in some coastal areas.

In addition, vanadium is – as is sulfur – a natural trace constituent of crude oil. During combustion, vanadium, together with sodium, can produce sticky, low-melting salts in a series of complex reactions, which may adhere to exhaust valves and turbochargers, causing high temperature corrosion. Apart from restricting the concentration of vanadium in residual fuels, there are operational (reduction of exhaust valve temperature) and metallurgical (specific alloys) solutions to avoid this type of corrosive wear.

Aluminum and silicon in fuel oils stem from disintegrated cat cracker catalyst. Of course these catalyst fines may only be present when cat cracked components are used as components. The catalyst fines may cause abrasive wear. Below the specified limit of 80 mg/kg they are safely reduced by centrifuging to non-critical levels.

The total sediment contains not only inorganic insolubles but also, after aging of the fuel, insoluble organic fuel components, such as agglomerated asphaltenes. As mentioned above, residual fuels contain significant amounts of asphaltenes, which are kept finely dispersed by aromatic diluents. Fuel stability tests are the main criteria of quality control procedures in refineries, and it is unlikely that any stability problem leading to the production of asphaltenic sludges would occur with fuels supplied from refineries where the in-house and international standards are followed. However, if a fuel is blended with paraffinic diluents or fresh asphaltenes are produced by heating (e.g., using electrical heaters with too high a skin temperature), it is possible that the asphaltene/aromatic ratio could be altered and the asphaltenes may agglomerate, adding considerably to sludge formation. Experience has shown that residual fuels with total sediments below the specified 0.1 % can be handled without any problems with the standard centrifuging equipment.

Another important quality criterion, which is not specified, is the fuel's specific energy. The specific energy can, however, be calculated with sufficient precision from the inspection properties of the fuel, if known. For details see ISO 8217. Increased density (i.e., higher carbon/hydrogen ratio), sulfur and, of course, water and ash content reduces the specific energy on a weight basis (in MJ/kg). However, fuels belonging to the high density/viscosity categories are usually priced lower, and their deficiency in the lower specific energy content is therefore generally more than offset by the price.

9.5 Outlook

While for several decades fuel research and engine development was geared to reducing the overall cost by improving efficiency and engine life with cheap fuels, the focus has now changed to improved environmental acceptability. In the area of marine diesel emissions, this is mainly the reduction of harmful exhaust gas constituents. Depending on the nature of these pollutants there are various options, i.e., changes in fuel composition, of the combustion process, or exhaust-gas aftertreatment. Various technologies are available and have partly been proven "on shore", such as selective catalytic reduction (SCR) exhaust-gas aftertreatment catalysts to reduce nitrogen oxides. Research in all three areas is ongoing, however, cost and practicability on board are some of the limiting factors. International standards on emissions taking into account the technical potentials with acceptable lead times or penalties, as already enforced in sensitive areas such as the Baltic Sea, are helping to neutralize the cost factor and will lead to environmentally improved ship propulsion systems.

Index

Handbook of Fuels: Energy Sources for Transportation. Edited by Barbara Elvers

ISBN: 978-3-527-30740-1